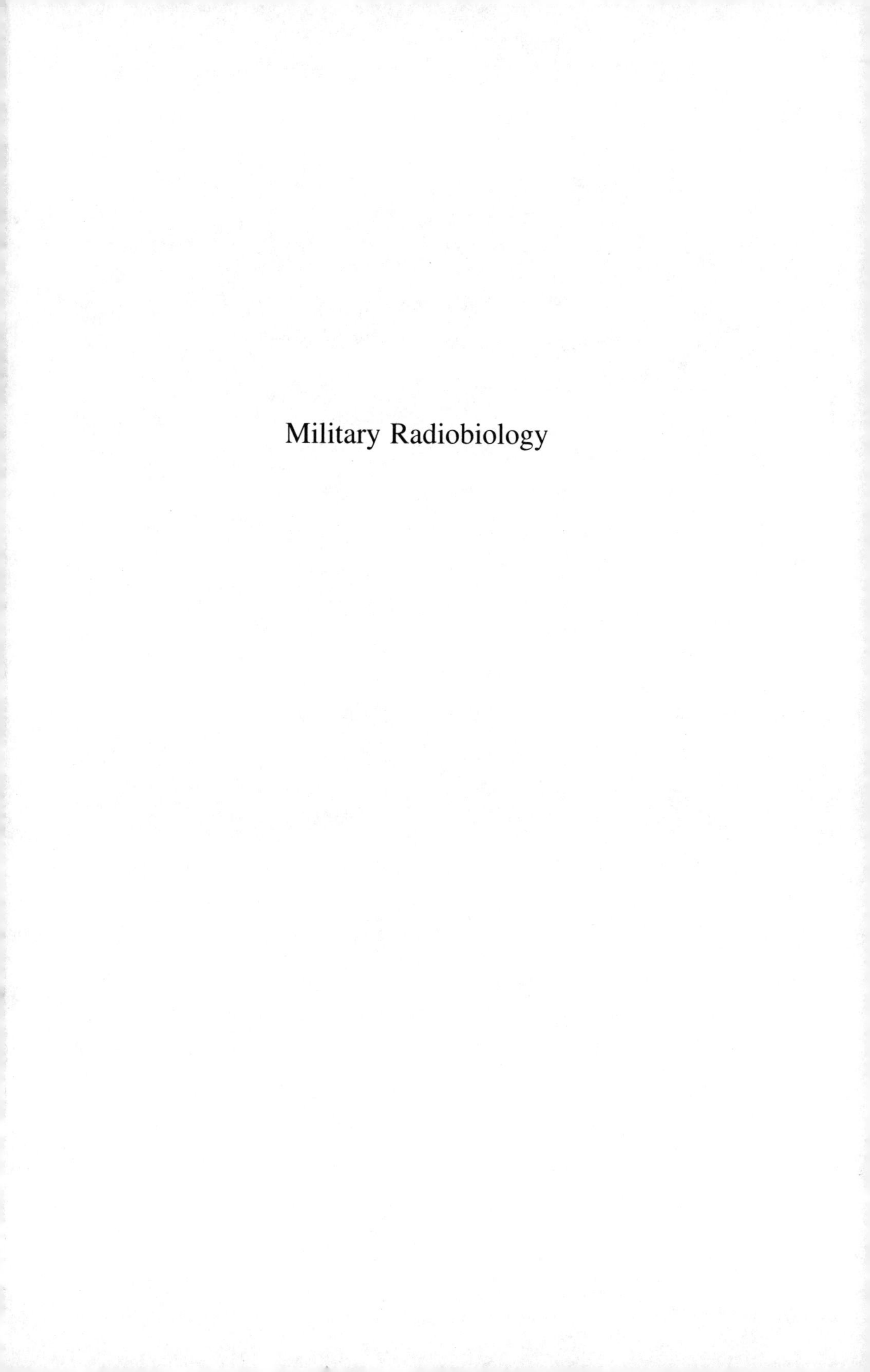

Military Radiobiology

Military Radiobiology

Edited by

JAMES J. CONKLIN
RICHARD I. WALKER

Armed Forces Radiobiology Research Institute
Bethesda, Maryland

1987

ACADEMIC PRESS, INC.
Harcourt Brace Jovanovich, Publishers
Orlando San Diego New York Austin
Boston London Sydney Tokyo Toronto

ACADEMIC PRESS, INC.
Orlando, Florida 32887

United Kingdom Edition published by
ACADEMIC PRESS INC. (LONDON) LTD.
24–28 Oval Road, London NW1 7DX

Library of Congress Cataloging in Publication Data

Military Radiobiology

Includes index.
1. Radiation—Physiological effect. 2. Nuclear weapons—Physiological effect. I. Conklin, James J. II. Walker, Richard I.
QP82.2.R3M53 1986 363.1'79 86-10759
ISBN 0–12–184050–6 (alk. paper)

PRINTED IN THE UNITED STATES OF AMERICA

87 88 89 90 9 8 7 6 5 4 3 2 1

Contents

4. Ionizing Radiations and Their Interactions with Matter

Joseph A. Sholtis, Jr.

5. Cellular Radiation Biology

Eugene V. Holahan, Jr.

6. Radiation Effects on the Lymphohematopoietic System: A Compromise in Immune Competency

Rodney L. Monroy

7. Effect of Ionizing Radiation on Gastrointestinal Physiology

Pamela J. Gunter-Smith

8. Postirradiation Cardiovascular Dysfunction

• 9. Acute Radiation Syndrome

10. The Combined Injury Syndrome

11. Mechanisms and Management of Infectious Complications of Combined Injury

12. Diagnosis, Triage, and Treatment of Casualties

James J. Conklin and Richard I. Walker

13. Internal Contamination with Medically Significant Radionuclides

Asaf Durakovic

14. Radioprotectants

Leo Giambarresi and Aaron J. Jacobs

15. Psychological Effects of Nuclear Warfare

G. Andrew Mickley

16. Effects of Ionizing Radiation on Behavior and the Brain

Walter A. Hunt

17. Nuclear Weapons Accident Response Procedure

Raymond L. Chaput

18. Management of Radiation Accidents

James J. Conklin and Rodney L. Monroy

CHAPTER 1

Military Radiobiology: A Perspective

RICHARD I. WALKER AND JAMES J. CONKLIN

Armed Forces Radiobiology Research Institute
Bethesda, Maryland 20814-5145

The year 1945 saw the detonation of three new weapons: one in a test in the desert of New Mexico and two over the cities of Hiroshima and Nagasaki, Japan. The detonations were so devastating and so powerful that they ended a war, and they changed the thinking of strategists and tacticians ever since. These weapons were the product of years of tremendous effort to harness the energy from discoveries in atomic physics. The weapons helped to end a war, but also began a new era from which there would be no turning back.

Along with the political consequences of the military use of nuclear energy came new medical challenges. In addition to blast and thermal shocks to living systems came a variety of radiations that profoundly affected the major organ systems of the body. Among these irradiations were packets of electromagnetic energy, such as gamma rays, X rays, and ionizing particulate radiation, including electrons, alpha particles, and neutrons.

When conventional injuries and radiation injury were superimposed, they resulted in a phenomenon known as combined injury, which greatly increases the mortality of victims. These medical problems are compounded by the fact that the sheer numbers of serious casualties that could be created by a nuclear detonation would far exceed the existing medical capabilities. For example, the single 12-kt (kiloton) weapon used at Hiroshima was responsible for 45,000 deaths and over 90,000 casualties on the first day after the explosion. At the same time, the existing medical facilities and personnel were all but eliminated by the detonation.

These problems have not diminished over the years, and their threat has increased by tremendous proportions. Only 4 years after the first United States detonation, the Russians detonated their first atomic weapon, and the arms race was on.

Atomic weapons began to undergo steady improvement in the early 1950s. In 1952 the first hydrogen bomb was used in a test to literally vaporize an island in

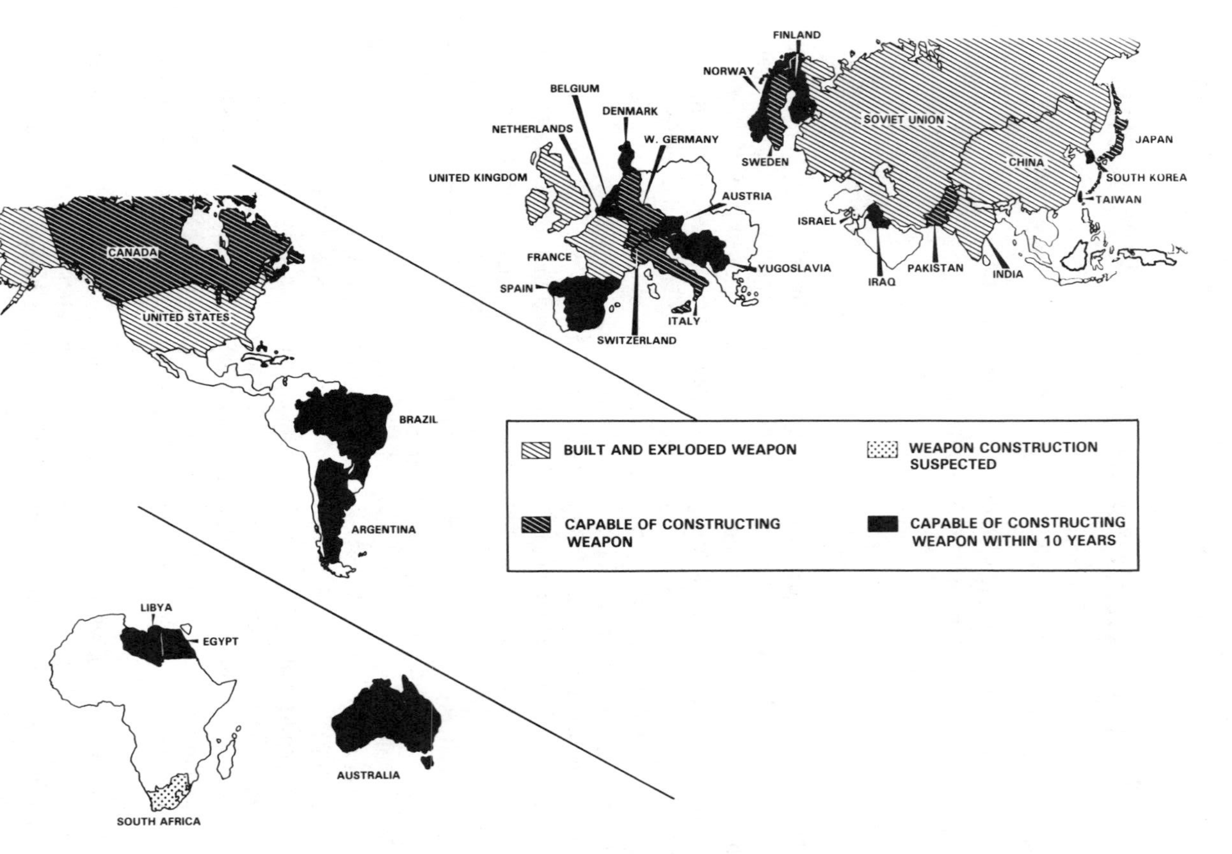

FIG. 1-1 Proliferation of nations with nuclear capabilities.

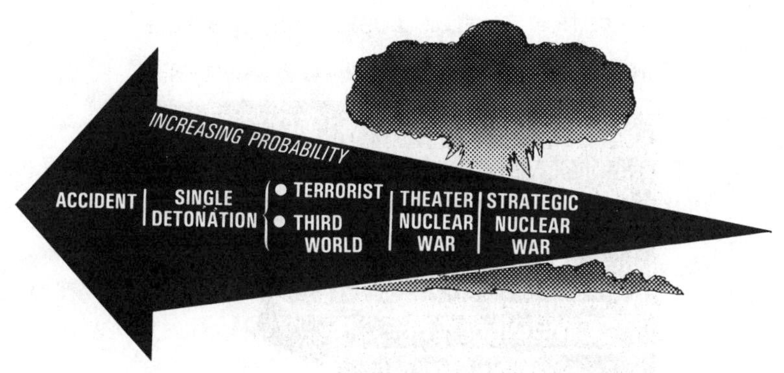

FIG. 1-2 Relationship between magnitude of nuclear incidents and probability of occurrence.

the Pacific. During this period the policy of massive retaliation was established, in which the United States threatened to unleash its entire nuclear arsenal against any nation that attacked the United States or an ally. But this policy lost credibility as the Soviet Union, under the leadership of Nikita Krushchev, sought nuclear supremacy over the United States.

As nuclear submarines, bombers, and missiles were developed by both superpowers, the suicidal concept of mutually assured destruction was reexamined. A flexible response was developed, which included only those weapon options thought absolutely necessary by the Kennedy administration. Today it is questionable whether any nuclear confrontation between the United States and the USSR can be controlled, and the nuclear capability of each country is overwhelming. One modern nuclear submarine carries more destructive firepower than 500 Hiroshima-sized bombs.

The awesome threat of a nuclear weapon confrontation between the superpowers has, we hope, given significant pause to the leaders of both nations. Unfortunately, the nuclear weapon threat grows continuously, as an increasing number of countries achieve nuclear weapons (Fig. 1-1). Perhaps it is even more likely that one of these nations or terrorist groups working within its boundaries will consider a nuclear option to be viable (Fig. 1-2). Until man turns away from violence as a solution to his problems, the possibility of nuclear destruction will have to be faced.

MEDICAL CONSEQUENCES

Military medical personnel must deal with the consequences of nuclear energy deposition and trauma in a living being. To do this, it will be necessary to

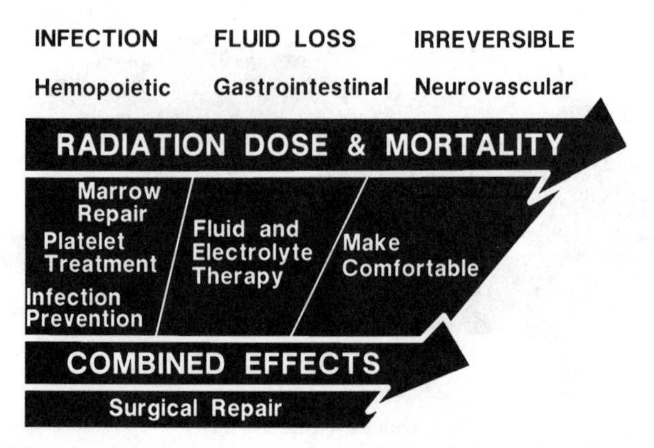

FIG. 1-3 Life-threatening syndromes associated with irradiation and combined injuries.

understand the postexposure effects in order to determine points of intervention. This is essential not only for humanitarian reasons but also so that the human component of operational systems can remain effective. When weapon systems are critically evaluated, it becomes obvious that the weak links in most systems are man and the man–machine interface. Further, an understanding of the effects of radiation and trauma will lead to better means of injury management to reduce the logistical drain that large numbers of nuclear weapon casualties will have on the limited medical resources. In addition to intervention to prevent or repair injuries related to nuclear weapons, knowledge of the medical consequences for military personnel is essential for contingency and targeting planning.

The medical problems confronting military radiobiology include target damage, which causes decrements in normal performance, physiological injury, and impairments of the immunological–hematological system that lead to life-threatening infectious complications. With the exception of performance decrements, these injuries are illustrated in Fig. 1-3. With increasing doses of radiation, one experiences what are known as the hemopoietic, gastrointestinal, and neurovascular syndromes. There is no realistic hope of managing injuries in the neurovascular range, but injuries associated with lower doses may be manageable. Lymphocytes and bone marrow cells are among the most sensitive to radiation, and their loss leads to thrombocytopenia (bleeding abnormalities) and increased susceptibility to opportunistic infections, which often arise from resident microorganisms. These problems may be controlled by new means to enhance marrow repair and/or insert marrow stem cells. Also, new antimicrobial and immunomodulating techniques may be used to prevent life-threatening infection until the regenerative process occurs in the marrow.

Loss of fluid and electrolytes is a consequence of radiation damage to the epithelial lining of the intestine. Among the means to offset intestinal injury is restoration of fluid and electrolytes. The replicating cells of the intestine, like

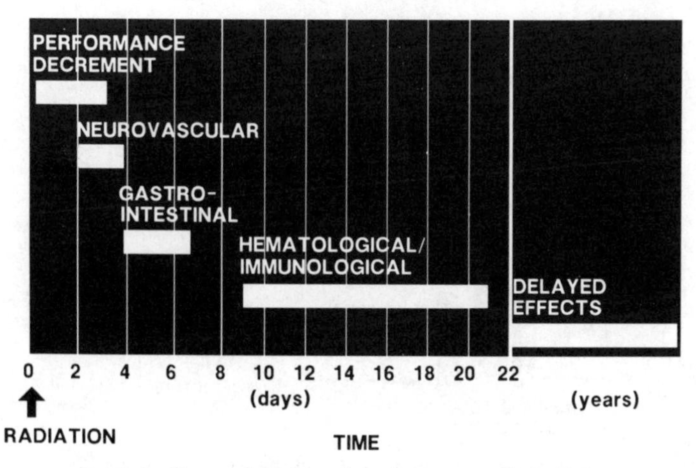

FIG. 1-5 Temporal relation of postexposure medical effects.

agement, another important facet of radiobiology is protection against the initial lesions following deposition of energy. Radiation induces the formation of reactive chemical products when it enters a biological system. For example, cellular water is converted to free radicals, which can form harmful peroxides and inflict cellular damage themselves. These reactive products as well as radiation energy itself can destroy the essential bonds between chemicals that make up the genetic material, cell membranes, and protein enzymes necessary for repair and regulatory processes. Much research is under way to develop natural and synthetic radioprotectants that eliminate reactive compounds, stabilize sensitive target sites, or enhance repair. If the dose response to deposited radiation energy can be reduced in this manner, the potential for recovery can be greatly enhanced.

The extent of damage caused in an unprotected cell depends on several physical factors in addition to sensitivity of the cell. These include dose and quality of the radiation, energy, and rate of delivery. The quality of the radiation can have major implications for the potency (relative biological effectiveness) of the exposure. For example, gamma photons are much more effective in producing performance decrements than are neutrons, but the neutrons have much greater effectiveness against intestinal epithelial cells. Although neutrons cause intestinal damage, gamma photons of similar dose may cause only hemopoietic injury.

With sufficient doses and qualities of radiation, cell injury causes the release of important circulating mediators, cell death, and, in those cells that do survive, delayed effects or subtle changes in cell physiology, which can sensitize against other insults (Fig. 1-6). Gastrointestinal damage occurs at the upper end of the intermediate dose range, but hematopoietic damage occurs at the lower end.

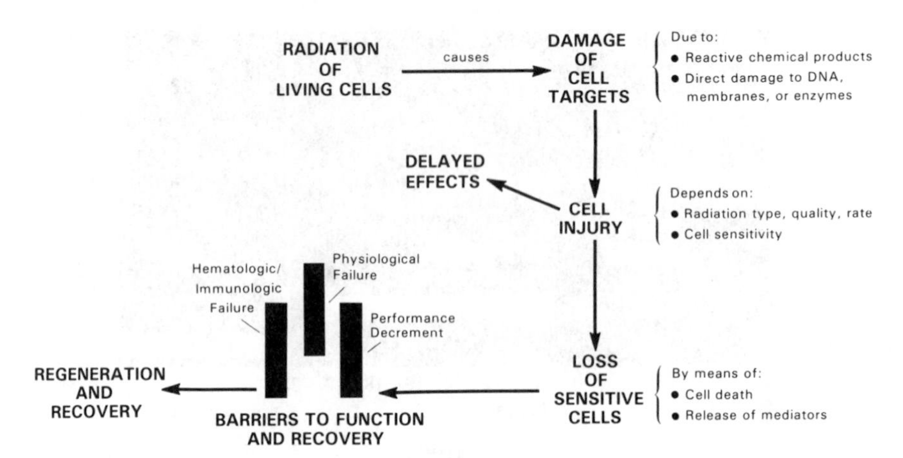

FIG. 1-6 Overview of medical consequences of irradiation.

However, subtle metabolic and physiologic deficiencies may occur in other organ systems. For example, at intermediate doses of radiation, cardiovascular changes are not apparent, but the radiation may cause subtle changes that sensitize the patient to sepsis or certain therapeutic drugs, which also alter cardiovascular physiology.

In summary, acute medical consequences affecting military personnel fall into two major classes: early events affecting performance and later more lethal events associated with single and combined injuries. If cells survive the radiation insult, they have the capability for repair. But the patient must survive fluid loss, infection, and bleeding defects until this can occur. Although no one can ever eliminate the incomprehensible destruction of human life associated with the use of nuclear weapons, significant medical advances can be achieved that will increase the performance and recovery of persons exposed to these weapons. Furthermore, these medical advances will go far toward improving the life and functioning of persons undergoing radiotherapy, trauma, accidental exposures, or a variety of other clinical situations. In the near future, the military battlefield will move into another dimension—space. Once outside the geomagnetic shield of the earth, military personnel will be exposed to a formidable array of new radiations. Among the new radiations will be high solar energy, solar particles and flares, and heavy nuclei from galactic cosmic rays. Associated stresses will be microgravity, vibration, and isolation. To protect man in these new environments will truly challenge our ingenuity.

CHAPTER 2

Physical Principles of Nuclear Weapons

VINCENT L. MCMANAMAN AND ERIC G. DAXON*

Armed Forces Radiobiology Research Institute
Bethesda, Maryland 20814-5145

The energy from nuclear weapons initially resides as potential energy within the nucleus of the atom. The origin of the nuclear energy is unrelated to the sources of the commonly known chemical, mechanical, thermal, etc., energies that derive from the molecular and macroscopic worlds. However, the energy that is released from the nucleus will ultimately be manifest as the familiar chemical, mechanical, thermal, etc., energies. This chapter describes the mechanisms by which this nuclear energy is stored within the nucleus, its release, and finally its conversion to those forces associated with nuclear weapons.

I. FUNDAMENTAL CONCEPTS

Energy is defined as the ability to do work or the storage or release of forces that can move objects. All nonnuclear energies can ultimately be traced to one common source—chemical potential energy. Fire, nonnuclear explosives, lightning, hydraulics, etc., all derive energy from that stored in molecular bonds or electron configurations of the atoms. Once this energy is released, it can be used to generate forces that can move objects.

Commonly used chemical units of energy, such as the erg, the electron volt, the calorie, etc., are too small to conveniently describe the energy associated with bombs and nuclear weapons. The energy equivalent to 1 ton of trinitrotoluene (TNT) (equivalent to 10^9 cal) is the standard unit. Nuclear yields are described in terms of kilotons (kt, thousands of tons of TNT) or megatons (MT,

*Present address: United States Army Environmental Hygiene Agency, Aberdeen Proving Ground, Aberdeen, Maryland 21010.

millions of tons of TNT). The detonation of 1 kt of TNT will generate 10^{12} cal. To release the same amount of energy from the nucleus, only 56 g of uranium-235 must be consumed. This means that pound for pound, nuclear weapons are approximately 10^8 times more energetic (and hence destructive) than conventional explosives.

II. MODEL OF THE NUCLEUS

The neutrons and protons with the nucleus are subject to the well-known forces of gravity and electromagnetism. However, if only these forces are present inside the nucleus, the neutrons and protons would fly apart due to the greater electrostatic repulsive force over the weaker gravitational attractive force. As a result, the presence of a short-range nuclear force of attraction accounts for the observed stability of nuclei. The stability of a nucleus is usually described in terms of how much energy is required to separate it into its constituent neutrons and protons. The more stable nuclei require greater energies to separate the constituent neutrons and protons, and hence are said to be bound tighter or to possess greater binding energies. Those nuclei that are more easily separated possess other binding energies. Nuclei have different binding energies, as shown in Fig. 2-1. In the figure, the vertical axis represents the quotient of the total binding energy (E_B) and the number of neutrons (N) and protons (Z), and the horizontal axis is the total number of neutrons and protons (A). The binding energies of the heaviest and lightest nuclei are lower than the binding energies in the midrange of weights.

A nucleus can transition from a state of lower binding energy to one of higher binding energy by giving up or losing energy. An analogy is the case of a ball in a well that contains different levels (Fig. 2-2).

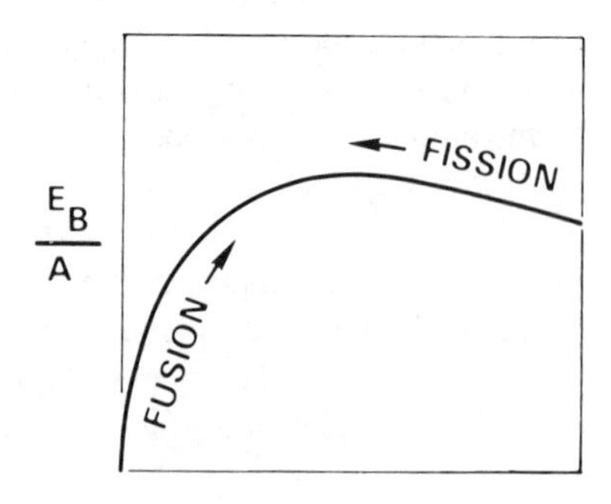

FIG. 2-1 Binding energy curve [$A = Z + N$ (total number of protons and neutrons)].

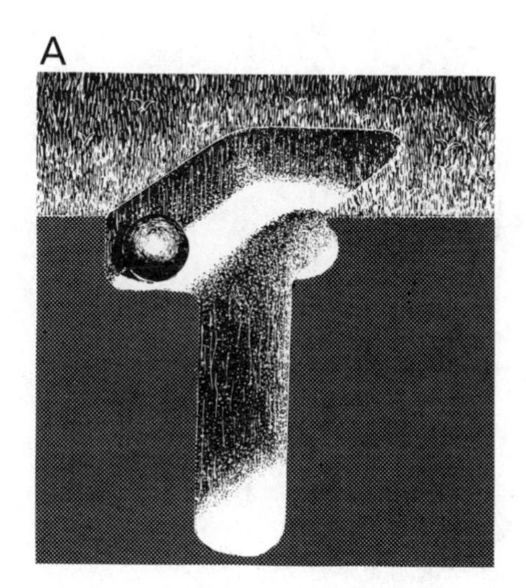

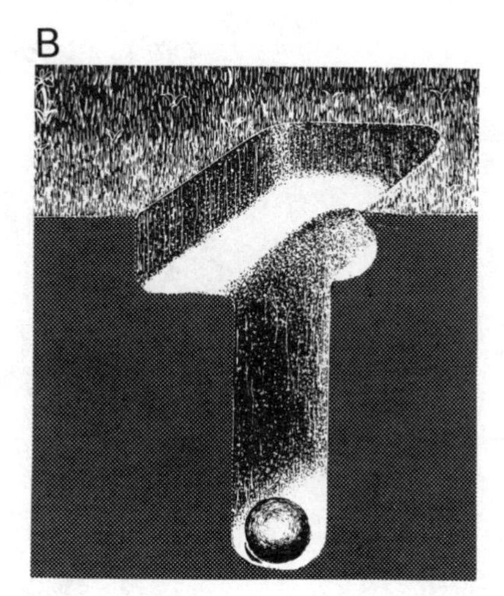

FIG. 2-2 Examples of binding energy.

The ball in Fig. 2-2A possesses a lower binding energy than in Fig. 2-2B because less energy is required to lift it completely out of the well. When the ball does drop to the lower level (thereby becoming more tightly bound and achieving a higher binding energy), it gives up energy that ultimately leaves the system as heat. As a rule, any system (a ball in a well or neutrons–protons in the nucleus) that transitions from a lesser to a greater binding energy must release energy. If the heaviest and lightest nuclei in Fig. 2-1 could be made to transition to nuclei in the midrange, then energy could be released. Indeed, heavy nuclei can be made to split (fission) into lighter nuclei that exist in the midrange of weights, with the release of a large amount of energy. Also light nuclei can be joined (fusion) to form nuclei in the midrange of weights, with the release of energy far greater than that required to cause the fusion.

III. FISSION

Certain isotopes, called fissionable materials, can be induced to fission by the absorption of neutrons. For example

$$^{235}_{92}\mathrm{U} + {}^{1}_{0}n \rightarrow {}^{94}_{38}\mathrm{Sr} + {}^{140}_{54}\mathrm{Xe} + 2\,{}^{1}_{0}n + \text{energy}$$

In this case, uranium-235 absorbs a neutron to form the compound nucleus uranium-236. This unstable compound nucleus then spontaneously splits (fissions) into two nuclei known as fission fragments and an average of two neutrons. It is accompanied by the release of energy.

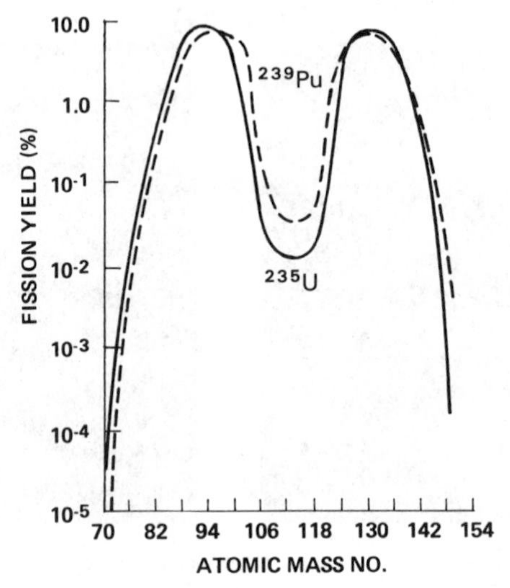

FIG. 2-3 Curve showing the relative fission yield of various isotopes versus their atomic mass numbers.

A competing reaction is

$$^{235}_{92}\mathrm{U} + {}^{1}_{0}n \rightarrow {}^{95}_{42}\mathrm{Mo} + {}^{134}_{50}\mathrm{Sm} + 2{}^{1}_{0}n + \text{energy}$$

with the formation of different fission fragments. Exactly how the uranium-236 nucleus fissions is probabilistic in nature, and because of this, over 400 normally radioactive fission fragments are produced when a fission weapon detonates. Figure 2-3 shows the relative probability of the production of a nucleus with a given total number of neutrons and protons (atomic mass number) from the fissioning of uranium-235 and plutonium-239 subsequent to neutron capture.

Radioactive fission fragments are unstable nuclei and achieve stability usually through the emission of radiations such as gamma rays (electromagnetic) or beta particles (electrons that are created and immediately ejected from the nucleus). The rate at which the radioactive fission fragments become stable is dependent on the specific element and its weight, and is characterized by the term half-life (the time for one half of the unstable nuclei to transition to the next more stable state). Table 2-1 shows some of the more important fission by-products of a fission weapon detonation, their half-lives, and modes of decay.

The energy of the fission reaction is partitioned as shown in Table 2-2. The fission fragments deposit their energy in the weapon itself. The energy ultimately is manifest as heat and a rapid rise in the temperature of the weapon and surrounding matter. It is this rapid temperature increase that gives rise to the blast and thermal effects of a nuclear explosion. A large portion of the prompt (or

TABLE 2-1

SIGNIFICANT FISSION FRAGMENTS

Element	Half-life	Primary mode of decay
^{89}Sr	50.5 days	β and γ
^{90}Sr	27.7 years	β^-
^{91}Y	57.5 days	β and γ
^{90}Y	64.2 hr	β^-
^{95}Zr	65 days	β and γ
^{95}Nb	35 days	β and γ
^{103}Ru	39.8 days	γ and β
^{106}Ru	1 year	β and γ
^{103m}Ru	57 min	γ
^{106}Rh	30 sec	β and γ
^{131}I	8.08 day	β and γ
^{137}Cs	26.6 years	β and γ
^{137m}Ba	2.6 min	γ
^{140}Ba	12.8 day	β and γ
^{140}La	40.22 hr	β and γ
^{141}Ce	33.1 days	β and γ
^{144}Ce	285 days	β and γ
^{143}Pr	13.76 days	β^-
^{144}Pr	17.27 min	β and γ
^{147}Pm	2.64 years	β^-
^{239}Pu	24.36 years	α and γ

TABLE 2-2

AVERAGE ENERGY PARTITION FROM THE FISSION OF URANIUM-235 BY THERMAL NEUTRONS

	Percentage
Kinetic energy of fission fragments	82.0
Prompt gamma rays	2.5
Prompt neutrons	3.5
Other decay products	12.0
Total	100

immediately emitted) gamma rays and neutrons escapes the weapon, and it is responsible for the prompt or initial nuclear radiation emitted when a weapon is detonated.

IV. CHAIN REACTION

If there are more uranium-235 nuclei in the vicinity of a fissioning nucleus, one or more of the prompt neutrons may be absorbed by these nuclei to cause fission, releasing more neutrons and energy. On the average, if fewer than one of the neutrons per fission event causes further fissioning, the process is not sustained and dies out. However, if there are enough uranium-235 nuclei present so that more than one neutron from each fission causes further fissioning, then the process continues. The number of fissions and amount of energy released then increase exponentially with each generation (Fig. 2-4).

The minimum amount of uranium-235 required to sustain such a chain reaction is called a critical mass. For masses less than the critical mass, too many neutrons escape without causing fission to achieve a sustaining reaction. By increasing the mass of uranium-235, a smaller fraction of neutrons escapes, and at the critical mass, a self-sustaining reaction is possible. These situations are depicted in Fig. 2-5. Obviously, mass is not the only factor influencing criticality. As the density of a material increases, interatomic spacing decreases, thus increasing the probability that a neutron traveling a given distance will interact with a nucleus (Fig. 2-6A). Impurities in the fissionable material (Fig. 2-6B) represent nonproductive or nonfissioning absorption centers and thus have

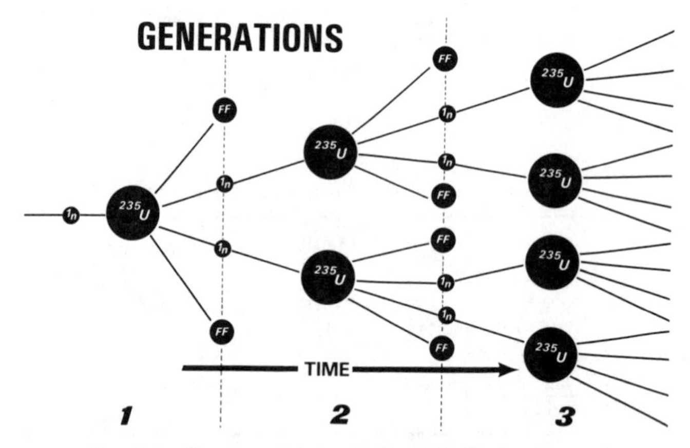

FIG. 2-4 Diagram of chain reaction. FF, Fission fragment.

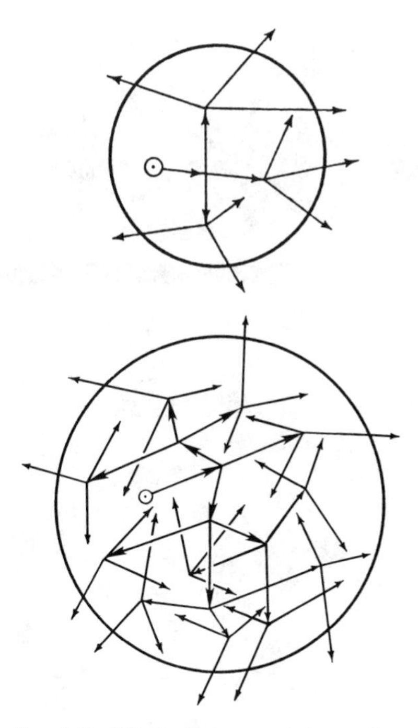

FIG. 2-5 Effect of mass on chain reaction.

the same effect as leakage. Figure 2-6C shows the effect of geometric shape on formation of critical mass. As the surface area of the fuel is decreased proportionally to its volume, there is less opportunity for neutrons to leak from the assembly. As the plane is folded into a cube and the cube is deformed into a sphere, the surface area gets smaller and smaller while the volume remains constant. The average distance that a neutron must travel to escape increases; hence the probability increases that the neuron will be absorbed to cause fission. In addition, if the mass is surrounded by a neutron reflector, criticality is achieved with less mass. There are other fissionable materials (plutonium is a notable example), and each requires a different critical mass to support a chain reaction. A mass smaller than the critical mass is subcritical; one larger than the critical mass is supercritical.

To construct a fission-type nuclear weapon, one starts with a configuration of fissionable material that is subcritical. Upon change of configuration, the material becomes supercritical and subsequently experiences a nuclear detonation. The simplest technique for accomplishing this is the gun assembly method, in which two subcritical masses are driven together at the desired time of detonation to form a supercritical mass. This technique was used in the Hiroshima bomb (Fig. 2-7).

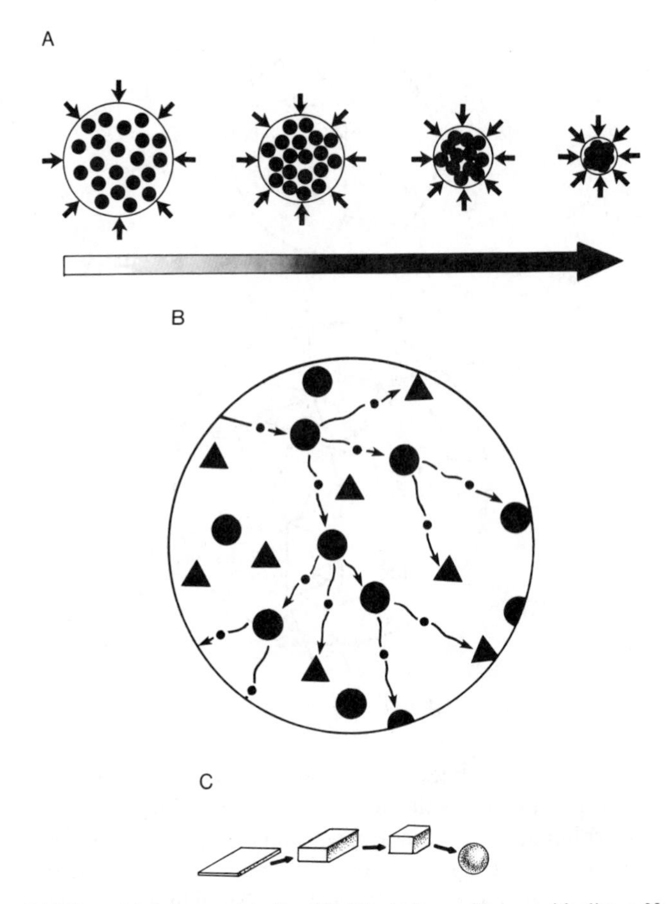

FIG. 2-6 (A) Effect of density on criticality. (B) Effect of impurities on criticality. •, Neutron; ●, uranium-235; ▲, impurity. (C) Effect of geometry on criticality.

Figure 2-8 depicts a more advanced design for a fission weapon. The fissionable material, plutonium, is in the form of a spherical shell surrounded by high explosives. Although subcritical in this configuration, the mass of plutonium becomes supercritical when compressed into a solid sphere upon detonation of the surrounding high explosives. The introduction of neutrons from a neutron source can then trigger a chain reaction and consequent nuclear detonation. This technique, called the implosion method, was used in the first nuclear detonation (Trinity shot at Alamogordo, New Mexico, in July 1945) and in the Nagasaki bomb.

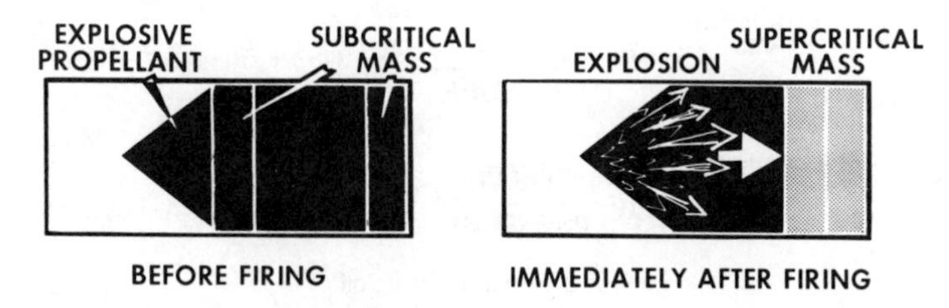

FIG. 2-7 Gun assembly nuclear device.

V. FUSION

The fusion of a deuterium nucleus (deuteron) and a tritium nucleus (triton) to form a helium nucleus, a neutron, and energy is shown in Fig. 2-9. In order for the fusion to take place, the deuteron and triton must first be given a large amount of energy to overcome the mutual electrostatic repulsion. This is achieved by heating the materials to very high temperatures ($\sim 10^8$ K). The pressure is needed to increase the density of the deuterium and tritium so that a significant number of fusions can take place before the weapon blows apart.

The high temperatures needed to ignite the fusion reaction in a thermonuclear weapon are obtained using a fission weapon. Figure 2-10 is a scheme of first a standard fission weapon and then a fusion or fission–fusion device. When the fusion weapon is detonated, the small fission weapon at the center of the device is ignited and the prompt neutrons, which are released, interact with the lithium to form tritium. The tremendous amount of energy that is deposited in the weapon rapidly heats the deuterium and tritium to the 10^8 K required to ignite the fusion reaction.

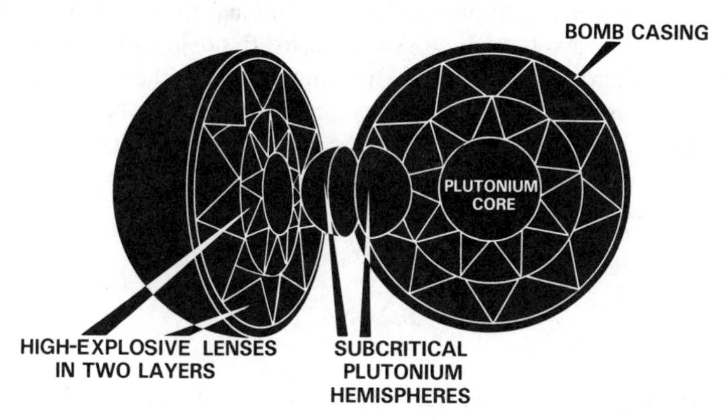

FIG. 2-8 Implosion weapon.

$$^{2}D + {}^{3}T \xrightarrow[\text{PRESSURE}]{\text{HEAT}} {}^{4}He + n^{1} + E$$

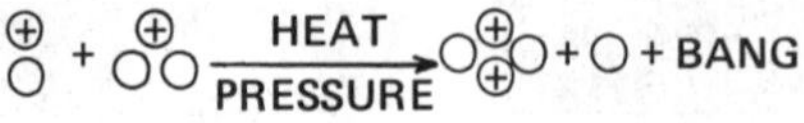

$$\oplus\!\bigcirc + \oplus\!\bigcirc\!\bigcirc \xrightarrow[\text{PRESSURE}]{\text{HEAT}} \bigcirc\!\oplus\!\oplus\!\bigcirc + \bigcirc + \text{BANG}$$

FIG. 2-9 Fusion reaction.

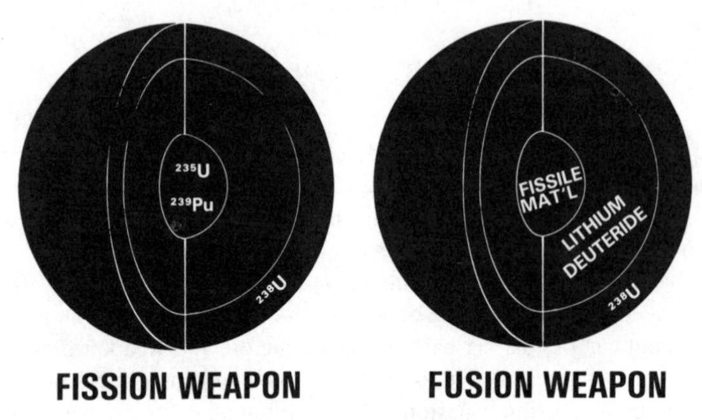

FIG. 2-10 Design of a nuclear weapon.

The uranium-238 tamper that surrounds both the fission and the fission–fusion weapon does several things to increase the yield of the weapon. First, it acts as a neutron reflector. Uranium-238 has a relatively high scattering cross section for neutrons, and it will scatter a significant number of the prompt neutrons back into the weapon, where the neutrons can cause additional fissioning. Further, uranium-238 has a relatively high cross section for fissioning when it absorbs an energetic fast neutron. This additional fissioning increases the yield of the weapon. Because of its high moment of inertia, the uranium-238 tamper tends to hold the weapon together longer. The longer the weapon is held together, the greater the number of fissions and fusions that will take place. The first thermonuclear weapon was detonated in November 1952 at Enewetak.

VI. DETONATION

The vast energy released by a nuclear detonation is transferred to the surrounding environment by four mechanisms: thermal radiation, blast, nuclear radiation, and electromagnetic pulse (EMP). The first three can cause massive casualties

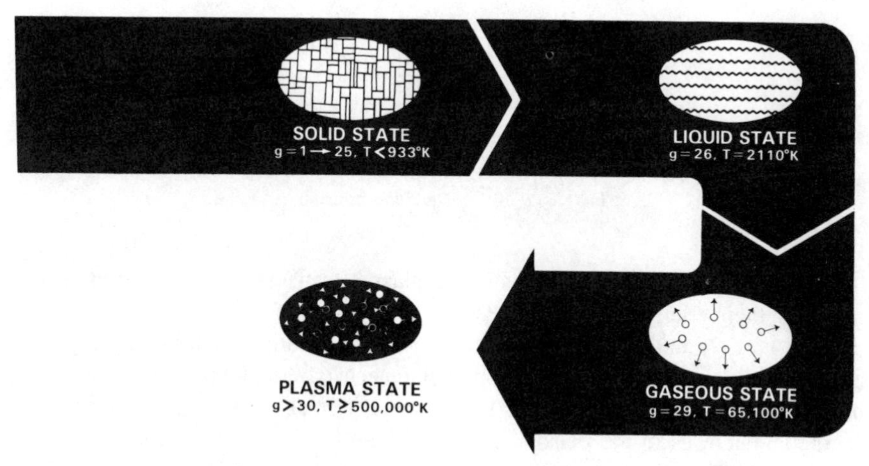

FIG. 2-11 Sequence of states in fissioning device. *g*, Generation; *T*, temperature.

and extensive damage to material and equipment. Electromagnetic pulse primarily affects electronics and communications equipment. Although each will be discussed in turn, their origins are interrelated and intimately connected to the physical phenomena taking place immediately after detonation. We will now examine the events taking place very early in the process of a nuclear explosion. For purposes of illustration, we will consider a 1-MT weapon, keeping in mind that the physics remains essentially the same for all nuclear detonations.

The sequence of events depicted in Fig. 2-11 takes place in a fraction of a microsecond in a typical fissioning device. The generation numbers and temperatures given are representative; in an actual device, they depend on the material and detailed construction of the weapon. Initially in a solid state, the bomb material remains in a solid state during the first 25 generations of the fission chain reaction, while the temperature increases roughly exponentially. As we have seen, more than 80% of the fission energy is in the form of kinetic energy of the fission fragments. These fission fragments collide with surrounding nuclei, transferring energy and thereby increasing the temperature of the material. During the next five generations, the bomb material passes through the liquid and gaseous phases as the temperature rises rapidly. After about 30 generations, the material exists in a plasma state, with atomic electrons completely stripped from their nuclei. Charged particles are flying about randomly at very high speed. These changes have occurred so rapidly that the material does not have time to expand significantly, leaving the volume and material density essentially unchanged. The longer this material is held together, the greater the number of generations of chain reaction that will occur, resulting in increased yield of the weapon.

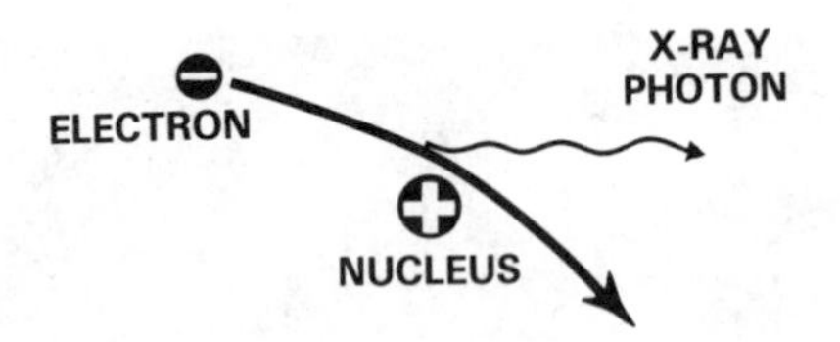

FIG. 2-12 Bremsstrahlung process.

The energetic, charged particles in the plasma, especially the electrons, lose much of their energy by the process of bremsstrahlung (Fig. 2-12). When a charged particle undergoes an acceleration or deceleration such as that experienced by an electron being acted upon by the electric field of a positively charged nucleus, it emits an X-ray photon. Due to the extremely large number of highly energetic electrons in the plasma, the magnitude of X radiation emitted in this manner is immense. These X rays travel outward at the speed of light, interact with and ionize the surrounding air, and heat it to millions of degrees. This layer of high-temperature air, called the radiation front, forms the initial surface of the fireball. As seen in Fig. 2-13, the radiation front develops rapidly in a fraction of a microsecond. The rate of expansion then drops suddenly when the energy from the nuclear reaction stops. During the next few microseconds, the hot air at the radiation front reradiates very rapidly into the undisturbed cold air, and the fireball grows by this process, called radiation diffusion.

Meanwhile, back at the point of detonation, the dense bomb material, in a plasma state, starts expanding at supersonic speed. This expansion compresses

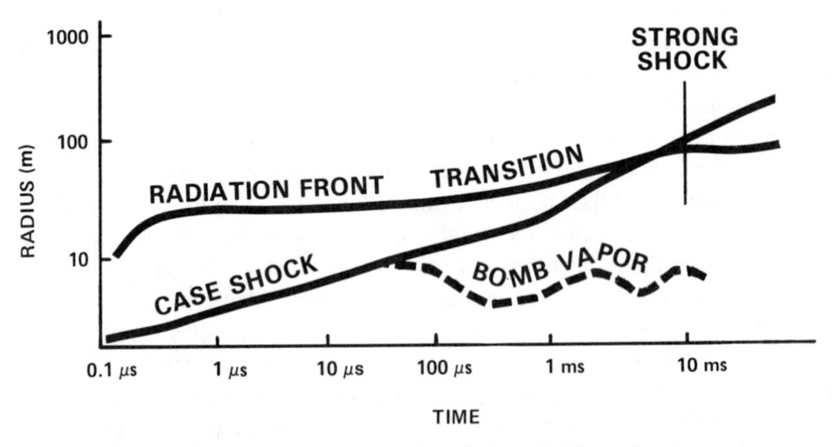

FIG. 2-13 Space–time plot of early fireball phase (1-MT, sea-level airburst).

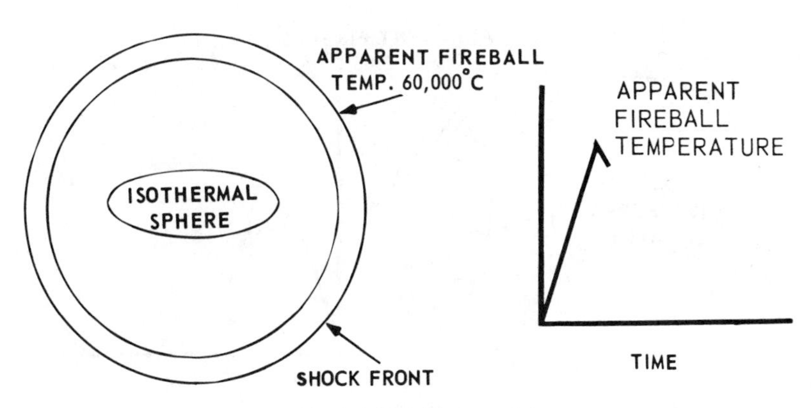

FIG. 2-14 A 1-MT airburst at 1.4 msec.

the surrounding air (which has already been ionized by X rays) and generates a shock wave within the fireball, called the case shock. As the case shock propagates rapidly outward within the fireball, it radiates thermal energy outward to the radiation front, causing the energy density (energy per volume) at the front to increase greatly. Consequently, at about 1 msec after detonation for a 1-MT burst, this fireball front itself starts to expand hydrodynamically (or materially). This generates a shock wave in the undisturbed air that moves supersonically outward away from the now slowly expanding radiation front. This outer, luminous shock front then forms the apparent edge of the fireball. At 1.4 msec after detonation (H + 1.4 msec) (Fig. 2-14), the fireball consists of two parts: the inner isothermal (or uniform temperature) sphere (sometimes called the true fireball) at a temperature of about 400,000°C, and the outer shock front at approximately 60,000°C. The shock front is opaque to the thermal radiation from the isothermal sphere, trapping it in the interior of the fireball. Therefore, the apparent fireball temperature is the temperature of the outer surface.

By H + 60 msec (Fig. 2-15), the shock front has traveled outward at supersonic speed, expanding and cooling to a temperature of about 2500°C. The isothermal sphere has also expanded and cooled, but remains much hotter than the shock front. When the expanding shock front cools below a few thousand degrees centigrade, it loses its luminosity, becoming transparent to the thermal radiation of the isothermal sphere.

As this thermal radiation escapes from the much hotter interior, the apparent fireball temperature rises to a new peak at about H + 1 sec. This phenomenon, known as the increasing optical depth concept, is illustrated in Fig. 2-16. As the isothermal sphere expands further and cools, the apparent temperature again decreases. By this time the transparent shock wave has broken away from the visible fireball and is traveling radially outward at supersonic speed.

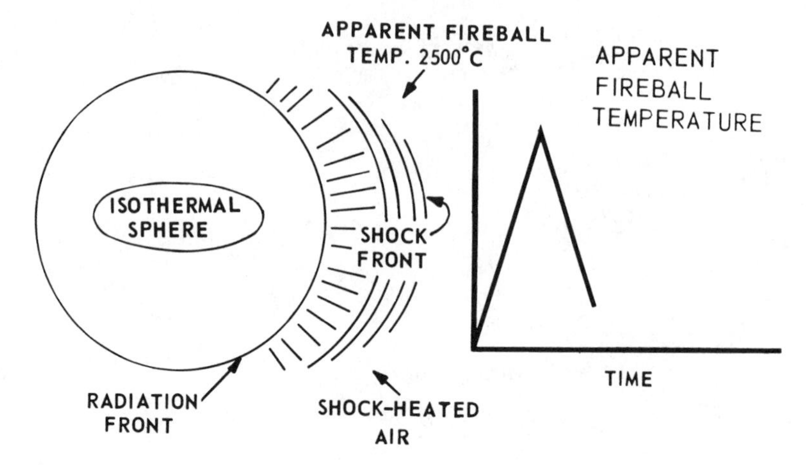

FIG. 2-15 A 1-MT airburst at 60 msec.

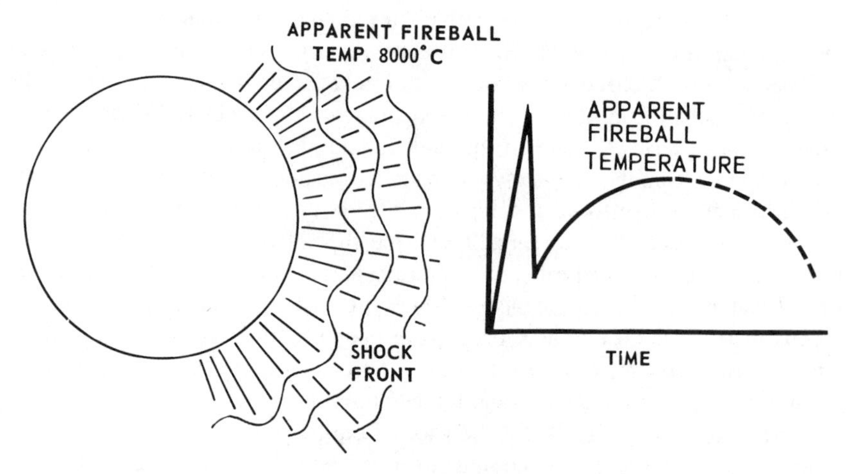

FIG. 2-16 A 1-MT airburst at 1 sec.

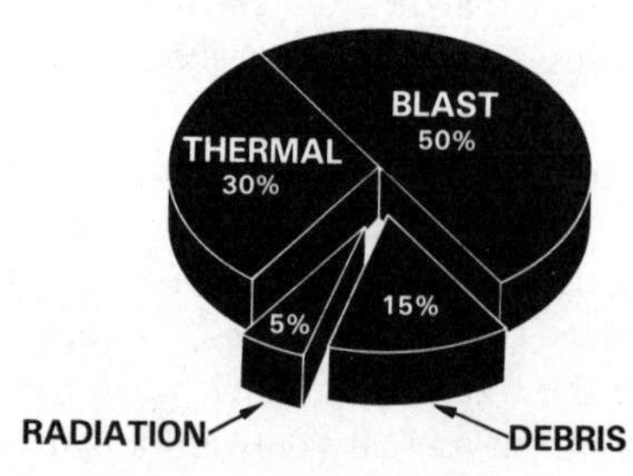

FIG. 2-17 Energy partition of a weapon detonated in the atmosphere.

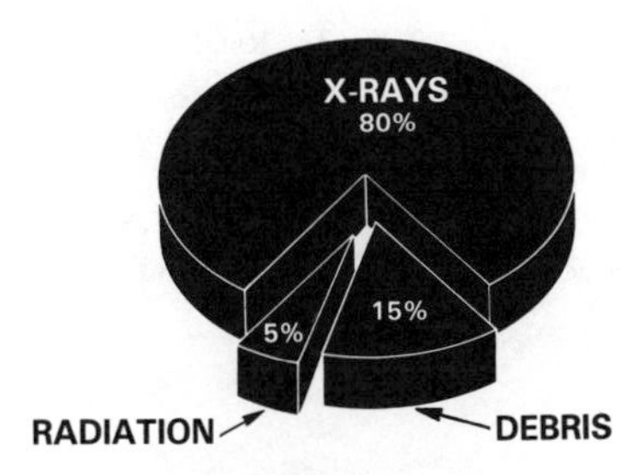

FIG. 2-18 Energy partition of a weapon detonated outside the atmosphere.

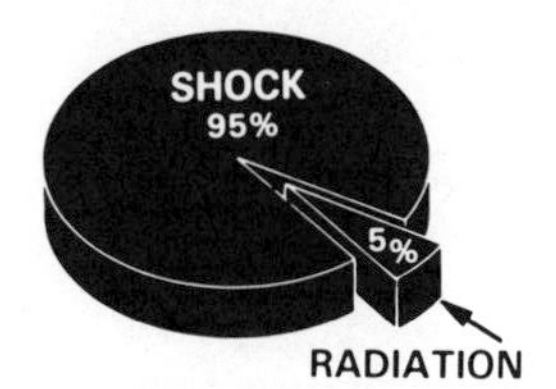

FIG. 2-19 Energy partition of a weapon detonated underground.

VII. PARTITION OF ENERGY

The height or depth at which the nuclear detonation takes place greatly affects the distribution of energy into the various channels of destruction. These channels are thermal radiation, blast, and nuclear radiation. In an endoatmospheric burst (a burst within the atmosphere), the partition of energy (seen in Fig. 2-17) is typical.

In a high-altitude burst, the lower atmospheric density will allow a smaller fraction of the energy to be carried away via blast. In fact, a burst outside the atmosphere (exoatmospheric) (Fig. 2-18) has no blast component since the majority of the energy is given off as (unabsorbed) X rays. On the other hand, in a subsurface burst (Fig. 2-19), the thermal component cannot be emitted due to the opacity of the surrounding material. Here the shock wave, in the form of ground shock, carries away 95% of the total energy. In an underwater detonation, the distribution of energy depends on the depth of the burst. A host of new phenomena (such as the formation of a radioactive "mist" or base surge and the generation of large-amplitude waves) are characteristic of shallow, underwater nuclear explosions.

SUGGESTED READINGS

Blanchard, C. H., *et al. Introduction to Modern Physics*. Prentice-Hall, Englewood Cliffs, New Jersey, 1969.

Brode, H. L. *Review of Nuclear Weapons Effects.* The Rand Corporation, Santa Monica, California, 1968.

Glasstone, S., ed. *The Effects of Nuclear Weapons.* Department of Defense and Department of Energy, Washington, D.C., 1977.

Lewis, K. N. The prompt and delayed effects of nuclear war. *Sci. Am.* **241,** 35–47 (1979).

York, H. F. The great test-ban debate. *Sci. Am.* **277,** 15–23 (1972).

CHAPTER 3

Nuclear Weapons Fallout

KENNETH P. FERLIC* AND VINCENT L. McMANAMAN

Armed Forces Radiobiology Research Institute
Bethesda, Maryland 20814-5145

The detonation of nuclear weapons usually generates vast quantities of atmospheric radioactive material (fallout) that ultimately falls out and is deposited over the surface of the earth. This chapter addresses the sources of fallout, the major factors that affect fallout, fallout patterns, radiological effects of fallout, and the management of fallout situations.

I. ORIGINS OF FALLOUT

A. General Description

The term "fallout" originated with the detonation of the first nuclear device. It was the Trinity shot at Alamogordo, New Mexico, on July 16, 1945 (Green, 1965). The term referred to the fact that the material "falling out" of the mushroom cloud produced by the explosion was radioactive. The process is one in which the radioactively contaminated dust and debris (which have been airborne and transported by the radioactive cloud) fall back to earth.

B. Formation of Fallout

Whenever a nuclear device is detonated, a fireball is produced. The fireball is extremely hot (~300,000°C), and all the atoms contained in the fireball (fuel, fission products, etc.) exist either in the gas or plasma state. Because of its heat, the fireball expands, engulfing and vaporizing the surrounding environment, and

*Present address: Division of Reactor Projects, Region I, United States Nuclear Regulatory Commission, King of Prussia, Pennsylvania 19406–1498.

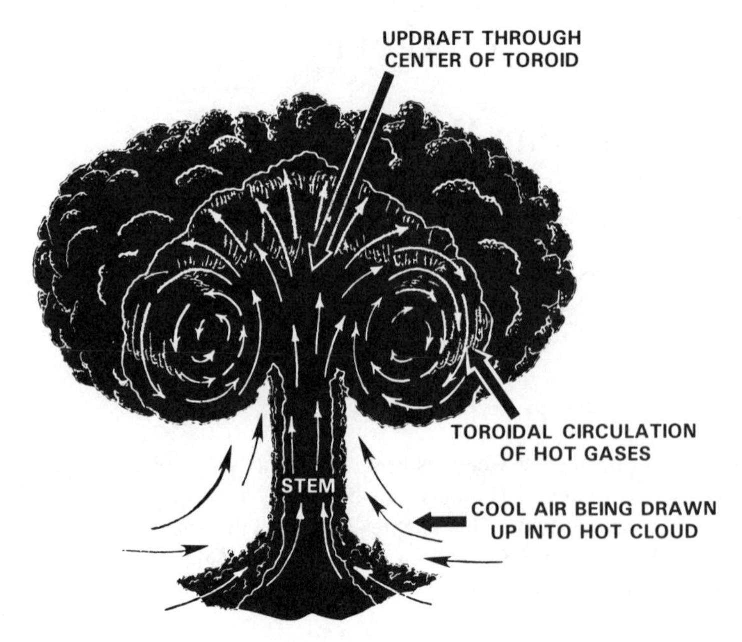

FIG. 3-1 Cloud formation. Toroidal circulation within the radiactive cloud from a nuclear explosion.

begins to rise at great speeds. As it rises, it creates a vacuum, which results in a tremendous updraft. As the surrounding atmosphere tries to fill the void created by the updraft, the afterwinds are produced. If the fireball is close enough to the ground, the strong updrafts result in the typical mushroom cloud (Fig. 3-1). If the detonation is high enough, the updrafts may not disturb the ground, and no stem or chimney may be seen. Any structure or other material near the fireball will be either engulfed or sucked up in the updraft. If the fireball touches the ground, the heat will vaporize that part of the surface, forming a crater (Fig. 3-2). The vaporized earth and ground will then be carried up into the cloud to whatever height the cloud ultimately obtains due to the upward movement of the updraft. The size of the crater depends on the height of the detonation, energy yield of the weapon, and nature of the soil.

If the prompt neutrons released during the fission process strike the ground, then an area of induced radioactivity is produced (Fig. 3-3). Depending on height of the fireball and strength of the updrafts, the radioactive ground material may also be taken up into the mushroom cloud, thereby increasing the total inventory of the radioactive materials in the cloud.

As the hot fireball and incorporated dirt and debris rise to higher altitudes, the vapors cool and condense to form a cloud containing solid particles. The cloud

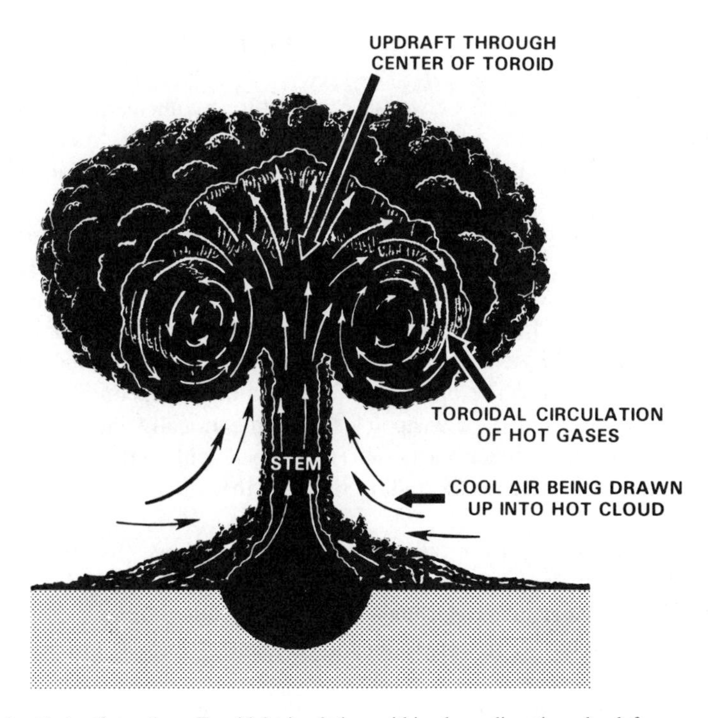

FIG. 3-2 Crater formation. Toroidal circulation within the radioactive cloud from a nuclear explosion.

reaches its maximum height in about 10 min, but continues to grow laterally, producing the mushroom shape (Glasston and Dolan, 1977). The fallout particles (ranging in size from 1 cm to less than 0.002 cm) are produced during this cooling and condensation phase. The radioactive residue in the cloud condenses and fuses with the earth particles in the cloud.

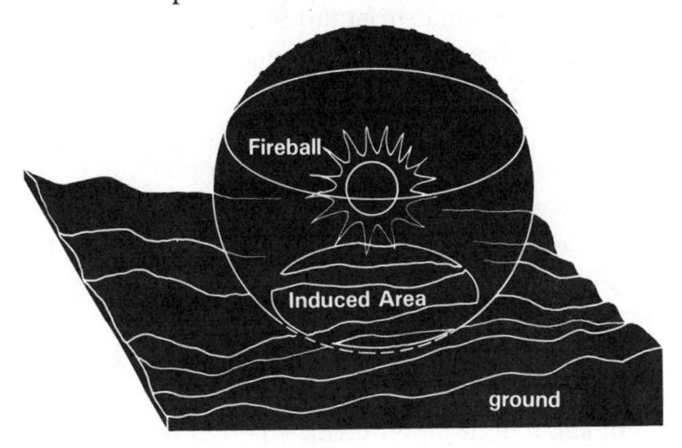

FIG. 3-3 Formation of an induced area.

C. Sources of Fallout Radioactivity

Some residual radiation will always be present in any nuclear detonation. In a standard fission weapon, residual radiation consists primarily of fission products, unspent fuel, and materials of induced radioactivity. When describing fusion weapons, the terms "clean" and "dirty" are often used to compare their radioactivity relative to that produced in an equivalent "normal" fusion weapon (approximately 50% fission yield) (Green, 1965). A normal fusion weapon is usually cleaner than a fission weapon of equivalent yield. A clean weapon is designed to yield less radioactivity than a normal fusion weapon. Only a small percentage of the total yield of a clean weapon comes from fissionable material (uranium or plutonium). The opposite is true of a dirty weapon.

The radioactive yield of a weapon can be intentionally increased through "salting," a process whereby a bomb is designed with certain elements that would substantially increase the radioactivity of the bomb debris produced on detonation.

In addition to the sources of radioactivity identified above (fission fragments, bomb materials of induced radioactivity, unspent fuel, and salting), ground elements of induced radioactivity may also enter the fallout cloud. Of these five sources of radioactivity, fission fragments are the greatest contributor to the fallout problem.

We may recall the fission process by examining two typical fission reactions

$$^{235}_{92}\mathrm{U} + {}^{1}_{0}n \rightarrow {}^{94}_{38}\mathrm{Sr} + {}^{140}_{54}\mathrm{Xe} + 2\,{}^{1}_{0}n + \mathrm{E}$$

$$^{235}_{92}\mathrm{U} + {}^{1}_{0}n \rightarrow {}^{94}_{42}\mathrm{Mo} + {}^{139}_{50}\mathrm{Sn} + 2\,{}^{1}_{0}n + \mathrm{E}$$

Exactly how the uranium nucleus fissions is probabilistic in nature, and because of this, over 400 normally radioactive fission fragments are produced when a fission weapon detonates (see Fig. 3-4). The most significant long-term fallout hazards arise from cesium-137 and strontium-90, both of which have an atomic mass that falls near the fission yield peaks, and consequently they and their precursors have a high probability of production.

Most of the fission fragments are highly unstable and decay very quickly. Most are beta-particle emitters, and many will also produce gamma radiation. Each fission fragment frequently undergoes three or four decays before stability occurs. Since each of these isotopes has a different half-life and since the many elements have different chemical properties (in particular, vaporization temperature), the resulting fallout problem is very complex. The actual fallout problem in a given area is determined greatly by how and when the various elements and isotopes return to earth. For example, the longer lived radioisotopes that last for many years become a long-term problem, whereas iodine-131 is an immediate problem because of its relatively short half-life (8 days).

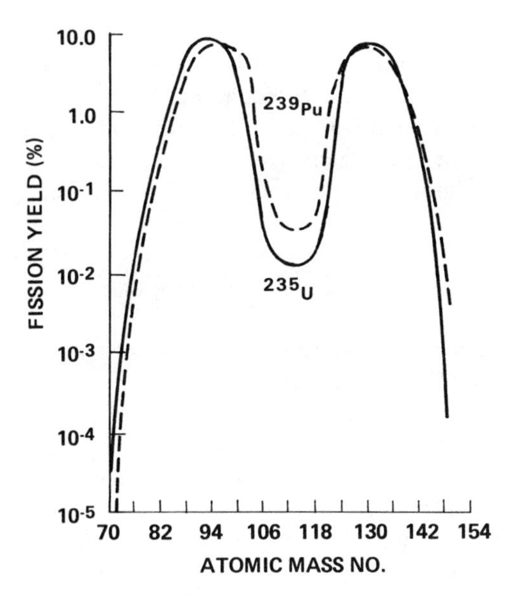

FIG. 3-4 Curve showing the relative fission yield of various isotopes versus their atomic mass numbers.

II. DESIGN AND DEPLOYMENT OF WEAPONS RELATED TO FALLOUT

A. AIRBURST

An airburst is a detonation in which the fireball does not touch the ground. As a result, no vaporization of the ground surface occurs. The fallout problem depends on the height of the burst and the nature of the terrain, because strong updrafts could occur as the fireball rises. Depending on the strength of these updrafts and afterwinds, varying amounts of dirt and debris can be taken up into the cloud. The full effect on the subsequent fallout problem depends on two factors: (1) the amount of dirt and debris carried into the fireball and (2) whether good mixing occurs when the fission fragments are still vaporized. Obviously, the closer the detonation is to the ground, the greater is its potential for hazardous early fallout. In general, fallout particles from an airburst (excluding a low airburst) are very small ($0.1–2 \times 10^{-5}$ m) (Glasston and Dolan, 1977) since the cloud does not contain large quantities of dirt and debris. The small particles can go to very high altitudes and, in the absence of snow or rain, early fallout is generally not significant (Glasston and Dolan, 1977). These particles remain airborne for long periods of time and decrease in overall activity through decay.

The particles become widely distributed, which reduces their concentration. The primary radiological hazard from an airburst is the long-term delayed fallout. Also, a radiological hazard may occur in the vicinity of ground zero as a result of neutron activation of ground materials. No early fallout occurred at Hiroshima and Nagasaki because both detonations were airbursts at 1670 and 1640 ft, respectively. These altitudes maximized the blast effects of the weapons.

B. Low Airburst and Near-Surface Burst

A low airburst and a near-surface burst are intermediate between the "ideal" airburst and the "ideal" surface burst. Although they may be strictly classified into one of the categories (e.g., the fireball in a low airburst would not touch the ground), other conditions may cause the fallout problem of one to resemble the fallout of the other. For example, a low airburst over the right surface with strong enough updrafts produces an unusual amount of early fallout because of good mixing between the fallout particles/vapors and the sucked-up dirt and debris. It would produce a fallout problem similar to that of a near-surface burst.

C. Surface Burst

A surface burst is a detonation in which the fireball touches the earth and causes the surface to vaporize. In addition, very strong afterwinds cause large amounts of dust, dirt, and surface debris to be sucked up into the fireball at a very early stage. A high degree of mixing occurs in the early stage of the fireball and cloud growth. As the fission products condense, they become fused with the foreign matter sucked up into the cloud. Highly radioactive, large particles (2×10^{-5} m to 1 cm) are formed (Glasston and Dolan, 1977). Larger particles (greater than 1 cm), formed later with larger unvaporized or incompletely vaporized foreign material, return very quickly to earth near ground zero. The final composition of the cloud depends on the nature of the surface materials and the extent to which the fireball contacted the surface. The early fallout potential from a surface burst is extremely high (40–70%) because of the large particles (Glasston and Dolan, 1977). A 60% early fallout fraction is normally assumed (Glasston and Dolan, 1977). The delayed fallout fraction that depends largely on conditions is a corresponding 30–60% of the total fallout problem.

It is of interest to note that a water surface burst results in almost the reverse fallout problem, i.e., early fallout is 20–30% and delayed fallout is 70–80% (Glasston and Dolan, 1977). This reversal of the early fallout and delayed fallout is due to the size of particles. In water surface bursts, development of the fireball is the same as in bursts over land, except that sea salts and water are sucked into the cloud. Because the vaporization point of water is 212°F (100°C), the fallout particles form only after the cloud has cooled substantially, resulting in very small particles. These small particles remain airborne for long periods of time, thereby causing a problem of long-term fallout.

D. Subsurface Burst and Deep Underground Burst

A true subsurface burst has no venting and so produces no surface contamination. In these cases the entire explosive energy yield results in shock and radiation, and any radioactivity formed is confined to the region near the point of detonation (except for some possibly escaping gases). On land, a subsurface crater may be formed, but it is not radioactive like the crater formed by a surface burst.

E. Shallow Subsurface Burst and Underground Burst

A shallow, subsurface land burst is one in which the fireball does not emerge from the ground; it would probably be used for cratering. For example, this device could be buried in a narrow mountain pass at a depth that would maximize the size of the resulting crater in the mountain valley. This would prevent the opposing forces from advancing, retreating, or delaying their movement. Since contact with the earth is present early in the formation of the underground equivalent of the fireball, a high degree of mixing of materials occurs. High levels of contamination will be found in the crater and surrounding areas. Depending on meteorological conditions, the small particles may remain suspended for some time and descend at great distances from the burst point. As the depth of the detonation becomes shallower and the fireball is formed and emerges from the surface, the effects become increasingly like those of a true surface burst.

Another important type of shallow subsurface burst is a shallow-water, subsurface burst, which could occur near a port facility or invasion beachhead. Several important facts were learned during the Baker shot [20 kt (kilotons)] at Bikini in 1946. (Bikini Baker was a test designed to observe the effects of a nuclear detonation on surface ships anchored near Bikini atoll.) Although the water was contaminated in the area surrounding the detonation, the attenuation capabilities of the ships would have offered sufficient protection to crew members to traverse the contaminated area while receiving only a small dose (Glasston and Dolan, 1977). The only significant problem was the fallout, which consisted of both solid particles and a slurry of sea salt crystals in drops of water. This contamination was difficult to dislodge from the ships used in the test (Glasston and Dolan, 1977). If there had been any personnel on these ships, they would have been exposed to a considerable dose of radiation unless the fallout could have been removed immediately (Glasston and Dolan, 1977). As a result, the United States Navy instituted a water washdown of ships (Glasston and Dolan, 1977). This test and subsequent tests indicated that there is no simple system of predicting residual radiation for underwater bursts, as there is for land surface bursts (Glasston and Dolan, 1977).

III. YIELD AND ATMOSPHERIC CONDITIONS

After considering the formation of fallout particles and their insertion into the atmosphere, we need to evaluate the conditions that control the return of these particles to earth. In general, the controlling factors are the type of particles formed, the yield of the original weapon, and the atmospheric conditions in the region of detonation. In any detonation that produces a mushroom cloud, about 90% of the contaminants are in the upper portion or head of the cloud, and the remaining 10% are in the stem (Fig. 3-5) (Office of Civil Defense, 1968). The particles in the cloud will fall because of gravity. Their rate of fall depends on their aerodynamic characteristics (size, weight, shape, etc.) and on the particular characteristics of the air they are in. As stated previously, particles ranging in size from 0.01 to 1 cm in diameter normally arrive within 1 day, after traveling up to a few hundred miles (Glasston and Dolan, 1977). Delayed fallout varies in size from very fine sand to fine sand (0.01 cm) (Glasston and Dolan, 1977). Very fine sandlike particles are reasonably stable aerodynamically, and their rate of descent is determined by the air structure where the detonation occurs. Larger particles have a more unpredictable rate of descent. Although fallout sometimes occurs when the cloud cannot be seen, it is the larger, more unpredictable visible particles that present the most serious radiological hazard. They range in size from that of fine sand (0.01 cm) to that of marbles (1 cm) (Glasston and Dolan, 1977).

Generally, the larger, heavier particles leave the fallout cloud early and deposit in a region closer to the burst point. In a near-surface burst producing many large particles, the particles deposit early, producing a highly concentrated radioactive fallout pattern. On the other hand, if an airburst occurs, the particles are smaller and tend to settle at larger distances, thereby diluting the radioactivity by spreading it over a larger area. Differing wind strength causes a similar result. A strong wind will cause the particles to travel greater distances. A stronger wind will spread the initial concentration over a larger area, thereby reducing its concentration and the subsequent radiological hazard. The height of the cloud has a diluting effect on the concentrations of radioactivity in the fallout pattern. Since the higher particles take longer to fall, they can travel greater distances,

FIG. 3-5 Distribution of fallout in a mushroom cloud.

covering more area and thereby diluting the radiological hazard in the fallout pattern. In addition, the longer the particles take to fall, the more they decay and also reduce the final radiological hazard in the fallout field.

Figure 3-6 illustrates a complicating factor in predicting fallout. The variability of wind direction could completely reverse the expected fallout pattern. Since the speed and direction of winds vary with height in the atmosphere, the actual height of the cloud is a very important factor. In general, United States winds increase with altitude from the surface up to 30,000–40,000 ft. Above 30,000–40,000 ft, the winds decrease; at about 60,000–80,000 ft, they are relatively light. Therefore, the strongest winds at 40,000 ft would determine the general direction and length of a fallout area (Glasston and Dolan, 1977; Office of Civil Defense, 1968).

Rain also affects fallout distribution. Localized "hot spots" may be formed in essentially unpredictable patterns. Rain has a cleansing effect on deposited fallout on higher ground and elevated structures, but it causes increased concentrations of fallout in lower areas. Rain can also wash some of the fallout into the soil, which then acts as a radiation attenuator or absorber, thereby reducing some of the radiological problem.

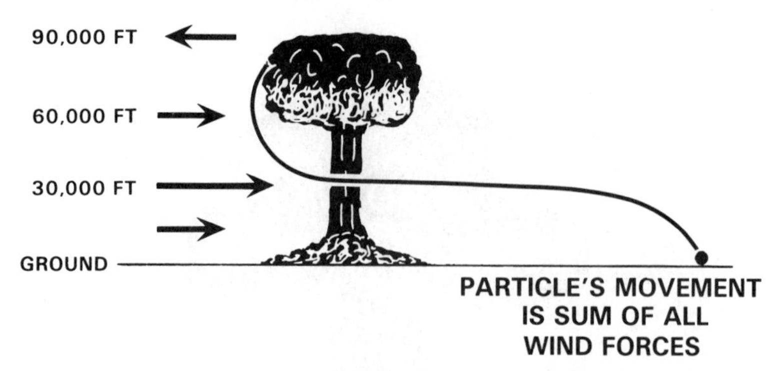

FIG. 3-6 Effect of variable wind on deposit of fallout.

IV. FALLOUT PATTERNS

In the previous sections the sources of fallout were identified, and the influence of the atmosphere on particle descent was discussed. This section addresses the actual deposition of fallout and the associated levels of external radiological hazard.

Since fallout gradually descends over a period of time, it is beneficial to look at a time sequence with ideal fallout patterns. Figure 3-7 (Glasston and Dolan, 1977) shows the dose rate contours from early fallout at 1, 6, and 18 hr postdetonation for a 2-MT (megaton) surface burst with a 1-MT fission yield and a 15 mph effective wind speed. As seen, a person 20 miles downwind from the explosion would find himself in a 3-rad/hr fallout field at 1 hr. This person would find the dose rate continuing to increase, so that at 6 hr after the burst, the dose rate 20 miles downwind would be over 100 rad/hr for this person. The dose rate would reach a maximum, either before or after 6 hr. It is at this time that the fallout has stopped at his location. Before this time of maximum dose rate, the fallout was not complete; so as accumulation of fallout increased, so did the dose rate. It is after this time of maximum dose rate that decay (according to the natural decay of the fission products) will be observable and dose rates will decrease. At 18 hr the dose rates will have decayed to near 30 rad/hr at 20 miles directly downwind from the detonation.

In addition to the dose rates, one can use the ideal fallout pattern to assess the total dose delivered during the time of deposition (or infinite dose). Figure 3-8 (Glasston and Dolan, 1977) uses the same detonation and time frame, but the total doses delivered at 1, 6, and 18 hr are given. As seen in Fig. 3-8, if a person at 20 miles directly downwind did not enter a fallout shelter, he would have received over 1000 rad at 18 hr. Although it is understood that the ideal fallout

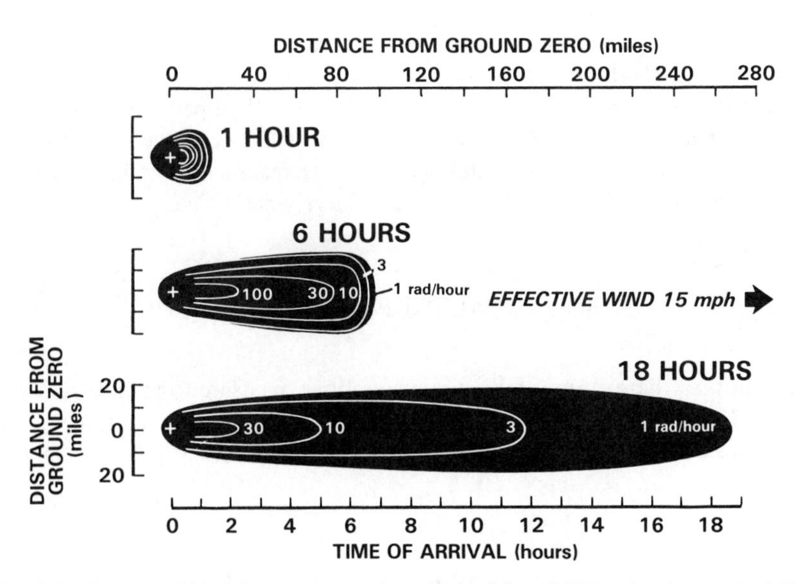

FIG. 3-7 Contours of ideal dose rate versus time of deposit for a 2-MT surface burst and a 1-MT fission yield.

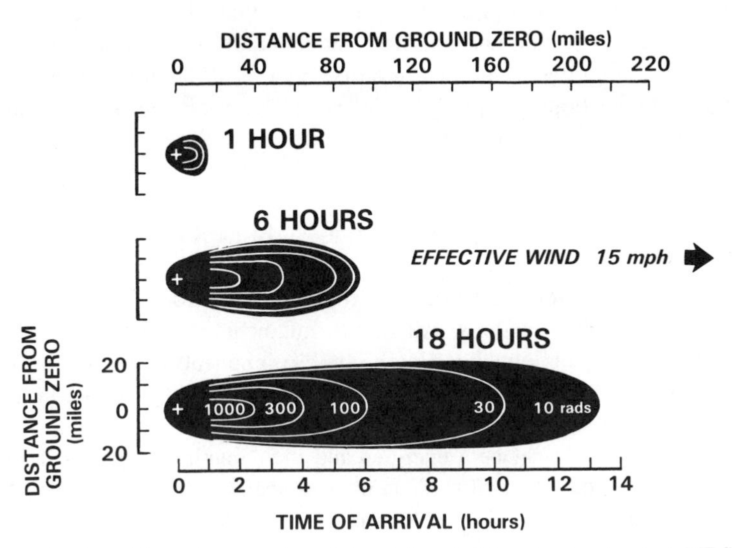

FIG. 3-8 Ideal total dose rate versus time of deposit for a 2-MT surface burst and a 1-MT fission yield.

pattern provides only average values, it does point out those regions in which fallout shelters should be seriously considered.

Other salient aspects derived from an analysis of fallout patterns include (1) a large percentage of the total dose possible will be delivered in the first 24 hr, (2) after the first few days, dose rates will have decreased substantially, and (3) fallout shelters will be needed for only a relatively short time.

V. RADIOLOGICAL PROPERTIES OF FALLOUT

The primary hazard from fallout is whole-body exposure to gamma radiation (Green, 1965). In addition, alpha and beta radiation will be present in the fallout field. These radiations present both external and internal hazards to man, and each will be addressed separately.

A. Protection from External Alpha Radiation

An alpha particle is a combination of two neutrons and two protons (a helium nucleus), which is emitted from a heavy nucleus at very great speeds. The range of an alpha particle in tissue is less than the thickness of the outer dead layer of the skin. Consequently, the alpha particles are readily stopped by the skin and pose no external radiation problem in the fallout field. However, care must be taken to prevent ingestion and internal contamination of alpha-particle emitters since they pose significant internal hazards. As a result the primary concern for external contamination by alpha particles is to keep it off the skin, clothing, or wounds to prevent incorporation into the body.

B. Protection from External Beta Radiation

Beta particles are very high-speed electrons emitted from the nucleus during certain nuclear transformations. All fission fragments are beta emitters, as are most of the induced radionuclides (soil, bomb components, etc.). All beta particles emitted during radioactive decay are not monoenergetic but are emitted in the form of a continuous energy spectrum.

Table 3-1 gives typical beta energies and their maximum ranges in air and tissue. In general, the range of the beta in air is about 12 ft (365 cm) per MeV of energy. Generally, 0.5 in. (1.27 cm) of most materials will stop all but the most energetic betas, and usually less than 0.5 in. will be required. The actual thickness of shielding that is required will decrease proportionally with the increase of a material's density.

TABLE 3-1

APPROXIMATE BETA RANGES IN AIR AND TISSUE

Energy (MeV)	Air (cm)	Tissue (cm)
0.1	13	0.14
0.5	155	0.18
1.0	380	0.46
2.0	840	0.96
3.0	1300	1.47

Although beta particles are not capable of producing external whole-body exposures, they can deposit very intense local irradiations. As seen in Table 3-1, the 0.1-MeV beta particle will penetrate well beyond the protective layer of the skin, which is 0.07 mm thick. The average maximum energy of the beta particles from the fission products is 1.2 MeV, and the absolute maximum energy probable can exceed 3 MeV (although rarely). Consequently, if the fallout particles with their fused fission fragments are allowed to remain in contact with the skin (or very near the skin separated by a very thin material), serious "beta burns" could occur. If a sizable fraction of the body should suffer serious skin damage from the beta radiation, the results would be similar to those from thermal burns, i.e., serious injury or death. The primary concerns from beta radiation in the fallout field will be twofold: (1) the prevention of beta burns and (2) the prevention of ingestion and internal contamination, which poses significant internal hazards.

An example of the demonstrated effects of beta burns was provided by the Bikini Bravo shot, which occurred in 1954. It was a thermonuclear device with an estimated 10-MT yield at a height of 7 ft above a coral reef. Upon detonation the device exceeded design specification and yielded 15–18 MT. An area of 7000 square miles was contaminated by fallout to such an extent that evacuation or protective measures were necessary to avoid death or serious radiation injury (Glasston and Dolan, 1977). The surface burst took large amounts of coral up into the fireball and formed lime-like flakes at high altitudes (Glasston and Dolan, 1977). Within 5 hr after the detonation, the radioactivity-contaminated coral ash began falling on the Marshall Islanders. Because the weather was hot and damp, the Marshall Islanders wore little clothing. Since they were unaware of the significance of the lime "snowflakes," appreciable amounts of fission products fell on their hair and skin and remained there for a considerable period of time. In addition, since the islanders did not wear shoes, their bare feet were continually contaminated from the fallout on the ground. After about 10 hr, the fallout cloud had thinned and was no longer visible. The visible particles were already deposited, and they presented the most serious hazard. The fallout was complete after 24 hr. Unaware of the fallout hazards, many inhabitants ate contaminated food and drank contaminated water from open containers for up to 2 days before they were evacuated. A total of 239 Marshallese were exposed.

Some Marshall Islanders bathed during the 2-day exposure period before evacuation, but others did not. Therefore, in general, optimal conditions existed for possible beta damage. In the group suffering the greatest exposure, 20% of them (13 persons) showed deep lesions, 70% (45 persons) superficial lesions, and 10% (6 persons) no lesions. Fifty-five percent (35 persons) showed some degree of epilation followed by a regrowth of hair (Glasston and Dolan, 1977).

The time sequence for those Marshall Islanders suffering from beta burns is as follows. During the first 24–48 hr, the more highly contaminated individuals experienced itching and burning of the skin. Those less contaminated experienced less itching and burning. Within 1 or 2 days, all skin symptoms had subsided and disappeared. After about 2 or 3 weeks, epilation and skin lesions were apparent on the contaminated areas of the body. No erythema was apparent as might have been expected, but it might have been obscured by the dark coloration of the Marshall Islanders' skin. The first evidence of skin damage was the increased pigmentation in the form of dark-colored patches and raised areas. These lesions developed on the exposed parts of the body (i.e., scalp, neck, shoulders, and depressions in forearms, feet, and limbs). The most frequently observed were epilations and skin lesions of the scalp, neck, and feet. Most lesions were superficial without blistering. No skin damage was observed under a covering of even a single layer of clothing. After 3–6 weeks, microscopic examination revealed that damage was most marked in the outer layers of the skin. This form of damage was due to the short range of the beta particles. The lesions formed dry scabs and then healed, leaving central depigmentation. Normal pigmentation spread outward in a few weeks. Regrowth of hair began in 9 weeks and was complete in 6 months. The more highly contaminated persons developed deeper lesions, usually on the feet and neck. They experienced mild burning, itching, and pain. The lesions were wet, weeping, and ulcerated, and became covered by a hard, dry scab. The majority of the lesions healed readily with the regular treatment for nonradiation skin lesions. Abnormal pigmentation existed for some time; in some cases, about a year passed before normal color was restored.

C. Protection from Fallout Gamma Radiation

The primary hazard from fallout is the external whole-body exposure to gamma radiation. Gamma radiation is a decay product from both the radioactive disintegration of the fission fragments and the induced radionuclides. The ability of a material to attenuate gammas basically depends on the electron density of the material (i.e., number of electrons per cubic centimeter). Consequently, mate-

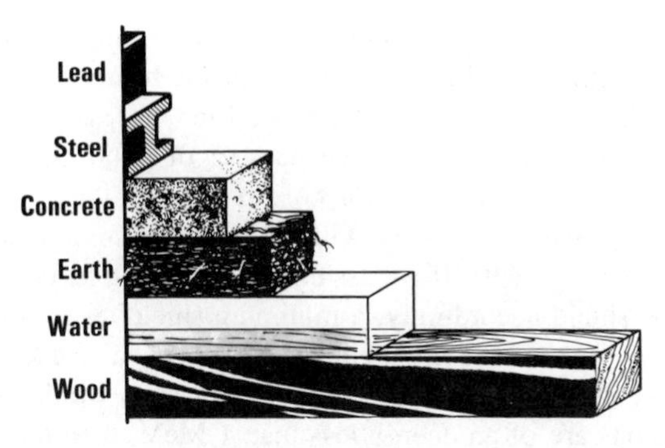

FIG. 3-9 Relative thickness of various materials to produce equivalent shielding.

rials with high mass density and/or high atomic number make the best gamma shields.

Figure 3-9 (Office of Civil Defense, 1968) compares the density and therefore relative effectiveness of various shields. As the density of a material increases, less thickness of that material is needed for effective shielding. Lead is one of the best shielding materials, but since it is not readily available when needed, something like iron or steel is the second-best practical choice.

Since gamma shields do not stop all the photons but rather attenuate some fraction of the initial beam, the concept of tenth thickness is used. Tenth thickness is the amount of a material needed to reduce the initial photon beam by a factor of 10. In addition, for broad beams, scattering occurs into a defined area as well as out of it. Since every photon that interacts with the shield is not necessarily removed from the beam, a little extra material is needed to reduce the total number of final photons (remaining initial photons plus secondary photons) to one tenth of the initial number of photons. Table 3-2 gives the tenth thickness for several materials for a narrow beam (having good geometry so that most of the

TABLE 3-2

TENTH THICKNESS FOR 1-MeV PHOTONS

Material	Narrow beam attenuation, no scatter (in.)	Point source tenth thickness, with scatter (in.)
Air	10,000	—
Water	13.0	24.0
Concrete	6.2	11.2
Aluminum	5.5	10.0
Iron	1.9	3.4
Lead	1.2	1.6

scattered photons leave the beam) and from a point source, which results in a broader "beam" with scattering into and out of the beam.

Since the photons in the fallout field are 1–2 MeV or less in energy and since the dose delivered is proportional to the number of photons, one can use the tenth-thickness values in Table 3-2 for construction of the emergency shield. Photons with energy less than 1 MeV will be attenuated to a greater degree than those at 1 MeV (see Fig. 3-9). If we assume that all photons present are 1 MeV and construct the shield accordingly, a minimum shielding factor would result. For example, if a flux of 1-MeV photons with a dose rate of 100 rad/hr is incident on a lead shield 1.6 in. thick, the final dose rate would be 10 rad/hr. If the incident photons are of an energy less than 1 MeV, then the final dose rate would be less than 10 rad/hr. Although the actual dose rate is not known, the upper limit would be identified. In an emergency situation, a good estimated dose rate may be all that is necessary.

Fallout will deposit like snow and become widely distributed on the horizontal surfaces, such as the ground, roofs, trees, buildings, etc. If a person stands in the center of a large, smooth, evenly contaminated plane, about 50% of the fallout gamma radiation reaching him would come from within a radius of 50 ft (15.2 m), 70% would come from within 150 ft (45.7 m), and 90% from within 500 ft (152 m) (Green, 1965). The remaining 10% would come from beyond 500 ft (Green, 1965). A real surface, such as asphalt, concrete, or the surface of a lawn, has some roughness to it. Therefore the gamma radiation from fallout on a real surface would be partially shielded, and the distances would not be so great. Half of the radiation would come from a circular area with a radius closer to 25 ft (7.6 m) than 50 ft surrounding the point of interest (Green, 1965). As ground roughness increases, this radius for 50% of the exposure would continue to decrease. It should be noted that the reduction of fallout contribution with distance due to ground roughness is more pronounced near the surface of the ground. The effect is reduced with increasing altitude (Green, 1965). If personnel cannot leave the fallout area, they can reduce their exposure by one-half by clearing an area of about 10 m in radius (Departments of the Army and Navy, 1968). In addition to clearing the area, it is beneficial to mound the dirt around the edges of the area (see Fig. 3-10). The dirt mound offers some additional protection by attenuating some of the radiation. Allowing for ground roughness, most of the exposure will probably be delivered within a radius of about 100 m rather than the 152 m (500 ft) for the smooth infinite plane.

Using the fact that most of the exposure will be delivered by the fallout in a radius of 100 m, we know that ground structures can aid greatly in reducing personnel exposure. In general, a person in an open, built-up city would receive about 20–70% less dose from fallout than in the absence of shielding (Glasston and Dolan, 1977). Figure 3-11 illustrates how this is possible. The personnel exposure is reduced by two methods: (1) the buildings themselves act as shield-

FIG. 3-10 Mounded dirt around a clearing of a 10-m radius.

ing material, and (2) the height of the building increases the distance between the fallout and personnel.

In a fashion similar to the above, a person standing against a building in the middle of a city block receives less radiation than at the intersection of two streets. Figure 3-12 diagrams this situation. In this case, the person makes the best use of the shielding offered by both the building and the distance from the fallout particles. Obviously, the best fallout protection is to remain inside the building, but if a person *must* go outside into the fallout field, he may minimize his exposure by the practical means discussed above.

Large structures can be used to reduce fallout exposure through a method known as geometric shielding. Geometric shielding, in contrast to barrier shielding, is the attenuation or reduction of the exposure to a person due to his location relative to the fallout field. Figure 3-13 diagrams the use of geometric shielding. (Figure 3-12 also exemplifies geometric shielding.) As seen in Fig. 3-13, although the building walls offer the same amount of shielding, the individuals in the center (×) of the large building receive less exposure because the fallout is kept farther away. In general, inside a building above ground, the center of the building will offer better fallout protection than will a location next to an outside wall. It is to be noted that the reverse is true below ground, i.e., in the basement of *low* buildings. This is because more radiation is scattered into the center of a basement than near the walls.

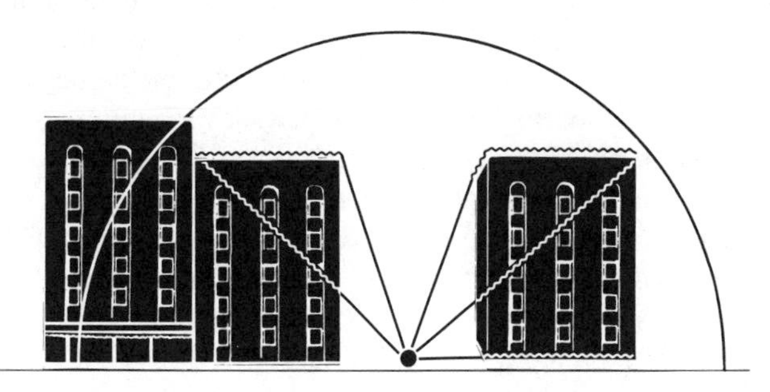

FIG. 3-11 Dose reduction offered by buildings in a city.

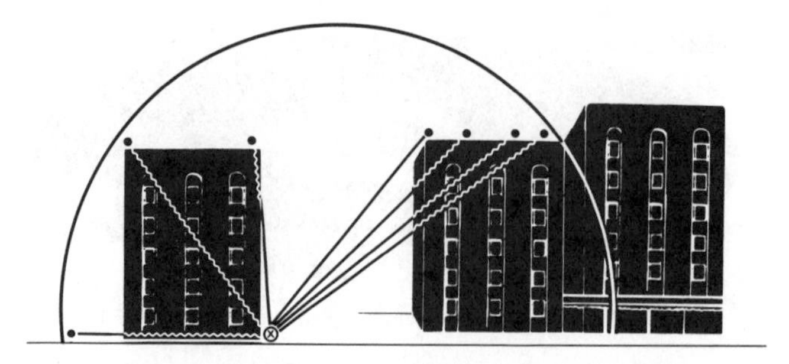

FIG. 3-12 Best use of protection offered by a city.

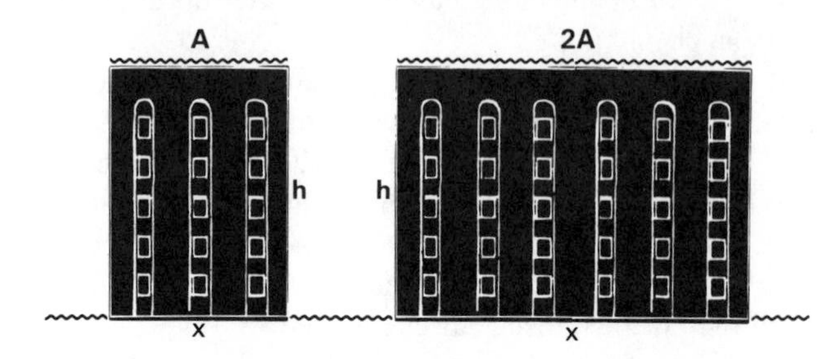

FIG. 3-13 Geometric shielding. A, width of building.

VI. INTERNAL HAZARDS FROM FALLOUT

Both early and delayed fallout are potential internal hazards. The internal hazards of early fallout are not as serious as the primary and secondary hazards of external whole-body gamma irradiation and beta burns. The primary hazard of delayed fallout is internal. Although many factors govern the potential fallout problem (type of weapon, yield, height of burst, height of cloud, distribution of radioactivity in the cloud, radioactivity associated with various particle sizes, rate of fall of particles, and meteorological conditions), the controlling factor for the long-term fallout problem and the subsequent problem of internal hazards is the length of time the particles stay in the air. The size of the particles formed and the height of ascent determine how long the particles remain in the air.

Global distribution of fallout occurs when the very fine particles reach very high altitudes. All bursts above ground can cause long-term fallout. Airbursts have approximately 100% delayed fallout (Glasston and Dolan, 1977). Delayed fallout from a surface burst depends on the type of surface under the detonation. Land surface bursts result in about 40% delayed fallout, whereas fallout from water surface bursts is approximately 70% delayed. For tactical or low-yield weapons (less than about 100 kt) detonated close to the ground, the fallout problem usually does not last longer than a few weeks (Glasston and Dolan, 1977).

A. General Concept of Internal Hazards

Fallout may enter the body and become an internal hazard. If this occurs, the concern for the type of radiation is the opposite of the concern for external exposure. Alpha particles, because of their relatively large size, produce localized regions of extremely high ionization that result in extensive and localized damage to tissue. Gamma radiation, on the other hand, because of its greater ability to penetrate tissue, is of relatively less concern. A photon travels much farther than an alpha particle before it interacts; at the time of interaction, it produces much less damage per unit path length than an alpha particle does. The result is a lower average dose. In addition, depending on the photon's energy, a high proportion of photons may leave the body without interacting in the tissue. Beta radiation is intermediate. Although the damage that beta radiation produces per unit path length is comparable to the damage a photon produces when it interacts, the beta radiation travels a relatively short distance in tissue, thus producing relatively high average doses.

Internal exposure causes great concern because radiation exposure of organs and tissues from an internal source is continuous exposure, and nuclides tend to collect in critical organs (e.g., iodine-131 in thyroid). The radionuclide is subject to depletion only by physical decay or by biological elimination. The actual internal dose rates are highly dependent on the circumstances surrounding the pathway by which the radionuclide becomes an internal hazard; as a result, they *may not be readily predictable.* A radionuclide becomes internal by inhalation, ingestion, or injection through a wound. The actual fate of the radionuclide depends on its chemical nature. Radioisotopes follow the same metabolic processes as the stable isotopes of the same element; for example, radioactive iodine goes to the thyroid.

Elements not normally found in the body behave like those with similar chemical properties that are normally present in the body. For example, strontium, barium, and cerium act like calcium by going to the bone. But since these elements are only chemical analogs, they will act differently and will deposit in tissues in which the original element may not be present; i.e., cerium, in addition to going to the bone, will go to the liver, spleen, and other tissues, and a greater percentage of strontium will deposit in the soft tissue relative to the percentage of calcium in soft tissue.

Any element that does not tend to concentrate in a particular part of the body is eliminated rapidly by natural processes. The probability of serious pathological changes caused by ingestion depends on the amount of radionuclide deposited, energy of radiation emitted by the radionuclide, type of radiation emitted, and length of time the source is in the body.

Of particular interest are the bone seekers: plutonium, strontium, cerium, and barium. Bone seekers are radionuclides that tend to concentrate in the bone. They are of serious concern because they can cause radiation damage to bone marrow, resulting in leukemia and other hematological abnormalities. They can also cause actual bone damage, producing bone necrosis and bone tumors.

B. Inhalation

At the time of nuclear detonation, particles of many different sizes are produced and become airborne. Early fallout particles range in size from 20 μm to 1 cm. Delayed fallout particles are smaller. Particles larger than 10 μm are the most hazardous. These particles do not reach the lungs because the nose is almost totally efficient in filtering particles larger that 10 μm and about 95% efficient for particles of 5 μm (Glasston and Dolan, 1977). Obviously, one would not want to let fallout particles accumulate in the nose and remain there, because damage from energetic betas could result. Table 3-3 gives the efficiency of various common household items and personal items that can be used as an emergency filter against aerosols with a particle size of 1–5 μm.

TABLE 3-3

Efficiency of Common Household Items Against Particle Sizes of 1–5 μm[a]

Item	Number of thicknesses	Approximate efficiency (%)
Handkerchief, man's, cotton	16	94
Toilet paper	3	91
Handkerchief, man's, cotton	8	90
Handkerchief, man's, crumpled	—	88
Bath towel, cotton terry cloth	2	85
Bath towel, cotton terry cloth	1	74
Shirt, cotton	2	66
Handkerchief, woman's, cotton	4	56
Slip, rayon	1	50
Dress material, cotton	1	48
Shirt, cotton	1	35
Handkerchief, man's, cotton	1	28

[a]Adapted from Table LXVII in *Evaluation of Radiation Emergencies and Accidents: Selected Criteria and Data.* International Atomic Energy Agency, Vienna, 1974, p. 110.

Soluble particles and insoluble particles entering the lungs behave in slightly different ways. Generally, 25% of the soluble airborne particulate matter is exhaled, 50% remains in the upper respiratory tract and is swallowed within 24 hr, and the remaining 25% is absorbed [International Commission of Radiation Protection (ICRP), 1968]. Twenty-five percent of the insoluble airborne particulate material is exhaled, 50% remains in the upper respiratory tract and is swallowed within 24 hr, 12.5% enters the deep respiratory tract and is swallowed within 24 hr, and the remaining 12.5% remains in the deep respiratory tract and is eliminated with a biological half-life of 120 days (ICRP, 1968). Since many of the contaminated particles of fallout are relatively insoluble, they are not transported to the blood (Glasston and Dolan, 1977). Particles remaining in the lungs are removed by cellular or lymphatic transport. Those removed by the lymphatic system accumulate in the tracheobronchial lymph nodes, thus causing intense localized radiation doses (Glasston and Dolan, 1977).

C. Ingestion

Ingestion of fallout particles occurs in two sources: contaminated mucous from the upper respiratory tract and contaminated food and water. Solubility of fallout material is the major factor in determining the distribution and thus the resultant dose within the body. The solubility varies, depending on (among other factors) the surface over which the detonation has occurred. Fallout material collected in soil samples at the Nevada Test Site is quite insoluble (Dunning, 1957). However, it is likely that activity actually present in drinking water is principally a soluble form (Dunning, 1957).

Like most fission products, uranium and plutonium are in the form of oxides and do not dissolve well (Glasston and Dolan, 1977). As oxides, strontium and barium are about 10% soluble; after entering the blood, they go to the bone (Glasston and Dolan, 1977). Iodine is soluble; it readily enters the blood and goes to the thyroid (Glasston and Dolan, 1977). Although large amounts of radionuclides pass through the kidneys, they do not greatly affect this organ (Glasston and Dolan, 1977). The large intestine receives a reasonable portion of the relative dose to body organs (Dunning, 1957). These doses to the large intestine occur because of the length of time the insoluble radionuclides remain in the large intestine while waiting to be excreted.

D. Injection

The amount, form, and subsequent effect of radionuclides entering the body through an injection or wound depend entirely on the situation. Action taken to

TABLE 3-4

MOST COMMON GROUND-INDUCED RADIONUCLIDES

Isotope	Radiation	Half-life
Sodium-22	Beta, gamma	15.0 hr
Chlorine-38	Beta, gamma	37.0 min
Manganese-56	Beta, gamma	2.6 hr
Aluminum-28	Beta	2.3 min
Silicon-31	Beta	2.6 hr

reduce the overall potential hazard from the injection should correspond to the normal methods of wound decontamination.

E. Isotopes of Concern

Although a wide range of radionuclides is produced in a nuclear denotation, only a few present particular problems. The relatively short-lived radioisotopes produced from sea and ground activation are listed in Table 3-4 (Glasston and Dolan, 1977). These isotopes are short-lived and pose no particular internal problems. Aluminum is a concern until about $\frac{1}{2}$ hr postdetonation, and manganese and silicon are a concern up to about 10–20 hr. Thereafter, sodium is the concern for up to a few days (Departments of the Army and Navy, 1968; Defense Civil Preparedness Agency, 1973).

Of the 36 elements produced by the fission process and by subsequent decay of fission fragments, only a few are of concern. Based on their potential hazard, these isotopes can be divided into three groups (Glasston and Dolan, 1977). Group I contains iodine-131, which is a problem for only the first few weeks because it has a relatively short half-life. The isotopes in Group II are strontium-89 and -90, cesium-137, and barium-140. These isotopes enter the stratosphere from moderate- and high-yield weapons. Due to their long half-lives, they persist for years and are the most significant problem of long-term fallout. Group III isotopes are cerium-144, yttrium-91, and other related rare earth elements. These isotopes are similar to Group II but are less significant.

Bomb materials can be activated. Of primary concern are the isotopes of zinc, copper, magnesium, and (to a lesser extent) iron (Glasston and Dolan, 1977), which are a concern only in early fallout. Other bomb materials include the unspent fuels uranium, plutonium, and tritium (from the lithium in the lithium deuteride). These isotopes remain and contribute to the problem of long-term fallout.

Two other isotopes are produced from the interactions of neutrons with the air. Carbon-14 is produced from the nitrogen-14 in air. Nitrogen-14 absorbs a neutron, and the subsequent compound nucleus (nitrogen-15) decays by emitting a proton and producing carbon-14. Tritium, to a minor extent, is also produced from the unstable nucleus of nitrogen-15. Nitrogen-15 can decay to stable carbon-12 by emitting a tritium nucleus.

F. Experience of Internal Hazards from Early Fallout

Evidence indicates that contamination of the Marshall Islanders was by ingestion instead of inhalation. They ate food and drank water from open, contaminated sources for up to 2 days before being removed from the islands. But only iodine, barium, and strontium isotopes and the rare earth elements were found to persist; all other elements were rapidly eliminated from the body. The body burden of radioactive materials among the more highly contaminated inhabitants was never large, and it decreased fairly rapidly in the course of 2 or 3 months. Activity of the strontium isotopes fell off more slowly. Although the Marshall Islanders lived for almost 2 days under conditions of maximum probability of contamination of food and drink and although they took few steps to protect themselves, the amount of internally deposited radionuclides was small.

The Marshall Islanders received whole-body gamma exposure up to 175 rem (Glasston and Dolan, 1977). The short-term effects from internal sources from early fallout are minor compared to those due to external radiation. However, delayed effects have been seen. Only one case of leukemia was reported, and no thyroid abnormalities were detected before the year 1963 (approximately 9 years after the detonation). But by 1966, 18 cases of thyroid abnormalities were reported. This number increased to 22 in 1969 and 28 in 1974. Most of the thyroid lesions occurred in children who were younger than 10 years old at the time of exposure (in 1954). Of those 28 persons with thyroid lesions, 3 developed malignancy, 2 suffered hypothyroidism, and all others developed benign nodules.

G. Potential Magnitude of Internal Hazards

Calculations indicate that fission products from detonation of thousands of megatons of yield would have to be in the stratosphere before delayed fallout would cause an average concentration in humans equal to the recommended maximum values for occupational workers (Glasston and Dolan, 1977). Thousands of megatons are typical of a large-scale nuclear attack.

VII. MANAGEMENT OF FALLOUT

Although actual fallout patterns are uncertain, fallout can be managed. A home with two or more stories and an average basement wall exposure of 2 ft or less provides sufficient protection to save 90% of the persons from fallout exposures that would otherwise prove fatal.

A. Detection of Fallout by the Physical Senses

Ideally, all radiation detection should be done with instruments. Although the thorough control of radiation requires instruments, the physical senses can evaluate the relative magnitude of the hazard. Much, if not all, of the heavy fallout observed during nuclear weapons testing was visible as individual particles falling and striking objects or as deposits accumulated on the surfaces of various objects (Green, 1965). Although fallout still occurs after the fallout clouds have thinned and cannot be seen, the most hazardous fallout is associated with visible particles (Glasston and Dolan, 1977).

During tests in the Pacific, the fallout was white because it consisted primarily of calcium oxides and carbonates from the coral islands (Green, 1965). At the Nevada Test Site, fallout was composed primarily of alluvial soil, so it was usually darker (Green, 1965). War fallout would probably be composed of a mixture of sharp-edged particles, irregular particles, and round particles with smooth surfaces (Green, 1965). Their color would depend on the explosion environment, but would probably be brown, gray, or black (Green, 1965).

Realizing that not all individuals would have instruments and that the most hazardous fallout is detectable by the senses, the Defense Civil Preparedness Agency (1973) has proposed that guidelines be issued to isolated persons. These guidelines (paraphrased unless directly quoted) would be as follows:

> If you know there has been a denotation, you will want to protect yourself from fallout by going to a shelter (basement, storm cellar, etc.). Fallout will not arrive immediately; it may take several hours. You will probably have time to protect your stock and equipment and to bring food and water into your shelter. If you do not have a radiation-detecting device, you can detect heavy fallout by
>
> (1) "Seeing fallout particles, fine soil-colored, some fused, bouncing or hitting a solid object, particularly visible on shining surfaces such as the hood or top of a car or truck. A white board or piece of white paper on a flat surface may serve as a visual detecting device."
>
> (2) "Seeing a dust cloud or general haze in the sky not associated with a dust storm."
>
> (3) "Feeling particles striking the nose or forehead or collecting on the hands and arms or in the eyes or between the teeth." This would result in irritation of the eyes, a gritty sensation on the lips and between the teeth, and a gritty feeling on the forehead, hands, and bare arms.

(4) "In the rain, after turning on the windshield wiper of your car seeing fallout particles in raindrops slide downward on the glass and pile up at the edge of the wiper stroke, like dust or snow. The particles generally move readily like sand, rather than tending to smear and stick to the glass like fine dust."

If fallout is detected, shelter should be sought immediately. Even if fallout is only suspected and not detected, shelter should be sought. If caught in fallout, the head should be covered with a hat, a piece of cloth, or a newspaper. All outer clothing should be buttoned or zipped. Clothing should be adjusted to cover as much skin as possible. Clothing should be brushed frequently.

In the field, unless otherwise required, use of a shelter should be automatic whenever a small-yield (tactical) burst has been observed, because this type of burst is very likely to produce early fallout (Departments of the Army and Navy, 1968). Shelter should be used until monitoring or passage of time proves that no fallout exists (Departments of the Army and Navy, 1968).

B. Decontamination of Patients

In a disaster such as nuclear detonation, there will be casualties, even if they are only those persons injured in rushing to a fallout shelter. In general, the radiological hazards to a contaminated patient and the attending medical personnel will be small. Medical or surgical treatment needed when a life is in danger should not be delayed because of possible contamination. A checkpoint should be established immediately after receiving a contaminated patient, and the patient should be surveyed for contamination. Ideally this checkpoint is located outside the treatment facility. If contamination does exist, removal of the outer clothing and shoes, in most instances, will remove 90–95% of it. Washing the exposed areas of skin increases the removal of contamination to 99%. If the hair cannot be decontaminated, it should be clipped closely. The skin should not be shaved since, depending on the type of contamination, its absorption may be enhanced. If a large number of patients are being stripped because of contamination, the attending personnel should be checked frequently or else they should change their clothes periodically, because it is inevitable that they will also become contaminated. Care should be taken not to contaminate survey instruments because they will then produce positive readings on patients who are not contaminated. No special soap, detergent, or acids are needed to remove fallout particles (Green, 1965).

Skin can be decontaminated by washing with mild soap and water or detergent and water for 2–3 min. If necessary, a soft brush with heavy lather and tepid water can be used, but care must be taken not to scratch or erode the skin. Each subsequent cleaning (if needed) should be checked for effectiveness by monitoring with a radiation detector. Cuts or breaks in the skin should be flushed with large volumes of running water as soon as possible. Wounds may be spread open to permit flushing. Mechanical cleaning or microsurgery can then be used to remove any remaining contamination. After each step, the wound should be checked until decontamination is complete.

Except in rare cases, external radioactive contamination of personnel is not a medical emergency, i.e., the life of a person does not depend on emergency procedures against the contamination within minutes or hours. While all reasonable measures must be taken to prevent the spread of contamination, the treatment for emergencies such as trauma, shock, and hemorrhage always takes precedence over decontamination. No patient should be denied therapy or medical attention because of contamination. It is important that necessary therapy not be hindered by monitoring of the patient for contamination.

In a mass-casualty situation such as could occur in a fallout field, it would not be possible to devote extensive periods of time to a few decontaminations. As stated above, the removal of contaminated clothing and the washing or wiping of exposed areas of the body are adequate. Although some contamination may remain, the risk incurred to the patient and attending medical personnel is negligible compared to the overall hazards and/or the inefficient use of medical personnel when a large number of casualties are present. After the emergency has subsided, any contamination remaining on the patients can be removed by the methods already described.

C. Postattack Food Sources

Gamma radiation from fallout does not damage food. Contamination is normally confined to the outer surface of sealed containers, so it is necessary only to wash the container before opening it. Unsealed food sources must be suspected of contamination and must be isolated until checked. Since only the outside of unsealed food becomes contaminated, the food can be made safe for eating by washing, peeling, or otherwise removing contaminated parts.

Depending on the location, many emergency sources of water may be available, such as hot water tanks, flush toilets, ice cube trays, or bottled water. If water is exposed, it must be considered contaminated. Specialized and normal water treatment can be used. Several 2-in. layers of sand, gravel, humus, coarse vegetation, and clay can be used as a filter to remove 90% of the dissolved radioactive materials (Office of Civil Defense, 1968). A 6-in. column of loose dirt gives reasonable results. In any case, the filtered water should be boiled or treated with iodine purification tablets or calcium hypochlorite to kill any biological contaminants.

Since the risk of internal contamination is relatively minor for short periods of time, no one should remain thirsty or hungry for fear of contaminated sources

(Green, 1965; Dunning, 1957; Sanders *et al.*, 1955). If it is necessary to eat contaminated food because noncontaminated food is not available, the least contaminated food should be eaten first. In terms of the overall impact of a nuclear attack, the problems of protecting food and water from radiation in the early postattack period are probably minor, compared to the problem of protecting them from bacterial contamination (Green, 1965). Bacterial contamination could result from the disruption of essential services such as gas and electricity, which are needed to preserve food.

In a postattack recovery, food sources growing in fields would be contaminated. Young plants can incorporate radionuclides from the soil, so they should be evaluated. Mature plants near harvest would probably not incorporate any significant amounts of radionuclides. If foodstuffs must be grown in contaminated soil, liming can be used to increase the concentration of calcium and thus decrease the uptake of strontium. In addition, in highly contaminated fields, food low in calcium requirements could be planted (e.g., potatoes, cereals, apples, tomatoes, peppers, sweet corn, squash, and cucumbers) (Office of Civil Defense, 1968). Foods requiring a high intake of calcium could then be planted in the areas of least contamination. Examples of foods with high calcium intake are lettuce, cabbage, kale, broccoli, spinach, celery, and collards (Office of Civil Defense, 1968).

D. Postattack Recovery

The three principal ways of reducing exposure of people in a contaminated area are to (1) shield against radiation by remaining in shelter, (2) evacuate from the area, and (3) decontaminate. However, any kind of practical countermeasure may be used.

Time spent outside a shelter should be kept to a minimum when the dose rates are high. In addition, as much protective clothing as practical should be worn. For example, boots should be worn and the cuffs of pants should be tied. Areas of high contamination such as puddles or dust should be avoided as much as possible. If winds sweep contaminated surfaces that are relatively smooth, the winds can quickly redistribute the fallout against curbs, buildings, or other obstructions, thus causing areas of high contamination. Puddles usually indicate areas of localized concentration caused by runoff. Disturbing a dusty area can cause resuspension of contaminated particles in the air, so a handkerchief or some other covering for the nose and mouth should be used. Unnecessary contact with contaminated surfaces should be avoided. Personnel dosimetry must be worn if available.

During postattack recovery, some areas may have higher dose rates than desired. In these areas it will be possible to leave the shelter, but in order to keep personnel exposures down, the shelter must function as a base camp. The recommended times for remaining in it are based on the radiation levels and the acceptable risks (doses).

In general, personnel in a fallout area should not leave a shelter during the first day or so after fallout arrives because radiation levels are high. If a shelter provides an appreciable amount of protection, it is generally better to remain in it and improve it rather than attempt to evacuate to an uncontaminated area. There are two important reasons for this: (1) personnel might receive excessive exposures while moving out of the area, and (2) some time might be needed to identify the location of safe areas and to check whether or not any habitable space is available in those areas. If evacuation on foot is attempted because no vehicles are available, some consideration should be given to using closely packed formation. A group of 20 persons can achieve an average dose reduction of 2 below the dose received if they were to move separately. The average dose reduction for a group of 60 persons could be a factor of 3 (Green, 1965).

REFERENCES

Defense Civil Preparedness Agency *DCPA Attack Environment Manual.* Department of Defense, Washington, D.C., 1973.

Department of the Army *Radiological and Disaster Recovery at Fixed Military Installations,* TM 5-225. Department of the Army, Washington, D.C., 1966.

Department of the Army *Nuclear Handbook for Medical Service Personnel,* TM 8-215. Department of the Army, Washington, D.C., 1969.

Department of the Army *Fallout Prediction,* FM 3-22. Department of the Army, Washington, D.C., 1973.

Hubner, K. F., and Fry, S. A. *The Medical Basis for Radiation Accident Preparedness,* Proc. REAC/TS Int. Conf. Elsevier/North-Holland, New York, 1980.

International Commission of Radiation Protection (ICRP) *Limits for Intakes of Radionuclides by Workers,* Report of Comm. II, Publ. 30. Pergamon, Oxford, 1978.

Kellogg, W. W. Atmospheric transport and close-in fallout of radioactive debris from atomic explosions. In *The Nature of Radioactive Fallout and Its Effect on Man.* Hearings before the Special Subcommittee on Radiation of the Joint Committee on Atomic Energy Congress of the Congress of the United States. Eighty-fifth Congress, May 27–29 and June 3, 1957.

National Council on Radiation Protection and Measurements *Management of Person Accidentally Contaminated With Radionuclides,* Rep. No. 65. NCRPM, Washington, D.C., 1980.

Office of Civil Defense *Shelter Design and Analysis. Vol. I. Fallout Prediction,* TR-20-(Vol. I). Department of Defense, Washington, D.C., 1964.

Office of Civil Defense *Fallout Shelters,* TR-39. Department of Defense, Washington, D.C., 1967.

Shleien, B. External radiation exposure to the offsite population from nuclear tests at the Nevada Test Site between 1951 and 1970. *Health Phys.* **41**(2) (1981).

U.S. Naval Radiological Defense Laboratory statement on prediction of fallout. In *The Nature of Radioactive Fallout and Its Effect on Man.* Hearings before the Special Subcommittee on Radiation of the Joint Committee on Atomic Energy Congress of the Congress of the United States. Eighty-fifth Congress, May 27–29 and June 3, 1957.

SUGGESTED READINGS

Defense Civil Preparedness Agency *Response to DCPA Questions on Fallout.* Prepared by the Subcommittee on Fallout Advisory Committee on Civil Defense, National Academy of Sciences, with notes and comments by J. C. Green. DCPA Res. Rep. 20. DCPA, Washington, DC., 1973.

Departments of the Army and Navy *Operational Aspects of Radiological Defense,* FM 3-12 (FM 11-5). Departments of the Army and Navy, Washington, D.C., 1968.

Dunning, G. M. Radiation from fallout and their effects. In *The Nature of Radioactive Fallout and Its Effect on Man.* Hearings before the Special Subcommittee on Radiation of the Joint Committee on Atomic Energy Congress of the Congress of the United States. Eighty-fifth Congress, May 27–29 and June 3, 1957.

Glasston, S., and Dolan, P. J. *The Effects of Nuclear Weapons.* Department of Defense and Department of Energy, Washington, D.C., 1977.

Green, J. C. *Fallout Radiation Exposure Control (An Introduction).* Post-Attack Research Division, Office of Civil Defense, Washington, D.C., 1965.

International Commission of Radiation Protection (ICRP) *Evaluation of Radiation Doses to Body Tissues from Internal Contamination Due to Occupational Exposure,* Report of Comm. IV, Publ. 10. Pergamon, Oxford, 1968.

Office of Civil Defense *Radiological Defense Textbook,* SM-11.22-2. Department of Defense, Washington, D.C., 1968.

Sanders, J. B., Placak, O. R., and Carter, M. W. Report of offsite radiological safety activities—Operation Tea Pot, Nevada Test Site, Spring 1955. In *The Nature of Radioactive Fallout and Its Effect on Man.* Hearings before the Special Subcommittee on Radiation of the Joint Committee on Atomic Energy Congress of the Congress of the United States. Eighty-fifth Congress, May 27–29 and June 3, 1957.

CHAPTER 4

Ionizing Radiations and Their Interactions with Matter

JOSEPH A. SHOLTIS, JR.*

Armed Forces Radiobiology Research Institute
Bethesda, Maryland 20814-5145

When a nuclear weapon detonates, a tremendous amount of energy is promptly released in the form of ionizing radiation. This radiation, composed of both electromagnetic and particulate forms, predominantly includes alpha particles, beta particles, positrons, neutrons, X rays, gamma rays, and fission fragments. The ways in which these individual radiation types interact with matter in the environment determine (along with the radiation amount, of course) the actual hazard that is presented. These interaction mechanisms also offer insights into methods that can be used to detect and measure the radiation as well as reduce the biological hazard. Therefore, study of these radiation types and how they interact with matter can be extremely valuable and practical if the basic principles are understood and used properly.

In this chapter a practical summary of protection and shielding is included as a distinct subsection for each radiation type discussed. However, in order to understand how such protective measures are determined as well as why and to what extent they are effective, deliberate study of this entire chapter (including the physics discussions) is necessary.

I. THE BASICS

A. Ionization and Excitation

Nuclear radiation ultimately results in ionization and excitation. Excitation, the less interesting of the two, simply involves raising the energy level of an

*Present address: Air Force Element, Headquarters, Department of Energy, Defense Energy Projects, Washington, D.C. 20545.

atom or molecule with which the radiation interacts without ionization occurring. Thus, excitation serves as a mechanism, albeit of secondary importance, for the dissipation of incident ionizing radiation energy. Ionization, on the other hand, is more effective in dissipating incident radiation energy, since more energy is generally required to strip off orbital electrons and impart to them kinetic energy than to simply excite them. Ionization, therefore, involves stripping one or more bound electrons from an atom or molecule, and is the predominant interaction mechanism for the attenuation (i.e., dissipation) of ionizing radiation.

B. Directly and Indirectly Ionizing Radiations

Ionizing radiations are defined as those types of particulate and electromagnetic radiations that interact with matter and either directly or indirectly form ion pairs. Consequently, ionizing radiations are divided into two general categories: directly ionizing and indirectly ionizing. Directly ionizing radiations are charged particles (electrons, protons, beta particles, positrons, alpha particles, etc.) that have sufficient kinetic energy to produce ionizations through direct coulombic interactions (or collisions) with the bound orbital electrons of an atom or molecule. Indirectly ionizing radiations are uncharged particles or photons (X rays, gamma rays, neutrons, etc.) that can liberate bound orbital electrons, but only indirectly, i.e., secondarily through a preliminary interaction. For photons, ionization does result; however, the interaction involves the entire atom or molecule, not just the bound orbital electrons. Similarly, for energetic neutrons, the interaction is with the nucleus, not its bound orbital electrons. In this particular case, a neutron can transfer kinetic energy to a nucleus sufficient to cause it to recoil as a positively charged nucleus (or ion) that has indeed lost one or more of its bound orbital electrons. For all indirectly ionizing radiations, the secondary charged particle liberated from the interaction (a secondary electron from a photon interaction or a positively charged recoil nucleus from an energetic neutron collision) goes on to cause a very large number of ionizations and excitations along its tract, as it is gradually brought to rest in the material. Thus, all but a very small fraction of the total number of ionizations and excitations from indirectly ionizing radiations are produced by the secondary particles. But regardless of the type of ionizing radiation, the final common event in all modes of absorption of ionizing radiation is the ejection or excitation of bound orbital electrons. This important fact should be remembered.

Directly and indirectly ionizing radiations exhibit marked differences in absorption. Beams of indirectly ionizing radiations can traverse considerable thicknesses of material without being completely absorbed. Such beams of uncharged

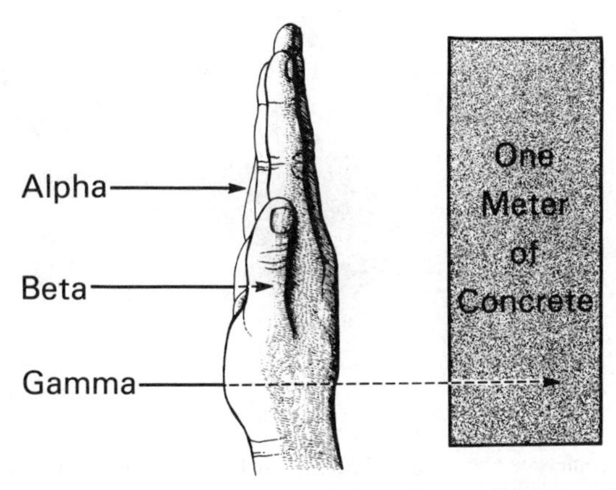

FIG. 4-1 Relative penetrating power of alpha, beta, and gamma radiation.

particles or photons are attenuated exponentially, whereas directly ionizing particles have a finite maximum range in materials (Fig. 4-1).

C. Prompt and Residual Radiation

The ionizing radiation emitted when a nuclear weapon is detonated is conventionally divided into two additional categories, both consisting of directly and indirectly ionizing radiations: (1) initial or prompt radiation and (2) residual radiation. Prompt radiation is defined as those ionizing radiations given off within the first 60 sec after the weapon detonation. Prompt ionizing radiations consist of neutrons, gamma rays, X rays, beta particles, the highly energetic and ionized fission fragments resulting from the fission and fusion processes themselves, and the ionizing radiations (gamma rays, alpha particles, beta particles, and occasionally even neutrons) emitted by decay of the shorter half-lived radioactive fission fragments and neutron-activated materials. Residual radiation is defined as those radiations emitted in time periods greater than 60 sec after the detonation. These radiations consist primarily of the gamma rays, beta particles, alpha particles, and (to a lesser extent) neutrons emitted by decay of the longer half-lived highly radioactive fission fragments and neutron-activated materials.

II. THE RADIATION TYPES

This section discusses alpha particles, beta particles, X rays, gamma rays, and neutrons. The characteristics, interaction mechanisms, and suggested methods for protection and shielding are treated separately for each.

FIG. 4-2 Nuclear emulsion photograph of alpha-particle tracks (enlarged 1000 diameters) emitted by a grain of alpha-emitting radioactive dust.

A. ALPHA PARTICLES

The following discussions regarding alpha particles are generally also true for fission fragments, which are energetic heavy charged particles with a very limited range and thus limited capability of penetration.

An alpha particle is simply the nucleus of a helium atom. That is, there are two neutrons and two protons bound together to form the nucleus, and no orbital electrons surround the nucleus. As a result, an alpha particle has a charge of +2 and is quite massive relative to the other types of ionizing radiation discussed in this section. Because alpha particles are massive and have a charge of +2, they have a very limited range in materials. When an alpha particle finally manages to capture two electrons and becomes a neutral helium atom, its ionization potential effectively drops to zero. But alpha particles do cause significant ionization, even though it is only over a very limited region (Fig. 4-2).

Alpha particles are typically emitted essentially monoenergetically by the decay of heavy radioactive isotopes, such as plutonium-239, uranium-235, in the energy range from ~4.5 to 5.0 MeV. Thus, they are a by-product of unconsumed nuclear fuel debris and heavy (i.e., high-Z) neutron-activated materials. Because alpha particles deposit all their energy over a very short distance, they are classified as a high linear energy transfer (LET) radiation. Linear energy transfer is a measure of the radiation energy lost per unit length of path through a biological material; typical units are kilo-electron-volts (keV) lost per micrometer.

B. PROTECTION AND SHIELDING AGAINST ALPHA RADIATION

Because alpha particles have such a short range in materials, external shielding is not necessary. A sheet of paper or less than ~1.0 cm of air or human skin is adequate to stop alpha radiation. Therefore, based on short distance alone, alpha radiation does not pose an external hazard. On the other hand, alpha radiation can pose a significant local internal hazard if alpha emitters are taken into the body via inhalation, ingestion, etc. As a result, solid foods contaminated with alpha emitters should be washed thoroughly (with uncontaminated liquids), and contaminated liquids should not be consumed. Also, contaminated wounds should be flushed and, if necessary, debrided. A self-contained breathing apparatus, gas mask, or at least fine-particulate mask should be used whenever airborne alpha radiation is present. Finally, contaminated surfaces should be washed.

Although alpha particles can cause activation of an absorbing material, they rarely do. Even then, this occurs only in certain materials. Thus, no problems should exist because of induced secondary radiation caused by alpha particles.

C. Beta Particles

Beta particles are identical to electrons except that they have a nuclear rather than electronic origin, and they typically have energies ranging from a few kilo-electron-volts to tens of mega-electron-volts. Beta particles from a nuclear detonation are produced primarily by decay of the highly radioactive fission fragments and neutron-activated substances. Beta particles cause ionizations primarily through direct, "billiard ball" collisions with the bound orbital electrons of a material. If a beta particle imparts enough energy to a struck bound electron, the atom or molecule involved will be ionized, and both the struck electron and the recoiling beta particle can go on to produce further ionizations (Fig. 4-3) and excitations. They do so by undergoing further collisions and generating more struck-free or excited electrons until the total energy of the incident beta particle is dissipated through such ionization and excitation reactions. Such struck-free electrons are called secondary electrons, and it is the subsequent interactions of these secondary particles that produce the majority of ionizations in a material. (Note that some of the energy of the struck-free electrons is lost due to bremsstrahlung radiation and is not available to cause ionization. Bremsstrahlung radiation is discussed in more detail in Section II,E.)

D. Protection and Shielding against Beta Radiation

Because beta particles are charged particles with some mass, they have a finite maximum range in materials, which is determined by the energy of the incident beta particle and the surrounding material. Generally, less than or approximately 1.0 in. of any material, including air, is adequate to stop beta particles. In fact, a

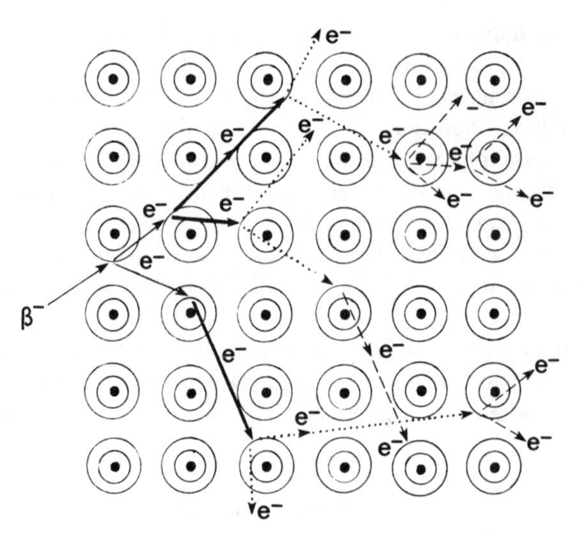

FIG. 4-3 Secondary electron production from beta radiation.

30–55% reduction in beta dose rate can be achieved simply by wearing protective clothing. Moreover, the thickness of material required to achieve a specified beta dose rate decreases as the density of the material increases. This makes sense, because beta particles interact with bound orbital electrons, and the effectiveness of a material against beta particles would be expected to increase as the concentration of bound orbital electrons increases, i.e., as the density increases.

Therefore, protection against beta radiation is a relatively simple affair. Both distance and/or the use of minor thicknesses of essentially any material provide an adequate means of protection against beta radiation. But it should be noted that beta particles can cause skin burns (beta burns) if skin-surface beta contamination is prolonged. Also, beta particles, like alpha particles, can pose a significant internal hazard if beta emitters are taken into the body. Therefore, the same precautions and protective measures identified for alpha radiation regarding internal hazards also apply to beta radiation.

Finally, very high-energy beta particles, like alpha particles, can cause activation of an absorbing material, but this rarely occurs since the beta energy threshold for activation is extremely high (~13 MeV). Although secondary activation is essentially of no concern, secondary bremsstrahlung (X-ray) radiation can be a concern because these secondary X rays have a much greater penetration capability. For more detail on this subject, see the discussion of X rays, specifically bremsstrahlung X rays (Section II,E,1).

E. X Rays and Gamma Rays

Both X rays and gamma rays are photons, or small quantized packets of pure electromagnetic energy just like light, but with a much higher nonvisible frequency. Although their origins are distinctly different and their associated energies are dependent on their origin, the ways in which they interact with matter are exactly the same. Specifically, a gamma ray has its origin from a nuclear interaction, whereas an X ray originates from electronic or charged particle interactions.

1. X-Ray Production

The very large number of X rays generated from a nuclear detonation are primarily produced by high-energy electrons and beta particles as they interact with their surroundings, very close to the detonation point. There are two basic interaction mechanisms that take place to generate these X rays: bremsstrahlung and characteristic bound inner electron ejection.

In characteristic bound inner electron ejection, a high-energy electron interacts with a bound electron from one of the inner electron shells of an atom, and imparts enough energy to the struck electron that it is ejected. This leaves a vacant inner orbital electron position that is quickly filled by an outer orbital electron. The filling of such inner vacancies is virtually assured because an inner orbital electron position is more stable than an outer orbital position. That is, an inner orbital electron is bound more securely by the nucleus than is an outer orbital electron. Therefore, different electron shells have distinctly different but fixed associated energy levels, and although it is more difficult to eject an inner orbital electron than an outer shell electron, when such a situation occurs it does not persist. When an outer shell electron drops to fill a vacancy in an inner shell, the process is accompanied by the emission of an X-ray photon (Fig. 4-4). Not surprisingly, the energy of this X ray is exactly equal to the energy difference between the two electron shells involved. Because the electron-shell energy levels are characteristic of the atom involved (thus the energy of X rays produced by this process is unique to the target atom), the process is called characteristic X-ray production, and such X rays are referred to as characteristic X rays.

In a bremsstrahlung reaction (example in Fig. 4-5), a charged particle undergoes acceleration. That is, a change in its scalar speed or a change in its direction must occur. This acceleration of a charged particle results in the emission of electromagnetic radiation, called bremsstrahlung radiation. (Bremsstrahlung means "braking" in German, so the term is appropriate.) In a nuclear detonation, high-energy electrons and beta particles are slowed down or deflected by coulombic interaction forces between the incident electrons or beta particles and the orbital electrons, as well as the nuclei of adjacent atoms.

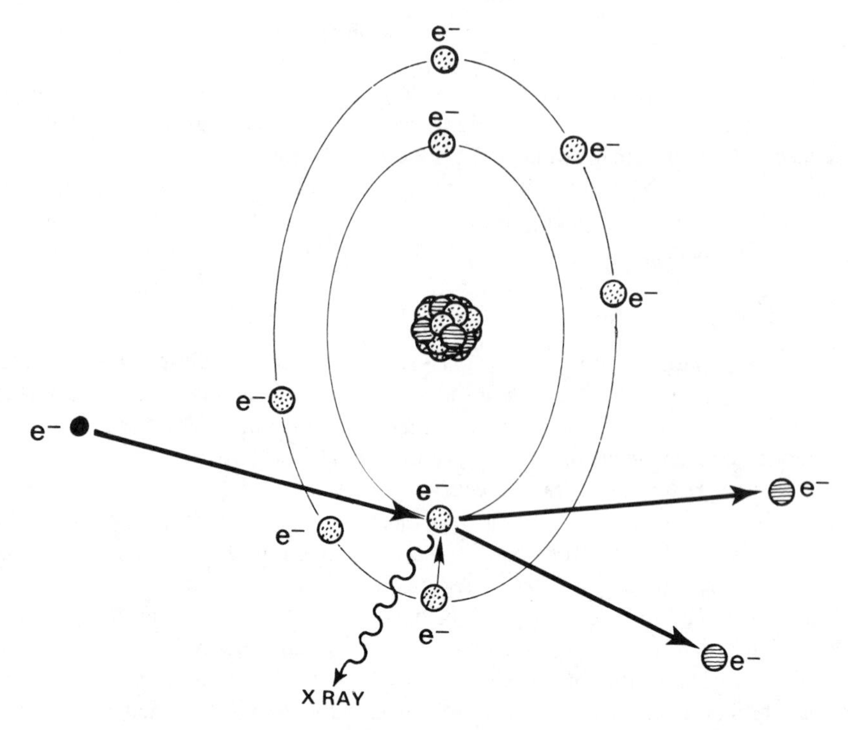

FIG. 4-4 Characteristic X-ray production. A high-energy electron interacts with an electron of an inner orbit and ejects it from the atom. An electron of a higher orbit fills the vacancy, and the energy difference is given off as an X-ray photon.

The energy of bremsstrahlung (X-ray photon) radiation depends on (1) the energy mass, and charge of the incident charged particle, (2) the Z value, i.e., the total number of protons in the nucleus, of the target medium, and (3) how closely the charged particle actually passes a nucleus in the medium. Because of these variables, the X rays generated by this process are not monoenergetic; they can have energies ranging from a minimum of near zero to a maximum equal to the energy of the impinging charged particle. Figure 4-6 shows a typical X-ray

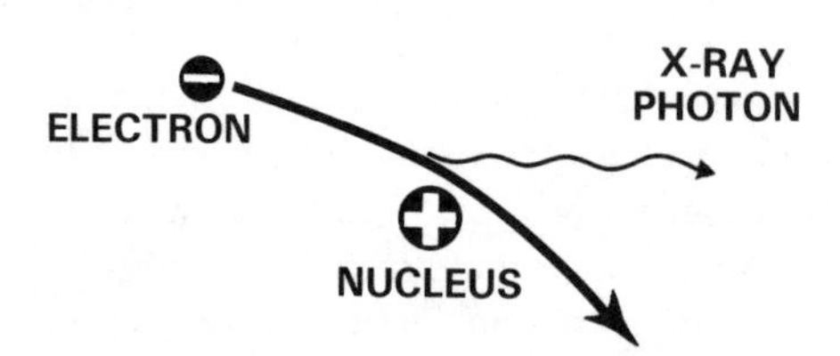

FIG. 4-5 Example of bremsstrahlung (X-ray) radiation production.

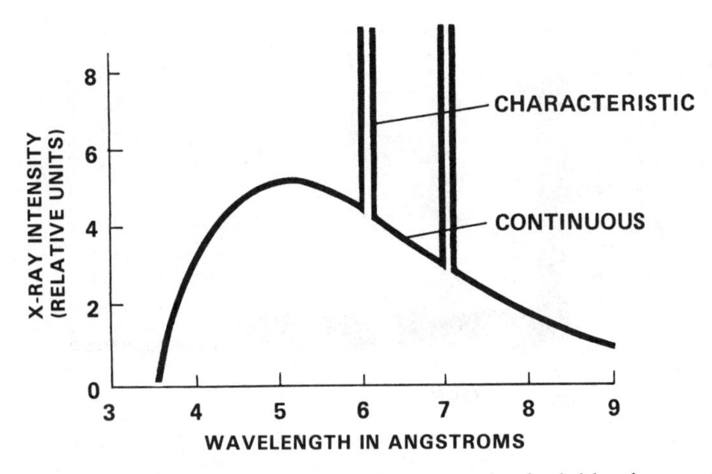

FIG. 4-6 X-Ray energy spectrum for a molybdenum target bombarded by electrons showing continuous and characteristic X-ray production.

energy spectrum with continuous energies (resulting from bremsstrahlung processes) and characteristic energies (resulting from characteristic bound inner electron ejection).

2. Gamma Ray Production

The large number of gamma rays emitted from a nuclear detonation are primarily generated from the fission process itself; they are also produced secondarily by the decay of neutron-activated or excited nuclides. These gamma rays can have energies that range from a few kilo-electron-volts to tens of mega-electron-volts. Figure 4-7 shows the energy spectrum for initial (prompt) gamma radiation at a distance of 2000 yards from a 20-kt (kiloton) nuclear explosion. Figure 4-8 illustrates the time-dependent nature and sources of gamma-ray production (per kiloton of yield) for atmospheric and exoatmospheric nuclear detonations. Two gamma-ray production sources identified in Fig. 4-8 (inelastic scattering in air and nitrogen capture in air) arise from neutron interactions (see Section II,G).

3. X-Ray and Gamma-Ray Interaction Mechanisms

Apart from their origins, X rays and gamma rays are the same physical phenomenon. Therefore, they will be considered as identical in the following discussion of interaction mechanisms. Both gamma rays and X rays interact with matter to cause ionizations through three mechanisms: the photoelectric effect, Compton scattering, and pair production.

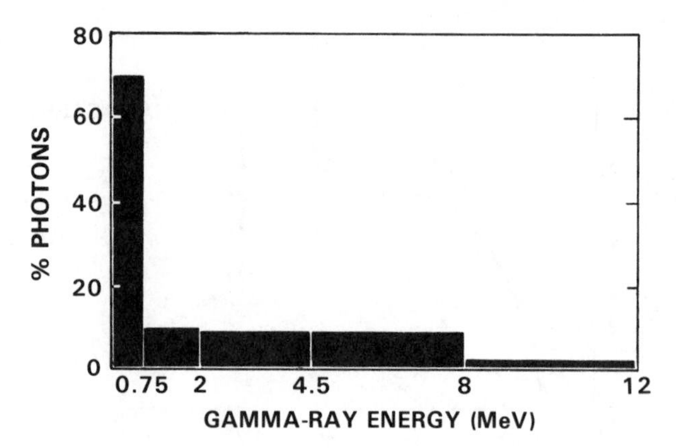

FIG. 4-7 Prompt gamma radiation spectrum (2000 yards from 20-kt explosion).

a. Photoelectric Effect. In a photoelectric interaction, a low-energy photon transfers all of its energy to a tightly bound inner electron of an atom (Fig. 4-9). The end result of this reaction is the total absorption or disappearance of the incident photon and the production of a high-energy electron (which will cause further ionizations in the same manner discussed for beta particles) and a soft X ray. The energy of the photoelectron is equal to the energy of the incident photon minus the energy expended in freeing the electron from the attractive force of its nucleus.

b. Compton Scattering. In Compton scattering, a relatively high-energy photon transfers a portion of its energy to one of the more loosely held outer orbit

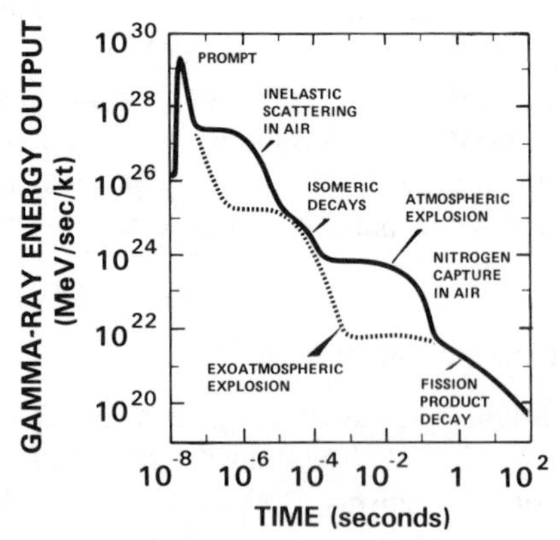

FIG. 4-8 Time-dependent gamma-ray energy output per kiloton of yield for atmospheric and exoatmospheric explosions.

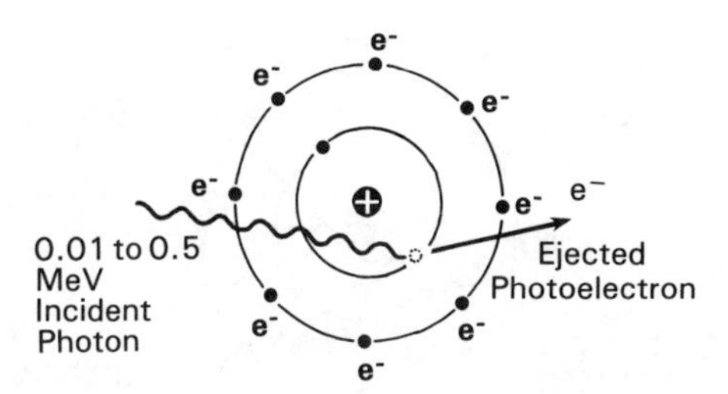

FIG. 4-9 Photoelectric effect.

electrons of an atom (Fig. 4-10). The end products of this interaction are a scattered, less energetic secondary photon and a high-energy Compton electron, both of which can go on to cause further ionizations. The energy of the scattered photon is equal to the incident photon energy minus the energy of the Compton electron minus the binding energy of the struck electron.

c. Pair Production. For high-energy photons ($E > 1.02$ MeV), a third type of interaction becomes increasingly important: pair production. In pair production (Fig. 4-11), the photon interacts with the electrostatic field of a nucleus and is converted into an electron and a positron. (A positron is a particle that has the same mass as an electron but is positive in charge instead of negative.) A portion of the incident photon energy, 1.02 MeV, is required to create the resulting electron and positron, since they have a rest mass that totals 1.02 MeV. The remainder of the incident photon's energy goes into providing kinetic energy to the electron and the positron. The positron will very likely react with an electron; they will annihilate each other and produce one or more photons with a collective energy equal to the rest mass of the electron and positron converted to energy

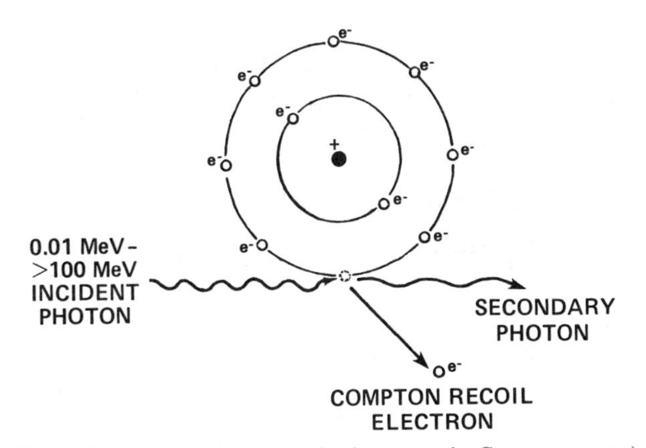

FIG. 4-10 Compton scattering (also known as the Compton process).

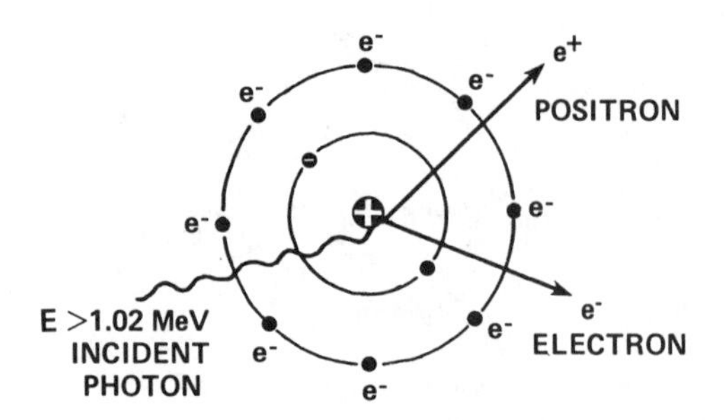

FIG. 4-11 Pair production.

plus the kinetic energy of the positron and the electron (Fig. 4-12). (Note that this particular reaction, termed annihilation, is the antithesis of pair production.) The electron emitted from the original pair production reaction can go on to produce further ionizations in the same manner as beta particles.

Exactly which of these three ionization mechanisms will occur when a photon interacts with an atom is a function of the Z of the target nucleus and the energy of the incident photon. Figure 4-13 is a graph depicting which interaction mechanism has the highest probability of occurring as a function of the energy of the incident photon and the Z value of the target absorber material. Photoelectric interactions are predominant for low-energy photons and high-Z absorbers. Pair production, on the other hand, is a major interaction mechanism for high-energy photons and high-Z absorbers, whereas Compton scattering interactions predominate the region of intermediate photon energies for all absorbers. Figure 4-14 illustrates the relative importance of each of these interaction mechanisms in aluminum, as an example.

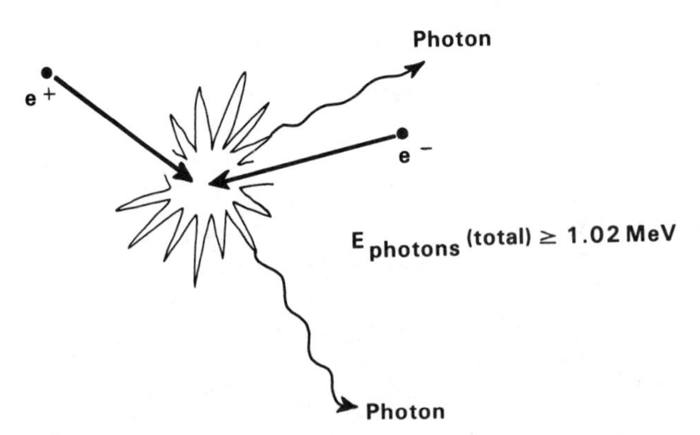

FIG. 4-12 Pair annihilation (the antithesis of pair production).

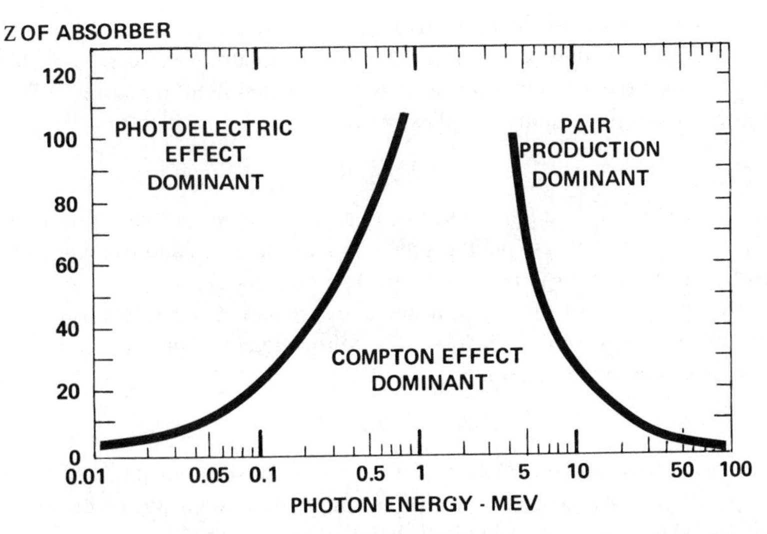

FIG. 4-13 Ranges of energy and ranges of Z of absorber where the various photon interaction mechanisms are dominant. (From Evans, 1972.)

F. PROTECTION AND SHIELDING AGAINST X-RAY AND GAMMA RADIATION

Because X rays and gamma rays are photons of pure electromagnetic radiation without mass or charge, they are highly penetrating. Protection against them can be accomplished in two ways: distance and shielding. Distance attenuation is achieved by placing as much distance as possible between oneself and the photon

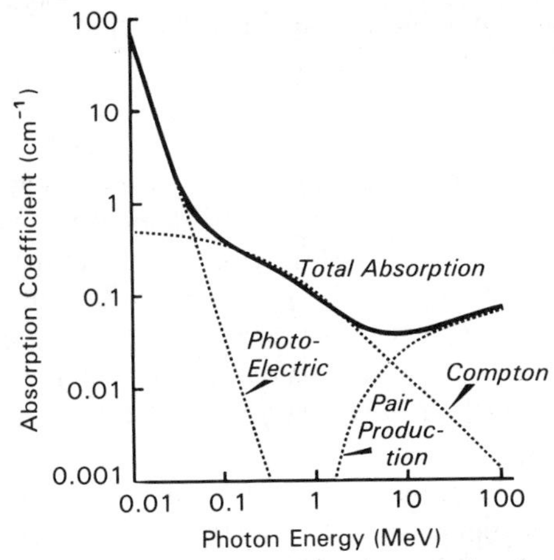

FIG. 4-14 Absorption coefficients in aluminum for various energies of gamma radiation. Relative contributions of the photoelectric effect, Compton process, and pair production are indicated.

source point. This form of protection takes advantage of the fact that photon intensities and thus dose rates drop off with the square of the distance from an isotropic source point. (Isotropic simply means equal in all directions.) Specifically, the following equation applies:

$$I(r) = S_0/(4\pi r^2)$$

where S_0 is the isotropic photon source strength (photons per unit time), r is the distance from the isotropic photon source point of interest, and $I(r)$ is the photon intensity at distance r (photons per unit time per unit area).

Unfortunately, the value of S_0 in the above equation is rarely known, which limits the usefulness of the equation. The following equation does not have such problems and so is more useful:

$$I(r_2) = I(r_1)r_1^2/r_2^2$$

where $I(r_1)$ is the photon intensity, or dose rate, at some distance r_1 from the isotropic photon source point and $I(r_2)$ is the photon intensity, or dose rate, at some distance r_2 from the isotropic photon source point.

This equation simply says that the intensity, or dose rate, at some distance, r_2, from an isotropic point photon source is equal to a known intensity, or dose rate, at some distance r_1 times the ratio of the squares of the distances. Interestingly, this equation holds true for any isotropic point source, not just X rays and gamma rays, as long as it is remembered that shielding attenuation is neglected.

For shielding attenuation, protection is achieved by placing material in the path of the photons so that photoelectric, Compton, and pair production interactions occur more readily to reduce the incident radiation intensity. If buildup is neglected (buildup is typically a second-order effect, so neglecting it is normally a valid approximation), it turns out that X-ray and gamma-ray beams are attenuated exponentially in a material according to

$$I(x) = I_0 \exp(-\mu x)$$

where I_0 is the incident photon beam intensity, or dose rate, at the entrance or front surface of the shield material; $I(x)$ is the photon beam intensity, or dose rate, at some distance x into the shield material (note here that x can be less than or equal to the shield thickness); and μ is the linear attenuation coefficient of the shield material for incident photons at a specific energy. (μ is the probability, per unit distance of travel in the shield material, that a photon will have an interaction. μ values are available in tabulated form, for various materials and photon energies, in numerous texts.)

It is important to note that this basic equation holds for any material subjected to a beam of X rays or gamma rays; the only thing that changes is the value of μ, which is determined by the material involved and the incident photon beam energy. This fact should become very clear if we solve this equation for the thickness of any shield material that will reduce the photon beam intensity, or dose rate, by a factor of 10; this is called the tenth thickness, $x_{1/10}$. First, the original equation is rewritten

$$I(x) = I_0 \exp(-\mu x)$$

Next, it is necessary to establish by definition that when $I(x)$ becomes $I_0/10$, then x becomes $x_{1/10}$. These values are substituted into the original equation, yielding

$$I_0/10 = I_0 \exp(-\mu x_{1/10})$$

Finally, solving for $x_{1/10}$ yields

$$x_{1/10} = \ln(10)/\mu$$

or

$$x_{1/10} = 2.303/\mu$$

This proves that the tenth thickness is determined only by the value of μ, the linear attenuation coefficient, which in turn is determined only by the shield material involved and the incident photon beam energy. Therefore, by obtaining the appropriate value of μ (which is readily available in most nuclear and health physics texts), the tenth thickness can be determined for any shield material. Based on this information, the necessary number of shield thicknesses can be provided to reduce the radiation intensity, or dose rate, to the desired level. When doing this, it must be remembered that a tenth thickness is the amount of a specific material required to reduce the photon beam intensity, or dose rate, by a factor of 10. That is, the application of two tenth thicknesses would reduce the radiation intensity, or dose rate, by a factor of 100, etc. Alternatively, the original basic equation can be solved for any desired shield thickness x if I_0 is known and $I(x)$ is specified or the $I_0/I(x)$ ratio desired is specified.

Sometimes μ, the linear attenuation coefficient, is not provided; instead μ_m, the mass attenuation coefficient, is listed. In those instances, it is important to remember that μ_m times the material density equals μ.

Values for μ and μ_m increase significantly with density from material to material. This supports the earlier claim made about increased shield efficiency with density because as μ or μ_m increases, $x_{1/10}$ will decrease; consequently, less shield material thickness is required to obtain the same effect. However, for a given shield material, μ or μ_m generally decreases with increasing incident photon energy. This indicates that the more energetic the incident photon beam, the greater the shield thickness required. These effects are depicted in Fig. 4-15.

Table 4-1 provides a comparison of the relative shielding capabilities of various materials against X-ray and gamma radiation. In particular, it lists the tenth thicknesses for lead, steel, aluminum, concrete, earth, and water against 1.0-MeV gamma photons.

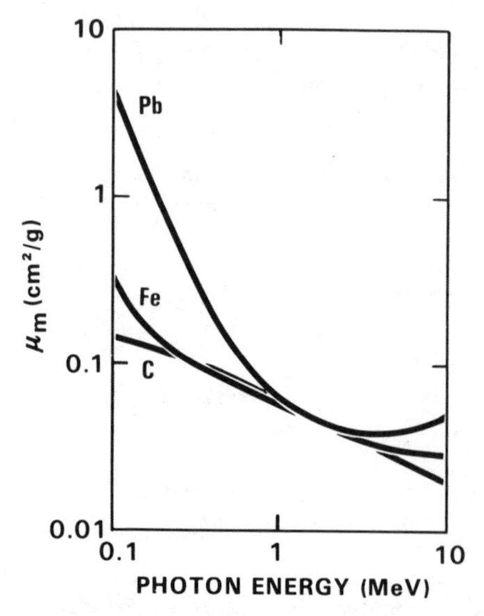

FIG. 4-15 Mass attenuation coefficients for various materials and photon energies.

In summary, X rays and gamma rays are high-energy photons of pure electromagnetic radiation that have no mass or charge and thus are highly penetrating. As a result, they can pose a significant external hazard even at great distances. Protection against X rays and gamma rays can be achieved by distance, shielding, or a combination of both. (Protection can also be gained by limiting the exposure time to the X-ray or gamma radiation source. However, this presumes that a safe, shielded haven is available. Thus, limiting exposure time also

TABLE 4-1

TENTH THICKNESSES FOR VARIOUS MATERIALS AGAINST A BEAM OF 1.0-MeV PHOTONS

Material	Tenth thickness (inches)
Lead	1.2
Steel	2.0
Aluminum	4.6
Concrete	5.7
Earth (dirt)	11.0
Water	14.0
Air	~10,700 (~890 ft)

must involve shielding.) Distance provides a $1/r^2$ reduction in dose rate, while shielding yields an exponential dose rate reduction with increasing thickness that is more pronounced as shield density increases. Accordingly, the recommended techniques for protection against X rays and gamma rays are to use separation distance to the extent possible and then to use shield material thicknesses that are necessary to further reduce the radiation field to an acceptable level. In doing so, remember that it is advantageous to use the densest material(s) available for shielding.

Finally, it is important to note that X-ray and gamma-ray photons normally do not cause activation of substances (except for certain materials and then typically only at very high incident photon energies, i.e., $\gtrsim$10 MeV). Therefore, secondary radiation caused by the absorption of X rays and gamma rays should not be a concern. The major biological concern for X-ray and gamma radiation is the total amount of external incident photon radiation energy that is absorbed. Appropriate measures must be taken to limit this absorbed dose over time to an acceptable level.

G. Neutrons

1. Neutron Production

The very large number of neutrons emitted from a nuclear detonation arise almost exclusively from the fission and fusion processes themselves. A small secondary neutron component is produced later as a result of delayed neutron emission, but only by a very limited number of neutron-activated substances and fission fragments. Consequently, delayed neutrons from nuclear weapons detonations are not significant, compared to the neutrons generated promptly from the fission and fusion processes. Neutrons generated by the fusion process are monoenergetic. That is, they all possess the same amount of kinetic energy. For example, the neutron produced from a deuterium–tritium fusion reaction has an energy of 14.1 MeV and is extremely penetrating. (This neutron has an effective velocity of about 80 million miles per hour.) In contrast, neutrons generated by the fission process are polyenergetic over a spectrum of energies from essentially zero (~1 eV) up to approximately 10 MeV. The exact energy of a given fission neutron is probabilistic in nature, depending on the type of fissile atom that fissioned, the number and atomic weights of the fission fragments produced, etc. Figure 4-16 depicts the fission spectrum for uranium-235. It shows the relative probability that a given fission neutron will have a given energy as a result of a uranium-235 fission event. Figure 4-17 illustrates the output neutron spectra from fission and fission–fusion (i.e., thermonuclear) weapons for comparison.

Although it is impossible to predict the exact energy for a particular fission

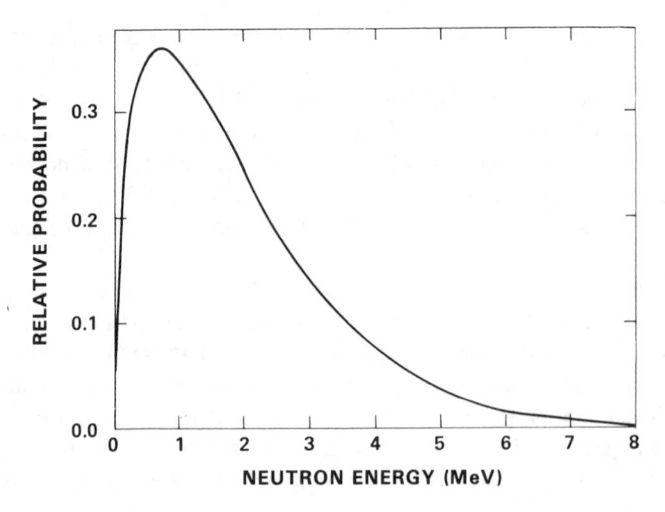

FIG. 4-16 Uranium-235 fission spectrum (sometimes referred to as the Watts fission spectrum).

neutron, it is possible to determine the most probable energy and mean energy of fission neutrons considered collectively. For fission of uranium-235, these two energy values are about 0.7 and 2.0 MeV, respectively, with the vast majority of fission neutrons in the energy range between 0.1 and 10 MeV. Therefore, fission neutrons in general can be considered fairly energetic to very energetic.

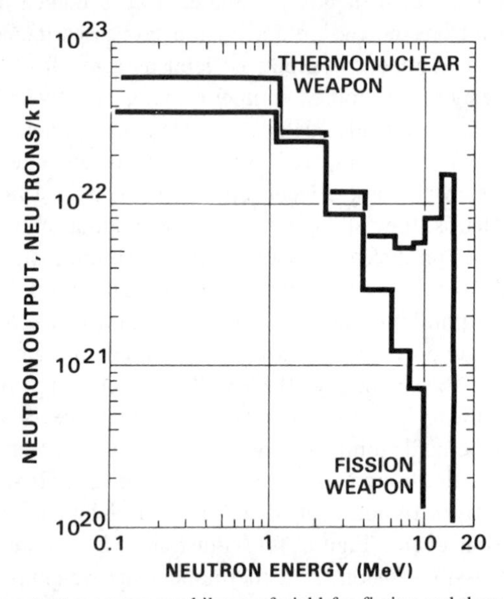

FIG. 4-17 Neutron output spectrum per kiloton of yield for fission and thermonuclear devices.

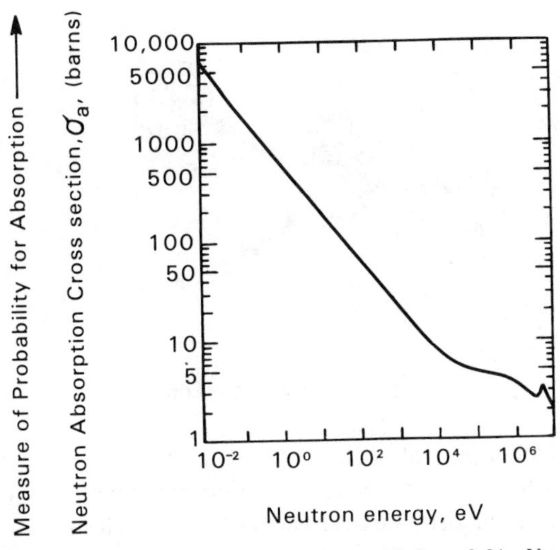

FIG. 4-18 Neutron absorption cross section for boron-10 from 0.01 eV to 10 MeV. (From Hughes, 1958.)

2. Neutron Interaction Mechanisms

Neutrons have a mass approximately equal to that of a proton, but they have no charge. As a result, they cannot interact directly with bound electrons to cause their excitation or ionization. Instead, neutrons cause ionizations in matter indirectly through three different interactions with nuclei of atoms: elastic scattering, inelastic scattering, and absorption. Determining exactly which of these interaction types will take place when a given neutron encounters the nucleus of a given atom is a complex function of the target nucleus and the energy of the incident neutron. A measure of the probability that any particular one of these interactions will take place is called the cross section for that particular reaction type, generally denoted by σ and specifically termed the microscopic cross section. Of the three possible neutron interaction mechanisms, only absorption results in the disappearance or death of the original neutron. Moreover, neutron absorption often requires that neutrons be slowed down considerably (on the order of five to six orders of magnitude) in energy before absorption is likely to occur. This is illustrated in Fig. 4-18, a typical microscopic neutron absorption cross-section curve, with cross-section values plotted as a function of neutron energy.

From Fig. 4-18 it is apparent that the cross section and thus a measure of the probability for neutron absorption is enhanced to a greater and greater degree as the neutron energy is degraded more and more. Intuitively this makes sense, but more importantly, this indicates that it would be advantageous to slow down

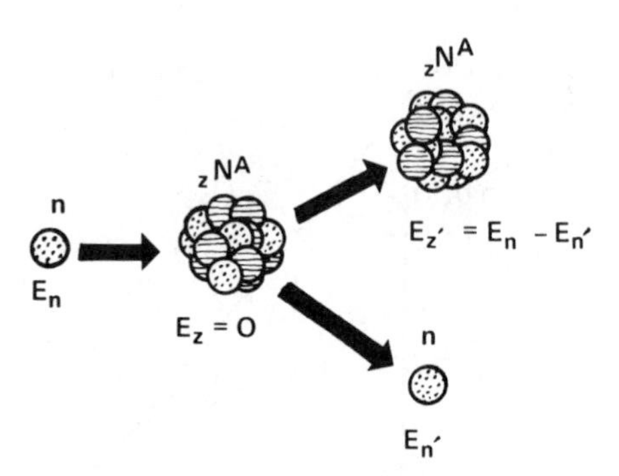

FIG. 4-19 Elastic scattering.

neutrons to the extent possible before attempting their absorption. This process of slowing down neutrons is termed moderation or thermalization; fortunately, it is not only possible but also practical via elastic and inelastic neutron scattering.

a. Elastic Scattering. Elastic scattering is the primary interaction mechanism for neutron moderation; it is also the primary way that neutrons indirectly induce ionizations in matter. Elastic scattering (Fig. 4-19) is like the familiar billiard-ball scatter collision. In elastic scattering, a neutron (n) of energy (E_n) "collides" with and transfers some of its kinetic energy to a struck nucleus ($_ZN^A$), which is assumed to have an initial energy (E_Z) equal to zero and a final energy ($E_{Z'}$) greater than zero, while the original neutron goes off at an energy ($E_{n'}$) that is less than E_n. (It should be kept in mind that mass, kinetic energy, momentum, and total energy are all conserved for elastic scattering.) If the energy transferred to the struck nucleus ($E_{Z'}$) is sufficient, the atom will be torn free of its chemical bonds or displaced from its lattice position and will begin traveling through the surrounding material at a velocity greater than the orbital velocities of its electrons. As a result, the now quickly moving struck atom will lose or be stripped of one or more of its electrons, thus becoming an energetic positively charged ion. This energetic ion will go on to cause further ionizations in the surrounding material by interacting with, stripping, and capturing electrons from other atoms (Fig. 4-20). As the swiftly moving ion captures electrons and even when it finally becomes an energetic electrically neutral atom, it may become reionized, depending on its remaining kinetic energy. This ionization and recombination process continues until the energy of the initially struck atom (or positively charged ion) has been depleted to the point where no further electron stripping can take place. At this point, the struck nucleus will deplete the remainder of its energy through nonionizing collisions with other nuclei before finally coming to rest.

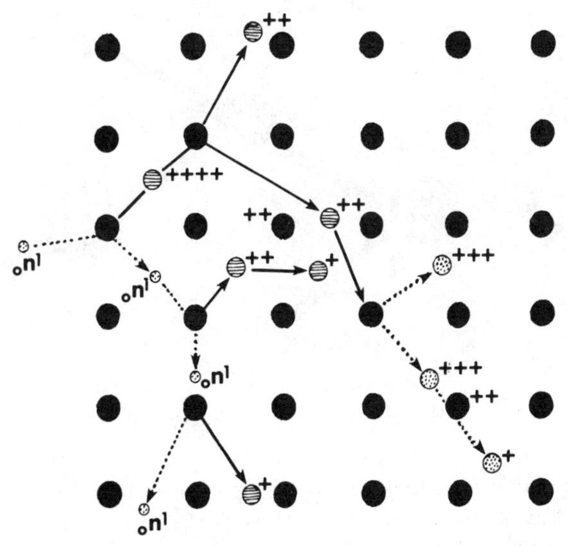

FIG. 4-20 Neutron elastic-scattering process and resultant ionization produced in a material.

The maximum amount of energy that a neutron can transfer to a nucleus depends on the mass of the struck nucleus. The larger or heavier the struck nucleus, the smaller the maximum energy transfer. This is analogous to throwing a Ping–Pong ball against a stationary bowling ball; the Ping–Pong ball bounces off with almost all of its original energy. Conversely, the smaller or lighter the target nucleus, the greater the energy that can be transferred. In fact, if the neutron and the target nucleus have the same mass (e.g., a neutron collides with a hydrogen nucleus), the neutron can transfer all of its energy to the struck nucleus. For example, if one billiard ball strikes another stationary billiard ball directly, the striking ball can transfer all of its energy to the struck ball and thus stop. Because neutrons can potentially transfer all of their energy to a hydrogen nucleus in one collision, hydrogenous materials are very effective at slowing down and even stopping neutrons. In general, elastic scattering is predominant and most efficient for low-Z target materials. Therefore, if the objective is to moderate neutrons so that they can be more readily absorbed, then low-Z materials should be used, with hydrogenous materials standing out as the best choice. But it should be remembered that neutrons are extremely efficient in secondarily producing a large number of ionizations in low-Z materials, including hydrogenous materials such as water or biological tissue, by the elastic-scattering process. Therefore, absorption of secondary electrons, in addition to the incident neutrons, must be provided.

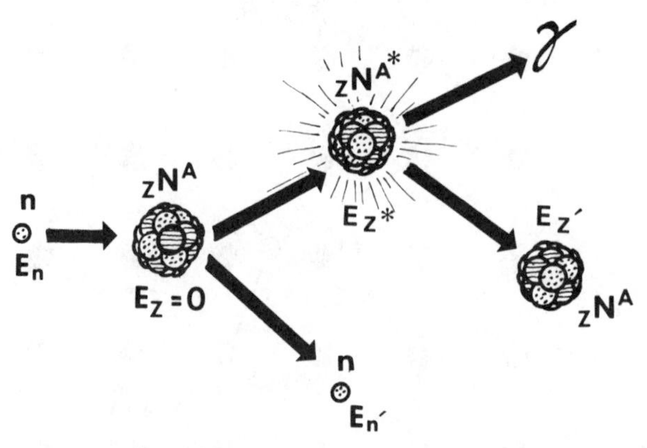

FIG. 4-21 Inelastic neutron scattering.

b. Inelastic Scattering. Inelastic scattering is more or less like elastic scattering except that excitation of the struck nucleus also occurs. This excited struck nucleus usually deexcites subsequently by emitting a gamma photon (Fig. 4-21).

For this type of neutron interaction, momentum and kinetic energy are not conserved, since part of the incident neutron's kinetic energy goes into excitation of the struck nucleus. Also, degradation of the incident neutron's energy is typically not as great for inelastic scattering as for elastic scattering, primarily because inelastic scattering is predominant for high-Z target materials. In fact, because of this, it is possible for a neutron to actually recoil with a greater kinetic energy than it originally had. In general, though, the neutron typically comes off at a somewhat degraded energy, and the struck nucleus acts very much like a struck nucleus for elastic scattering except that it is excited and usually deexcites by the emission of a gamma photon.

Because inelastic scattering results in the secondary production of penetrating gamma photons, is intrinsically less efficient in the moderation of neutrons, and typically requires high-Z materials, it is generally not preferred or recommended if elastic scattering can be used instead. Nevertheless, it is an interaction mechanism that can occur, particularly in high-Z materials; if it does, special considerations for the absorption of the secondary gamma photons produced are in order.

c. Absorption. Fission is a special case of neutron absorption. Thus, absorption can be subdivided into two more-specific categories: fission and capture.

In an absorption reaction, the incident neutron is totally absorbed by the struck nucleus, forming an excited compound nucleus with a mass number one greater than before. This compound nucleus is unstable and usually deexcites to a stable state by decay emission of a photon, or a charged particle (proton, alpha particle, beta particle, etc.), or even another neutron. In those cases where decay emission does not occur, fission does. However, fission is limited to a very select few target nuclides, and typically they must be struck by neutrons within a particular energy range. Hereafter in this chapter, fission as a special case of absorption is no longer discussed, and capture is considered to be synonymous with absorption.

The decay products from neutron capture interactions (alphas, betas, gammas, protons, and neutrons) normally are emitted with a substantial amount of kinetic energy (keV–MeV range), and they ultimately will cause significant ionizations in the surrounding material. The penetration capability and the techniques used to absorb these secondary radiations obviously depend on the radiation type involved.

Absorption is the primary interaction mechanism resulting in ionization of matter by thermal or low-energy neutrons (neutrons with energies $\lesssim 0.1$ eV). It also can play a relatively significant role in ionizations by very high-energy neutrons (greater than ~ 15 MeV). In hydrogenous materials such as water or biological tissue, a significant number of ionizations can be caused by the low-energy neutron capture reaction, shown below:

$$ {}_1H^1 + {}_0n^1 \rightarrow [{}_1H^2]* \rightarrow {}_1H^2 + \gamma(2.2 \text{ MeV}) $$

In this particular (n, γ) reaction, a very low-energy (thermal) neutron is absorbed by a hydrogen nucleus, forming an excited deuterium compound nucleus $[{}_1H^2]*$, which then deexcites, forming a stable deuterium nucleus ${}_1H^2$ and releasing a 2.2-MeV gamma photon. The 2.2-MeV photons produced from this reaction, as a result of many thermal neutron absorptions by hydrogen, will cause the vast majority of ionizations in the surrounding material.

Absorption is the only mechanism available to eliminate neutrons. As mentioned earlier, the probability that neutron absorption will occur in a material increases dramatically as the neutron energy decreases; this is generally true. However, some materials inherently have a much stronger affinity for neutron absorption than others, but they too exhibit a marked decrease for absorption with increasing neutron energy. Unfortunately, no ready rules of thumb can be used to predict these strong neutron-absorbing materials; one just has to remember some of them or rely on a nuclear engineering or physics text to find them. In general, at neutron energies above ~ 1 eV, most materials other than hydrogenous materials exhibit about the same neutron absorption affinity or cross section. But at low-neutron energies, certain materials stand out as extremely effective neutron absorbers. Such materials include boron, cadmium, gadolinium, europium, dysprosium, indium, iridium, mercury, samarium, and xenon. Materials like aluminum, iron, lead, and other readily available high-Z or known efficient photon attenuation materials offer little neutron absorption capability, particularly compared to the materials listed above. This is an extremely important point to remember for shielding purposes.

H. Protection and Shielding against Neutrons

When an unwanted neutron field exists, it should be assumed that the neutrons have not been moderated. That is, in all cases the first step in protecting against neutron exposure should be to provide for adequate neutron moderation. Thus, the recommended procedure for shielding against neutrons is to first moderate them to low energy (preferably via elastic scattering), then absorb them, and finally provide adequate shielding against secondary radiations produced (particularly gamma photons). The first step, moderation, is relatively easy because many low-Z materials (including hydrogenous materials) are readily available. They include such things as water, wood, plastics, concrete, water-laden earth, cardboard, paper, and even cloth when enough is provided. The second step, absorption, is more difficult since the best absorbers of low-energy neutrons (identified in the previous section) are not typically available. In their absence, additional quantities and thicknesses of concrete, wood, plastics, water-laden earth, or water alone can be used. The third and last step is identical to those identified for protection and shielding against X rays and gamma rays: the use of dense high-Z materials. This last step presumes that photons will be the predominant secondary radiation produced, and this is generally true. Even when it is not true, if any appreciable photons are (or can be) produced secondarily, then photon shielding should and must be provided. (In general, such photon shielding will afford adequate protection against any secondary radiation from incident neutrons, regardless of the type of secondary radiation generated.)

III. UNITS OF IONIZING RADIATION

A. Units of Activity

1. The Curie

A curie (Ci) is a unit of activity, i.e., the characteristic rate of decay of a radioactive substance where 1.0 Ci is equal to 3.7×10^{10} disintegrations/sec. This number of disintegrations per second is approximately equal to the activity or rate of decay of 1.0 g of radium.

All radioactive substances decay exponentially in time according to

$$A(t) = A_0 \exp(-\lambda t)$$

where A_0 is the activity of a particular radioactive substance at some initial time t_0; $A(t)$ is the activity of the same radioactive substance at some given time t after t_0; and λ is the decay constant of the radioactive substance, in inverse units of time.

The decay constant λ varies from material to material. In reality, it is the probability per unit of time that a radioactive atom will decay. It is related to a radioactive material's half-life $t_{1/2}$ by

$$t_{1/2} = \ln(2)/\lambda = 0.693/\lambda$$

where the half-life $t_{1/2}$ is defined as the amount of time required for the activity of a radioactive substance to decrease by a factor of two.

2. The Becquerel

A becquerel (Bq) is a unit of activity where 1.0 Bq is equal to 1.0 disintegration/sec. Thus, 1.0 Ci = 3.7×10^{10} Bq, or conversely, 1.0 Bq $\cong 2.703 \times 10^{-11}$ Ci.

B. Units of Radiation Dose

1. The Roentgen

A roentgen (R) is a unit of exposure dose in air that applies only to X-ray and gamma radiation. It is a measure of charge liberation or ionization in air caused by only X rays or gamma rays. When air at standard temperature and pressure is exposed to 1.0 R of radiation, 2.08×10^9 ion pairs are created per cubic centimeter (cm^3). This may seem like a strange value for a unit of radiation dose, but 1.0 R is also equal to 1.0 esu (electrostatic unit) of charge buildup, of either sign, per cm^3 of air exposed to X rays or gamma rays at standard temperature and pressure conditions.

Reminder: The roentgen as a unit of exposure dose applies to only, and thus can be used for only, X rays or gamma rays in air.

2. The Rad and the Gray

The rad and the gray (Gy) are units of radiation absorbed dose for any radiation type and in any material. Radiation absorbed dose is the quotient of dE_D divided by dm, where dE_D is the differential energy deposited by ionizing radiation in a differential mass dm. In the centimeter-gram-second (cgs) system of measurement, the unit of radiation-absorbed dose is the rad. One rad corresponds to the deposition of 100 ergs of energy in a material per gram of that material by ionizing radiation. In the meter-kilogram-second (mks) and SI units of measurement, the unit of radiation-absorbed dose is the gray. One gray is equal to one joule of energy deposited in one kilogram of material by ionizing radiation. In summary

$$1 \text{ rad} = 100 \text{ ergs/g} = 0.01 \text{ Gy}$$

$$1 \text{ Gy} = 1 \text{ J/kg} = 100 \text{ rad}$$

The quantification of radiation-absorbed dose makes it possible to provide and compare in a meaningful way the physical description and results of the interaction of radiation with matter; it can usually be closely correlated to the chemical or biological effects produced in the irradiated object. The importance of radiation-absorbed dose is illustrated in the case of bone and adjacent soft tissue that are identically irradiated with 100-kV X rays, resulting in an absorbed dose in the bone that may be more than three times the absorbed dose in the adjacent soft tissue.

By the way, 1 R of X-ray or gamma-ray radiation is equivalent to about 0.9 rad in tissue. So it can be conservatively said that 1 R = 1 rad in tissue.

3. Kerma

Very elaborate calculations are required to determine the radiation-absorbed dose in tissue exposed to fast neutrons on the basis of the neutron fluence. For neutrons with energies between 10 keV and 10 MeV, elastic scattering is the most important process of energy transfer. The mean energy transferred in such an elastic-scattering collision depends on the cross section for elastic scattering in the medium and the mass of the recoil nucleus. Only the first collisions of the neutrons have to be taken into account if the thickness of the medium under consideration is appreciably smaller than the mean free path of neutrons in the medium; this is the case for human exposure to fast neutrons. In order to clearly and quantitatively describe this transfer of energy from the neutrons to directly ionizing particles in an irradiated material, a quantity called kerma (kinetic energy released in material) has been introduced. Kerma description is valid for all types of indirectly ionizing particles. The derma is defined as the quotient of dE_k divided by dm, where dE_k is the sum of the initial kinetic energies of all the charged particles liberated by indirectly ionizing particles in a small differential volume element of the specified material, and dm is the mass of the matter in that volume element. Kerma has the same basic units as absorbed dose, i.e., it is typically expressed in ergs per gram. Therefore, both quantities ultimately have the same unit of measurement, the rad. However, only when radiation equilibrium exists (i.e., when all volume elements are exposed to the same neutron or indirectly ionizing radiation field) are the quantities kerma and absorbed dose truly equal. Because of the short range of secondary charged particles resulting from fast neutrons, the kerma is (for practical purposes) equal to the absorbed dose received by a small amount of tissue (formerly called "first collision dose"). Table 4-2 shows the average kerma values per unit of neutron fluence in standard tissue for some common neutron energies; here fluence is expressed in neutrons per centimeter squared.

TABLE 4-2

KERMA PER FLUENCE OF NEUTRONS IN STANDARD TISSUE

Neutron energy (MeV)	Approximate kerma[a] (10^{-7} ergs/g/neutron/cm^2)
1	2.45
3	3.50
10	5.13
14	5.0–6.7

[a]Values are averages based on reports prepared by International Commission on Radiation Units, reports of National Commission on Radiation Protection, and published individual reports.

4. The Rem and the Sievert

The rem and the sievert are units of biological dose equivalent in man, relative to X-ray, and gamma radiation. The rem dose (rem at one time stood for roentgen equivalent man) makes use of the radiation absorbed dose (which applies to all types of radiation and materials) measured or determined in biological tissue for each type of radiation that is involved, then multiplies each such value by an appropriate quality factor (QF) to determine each radiation type's contribution relative to X-ray and gamma-ray radiation, and finally adds each contribution together. In other words,

$$\text{Dose(rem)} = \sum_{\text{all } i} \text{Dose(rad)}_i \times \text{QF}_i$$

where i denotes each of the different radiation types involved and $\Sigma_{\text{all } i}$ denotes the sum over all i of that which follows this symbol.

Rewriting this equation in more readily understandable form yields: Dose(rem) = $[\text{Dose(rad)}_\alpha \times \text{QF}_\alpha] + [\text{Dose(rad)}_\beta \times \text{QF}_\beta] + [\text{Dose(rad)}_{\text{X rays}} \times \text{QF}_{\text{X rays}}] + [\text{Dose(rad)}_\gamma \times \text{QF}_\gamma] + [\text{Dose(rad)}_n \times \text{QF}_n] + ,\ldots,$ until all the individual types of radiation involved are included.

This same formula can be used for dose rates simply by making appropriate dose rate substitutions wherever doses appear.

The sievert (Sv) is the equivalent SI biological exposure dose unit in man, where 1.0 Sv = 100 rem, or conversely, 1.0 rem = 0.01 Sv.

The next obvious question is, What is this thing called the quality factor (QF)?

TABLE 4-3

QUALITY FACTORS FOR VARIOUS TYPES OF RADIATION[a]

Type of radiation	Quality factor (QF)
X Rays and gamma rays	1
Beta particles ($E_{max} > 0.03$ MeV)	1[b]
Beta particles ($E_{max} < 0.03$ MeV)	1.7[b]
Alpha particles	10
Protons	10
Heavy recoil nuclei or ions	20
Fission fragments	20
Neutrons	
Energy not specified	10
0.025 eV to 1 KeV	2
10 KeV	2.5
100 KeV	7.5
500 KeV	11
1 MeV	11
2.5 MeV	9
5 MeV	8
7 MeV	7
10 MeV	6.5
14 MeV	7.5
20 MeV	8

[a]Based on NCRP Report Number 39.
[b]Recommended in ICRP Publication Number 9.

5. The Quality Factor

As already alluded to, the quality factor (QF) is used to determine the biological radiation dose equivalent in man. Its value is determined by the particular type of radiation involved and (in some cases) also the energy of that particular radiation type. It pertains to the damage caused in man by each particular type of radiation relative to X rays and gamma rays. Therefore, it provides a fairly simple means for taking into account the relative differences in the overall human biological effect of different radiation types. Quality factor values (sometimes called Q values) are somewhat arbitrarily chosen, average round-offs of whole-body relative biological effectiveness (RBE) values for man, depending on the type of radiation involved. Extreme care must be exercised when relating QF to RBE, because RBE is a specific term tied to a particular given end result and a particular incident radiation type (and sometimes energy as well).

In any event, quality factors for various radiation types and energies are provided in Table 4-3. These values can be used to determine rem or sievert doses or dose rates by using the equation provided in the previous section, provided that

TABLE 4-4

COMPARISON OF RADIATION-ABSORBED DOSE RATE TO BIOLOGICAL EXPOSURE DOSE RATE BY EXAMPLE FOR A SPECIFIED SITUATION

Radiation type (given)	Radiation-absorbed or exposure dose rate (given)	QF[a]	Rem dose rate (calculated)
X Rays	1 mR/hr ≅ 1 mrad/hr	1	1 mrem/hr
Gamma rays	1 mR/hr ≅ 1 mrad/hr	1	1 mrem/hr
Beta particles (E_{max} = 100 KeV)	1 mrad/hr	1	1 mrem/hr
Alpha particles	1 mrad/hr	10	10 mrem/hr
Neutrons (E_n not specified)	1 mrad/hr	10	10 mrem/hr
Totals	5 mrad/hr	—	23 mrem/hr

[a]See Table 4-2.

radiation absorbed doses, or dose rates as appropriate, are known for each radiation type involved.

With this information, rem doses (or dose rates) can be readily calculated if appropriate rad doses (or dose rates) are available or can be determined. For example, assume that a combined alpha, beta (E_{max} = 100 KeV), gamma, neutron (E_n not specified), and X-ray field exists at some point of interest as specified in Table 4-4; the question to be answered is, What rem dose rate exists at that point? This example problem indicates that, although the total radiation field specified represents an overall radiation-absorbed dose rate of just 5 mrad/hr, it is equivalent to a biological exposure dose rate of almost five times that value or 23 mrem/hr.

IV. RADIOBIOLOGY: A BRIEF INTRODUCTORY SYNOPSIS

This section attempts to provide some perspective to the foregoing discussions, particularly from a radiobiological point of view. First, one might ask, "What does all this have to do with human biological damage and radiation effects?" This chapter discusses the details of ionization, i.e., the process of stripping off electrons from an atom or molecule. Biologically, we are made up of cells, which in turn are made up of both simple and complex molecules. These molecules are composed of atoms, which are chemically bound to one another via ionic or covalent bonds. Electrons that intermingle with and are shared by the molecule's constituent atomic parts are the "glue" of these bonds. If radiation can cause ionization (and it does), then it can cause these bonds to break, resulting in the creation of free radicals. Such free radicals can themselves be toxic to the cell, or they can recombine into substances that are benign or also toxic to the cell. Consider water, one of the body's main components, for example. When water is chemically broken apart by ionizing radiation, H^+ and OH^- free radicals can result. These free radicals can be toxic to the cell. They can also recombine, either benignly back into water or forming other species like hydrogen peroxide (H_2O_2), which is also toxic to the cell. Moreover, the things within and of a cell (each with a specialized function) are also made up of molecules, which can be broken apart by ionizing radiation. The breaking apart of these molecules can cause a change in cell functioning, or a change in cell genetic makeup, or cell death. Taken collectively, this is why it is important to know and understand the different types of radiation, their characteristics, the ways in which they interact with matter on a microscopic level, and the ways in which protection and shielding can be provided against them.

One might also ask, "How much radiation is too much; or conversely, how much radiation can man be expected to tolerate acutely?" The remainder of this chapter attempts to provide some general insights on the effects of acute high radiation doses, and to put these into perspective with everyday radiation doses.

The majority of this volume is dedicated to radiobiology, from the level of the cell up to the organism, with a great deal of focus on the acute radiation syndrome. The following paragraphs serve as a quick overview, using the concept of $LD_{50/30}$.

$LD_{50/30}$ is defined as the lethal whole-body radiation dose to a large population that has been uniformly and acutely exposed so that, on the average, 50% of that exposed population will die within 30 days. Therefore, the concept of $LD_{50/30}$ is a probabilistic or statistical one. For man, $LD_{50/30}$ is somewhere between 250 and 600 rads without extraordinary medical intervention; a value of about 450 rads is the generally accepted and quoted $LD_{50/30}$ for man. An acute whole-body dose of 450 rads is indeed large, and most certainly must be avoided. On the other hand, national bodies [like the National Council on Radiation Protection and Measurements (NCRP) in its Report Number 39] suggest, as once-in-a-lifetime emergency acute dose criteria, the limits of 100 rem (whole-body) to save a life and 25 rem (whole-body) to preserve valuable or essential facilities and equipment, or to circumvent further population exposures in an emergency. Further, it is generally accepted that little or no deleterious human clinical effects can be detected in the near term below ~25 rem (whole-body), even when delivered acutely; almost everyone would agree with this statement for an acute whole-body dose below ~5 rem. At low doses (i.e., doses well below 5 rem), such discussions inevitably center on what is acceptable from a societal point of view versus from an individual's point of view. Levels of acceptability are debatable and often involve both tangible and intangible factors as well as real and perceived issues. It is hoped that the information in Table 4-5 can provide the reader with a useful point of

TABLE 4-5

COMPUTATION OF ONE'S ANNUAL RADIATION DOSE[a]

Condition	Common source of radiation	Your annual inventory (mrem)
Where you live	Location: Cosmic radiation dose at sea level	40
	Add 1 for every 100 ft of elevation above sea level	____
	Home construction: Wood 35; brick 75; concrete 50; stone 70 (can use a combination by averaging)	____
	Ground: U.S. average	15
What you eat, drink, and breathe	Air: U.S. average	5
	Water and food: U.S. average	25
How you live	Jet travel: Four times the number of 6,000-mile flights above ~12,000 ft	____
	Television	____
	Black and white: One times the average number of hours per day	____
	Color: Two times the average number of hours per day	
	Smoking: 125 times the average number of cigarettes consumed/smoked per day (note that this represents a local dose to the lung caused by polonium-210, an alpha emitter)	____
Medical	X-Ray diagnosis (local doses)	
	Chest X ray ____ ×50	____
	GI tract X ray ____ ×2000	____
	Dental X ray ____ ×15	____
	Total	____ mrem/year

[a]Compare your annual dose to the U.S. annual average of 125–170 mrem.

reference for formulating informed personal conclusions and judgments on this subject.

BIBLIOGRAPHY

Asimov, I., and Dobzhansky, T. *The Genetic Effects of Radiation* (Library of Congress Catalog Card No. 66-62747). U.S. Atomic Energy Commission, Office of Information Services, Washington, D.C., 1973.

Attix, F. H., and Roesch, W. C., eds. *Radiation Dosimetry*, Vol. 1. Academic Press, New York, 1968.

Brode, H. L., *Review of Nuclear Weapons Effects*. The Rand Corporation, Santa Monica, California, 1977.

Evans, R. D. *The Atomic Nucleus.* McGraw-Hill, New York, 1972.
Glasstone, S., ed. *The Effects of Nuclear Weapons.* Department of Defense and Department of Energy, Washington, D.C., 1964.
Henry, H. F. *Fundamentals of Radiation Protection.* Wiley, New York, 1968.
Hogerton, J. F. *The Atomic Energy Deskbook.* Reinhold Publishing Corp., New York, 1963.
Hughes, D. J., and Schwartz, R. B. *Neutron Cross Sections,* 2nd ed., BNL 325. Brookhaven National Laboratory, Upton, New York, 1958.
International Commission of Radiological Units *Radiation Quantities and Units,* ICRU Rep. 19. ICRU, Washington, D.C., 1971.
Lamarsh, J. R. *Introduction to Nuclear Engineering.* Addison-Wesley, Reading, Massachusetts, 1975.
Lapp, R. E., and Andrews, H. L. *Nuclear Radiation Physics,* 3rd ed. Prentice-Hall, Englewood Cliffs, New Jersey, 1963.
Lewis, K. N. The prompt and delayed effects of nuclear war. *Sci. Am.* **241,** 35–47 (1979).
Meyerhoff, W. E. *Elements of Nuclear Physics.* McGraw-Hill, New York, 1967.
National Council on Radiation Protection and Measurements *Basic Radiation Protection Criteria,* NCRP Rep. No. 39. NCRP, Washington, D.C., 1971.
Taylor, L. S. *Radiation Protection Standards.* CRC Press, Cleveland, Ohio, 1971. (Contains thorough review of history of radiation protection standards together with excerpts from the most important ICRP, ICRU, NCRP, and FRC documents.)
U.S. Atomic Energy Commission *Nuclear Terms: A Glossary,* 2nd ed. (Library of Congress Catalog Card No. 67-60141). Office of Information Services, USAEC, Washington, D.C., 1974.

CHAPTER 5

Cellular Radiation Biology

EUGENE V. HOLAHAN, JR.

Armed Forces Radiobiology Research Institute
Bethesda, Maryland 20814-5145

Radiobiology is the field of study that describes and explains the many changes that radiation produces in biological material, from the gross to the microscopic, the lethal to the nonlethal, and the effects observed immediately after irradiation to those seen only years later. The objectives of this chapter are to examine three questions: Why is exposure to ionizing radiation hazardous? Can we compare and relate different types of ionizing radiation in terms of the relative biological effects produced? Can the potentially lethal effects of ionizing radiation be reduced, and is there a theoretical and/or practical basis for chemically induced radioprotection in man?

I. RADIATION INTERACTIONS

The previous chapter discusses the physical principles of how ionizing radiation may be classified as directly or indirectly ionizing. Charged particles (e.g., electrons, protons, beta particles, alpha particles, or fission fragments) are referred to as directly ionizing radiations since they can directly disrupt (ionize) chemical bonds and thus produce chemical and biological changes. Electromagnetic radiation (e.g., X rays, gamma rays, and neutrons) are referred to as indirectly ionizing radiations since they do not themselves produce any chemical or biological change. Rather, they deposit kinetic energy into the material through which they pass, and produce secondary charged particles (electrons) that can induce subsequent chemical and biological changes.

Ionizing radiations transfer energy to biological material by two mechanisms: ionization and excitation. Although a range of energies may cause ionization, an average of 33 eV (electron volts) is deposited in a system for each ionization that occurs. Only 10–25 eV is needed to eject an electron from a molecule (ionize) while the excess energy is distributed over the entire molecule and subsequently removed from the molecule as oscillation energy. However, this excitation energy also may be concentrated at one bond and induce a breakage. Generally, ionizations are not produced singly, but as double or triple events, known as clusters. Based on the assumption that an average of three ionizations occur per cluster, the figure of 100 eV/primary ionization is often used when discussing energy transfer. Even though the amount of energy involved in ionization appears very small, it tends to be very efficient and extremely lethal. If 100 eV/cluster were deposited in a sphere 30 Å in diameter, it would increase the temperature (locally) from 37°C to approximately 80°C. Consequently, it is the distribution of the energy and not the total amount of deposited energy that is significant for cell inactivation.

II. RADIATION CHEMISTRY

It is important to distinguish between (1) a molecule that has received energy directly through ionization or excitation from the incident radiation and (2) a molecule that has received the energy by transfer from another molecule. These two processes are referred to as the direct effect of radiation and the indirect effect of radiation, respectively. The indirect effect is especially important in aqueous systems, in which a water molecule may be ionized and then may transfer its acquired energy to another molecule.

A. Formation of Ion Pairs

When ionizing radiations deposit kinetic energy in biological material, they create charged particles, which subsequently ionize molecules or atoms. When an electron (e^-) is ejected from a molecule, a positive ion (A^+) is created. This process may be described by the equation

$$A \rightarrow A^+ + e^-$$

The ions produced, A^+ and e^-, are known as ion pairs. The formation of the ion pair is a purely physical process, and it is the initial step to both the direct and indirect effects of radiation. Ion pairs have an extremely short lifetime, on the order of 10^{-18} to 10^{-16} sec. They then undergo one of many reactions to form free radicals.

B. Free Radicals

A free radical is an electronically neutral, highly reactive molecule with an unpaired electron in an outer orbital shell. Free radicals are almost always intermediaries between ion pairs and the final chemical products. The process of free radical formation can be exemplified with water. Since most biological systems are composed of about 70–75% water, most primary ionizations in living tissue will occur in water molecules. When a charged particle interacts with a water molecule, an electron is ejected, which results in an ionized molecule (H_2O^+)

$$H_2O \rightarrow H_2O^+ + e^-$$

The ionized water may then react with another water molecule to form the highly reactive hydroxy radical (OH·)

$$H_2O^+ + H_2O \rightarrow H^+ + OH\cdot$$

The unpaired electron (e^-) created by the ionization of the water molecule may recombine with the water ion or more probably be picked up by another water molecule to form an aqueous electron (H_2O^-). This aqueous electron most probably will decompose in the presence of water to form a hydroxy ion (OH^-) and a hydrogen radical (H·)

$$H_2O^- + H_2O \rightarrow OH^- + H\cdot$$

Since the hydroxy ion (OH^-) and the hydrogen ion (H^+) do not contain an excess of energy, they will combine to reform a water molecule. Hence, the net result from the irradiation of water will be:

$$H_2O - H\cdot + OH\cdot$$

The free radicals, H· and OH·, are highly reactive and have a lifetime on the order of 10^{-9} to 10^{-11} sec. These radicals are of considerable energy and can break chemical bonds. Furthermore, these free radicals can rejoin each other by sharing their unpaired electrons in a chemical bond, thus forming H_2O, H_2, or H_2O_2 (hydrogen peroxide), or they can react with other ordinary molecules. It is thought that the oxygen effect operates at this level. Oxygen is an excellent electron accepter and can combine with the hydrogen radical (H·) to form a peroxyl radical

$$H\cdot + O_2 \rightarrow HO_2^{\cdot}$$

This product also is very reactive, and like other free radicals, its production represents a new molecule to the biological system. It is a less powerful oxidizing agent than OH· or H_2O_2, and has a longer life and is therefore capable of diffusing farther from its site of formation (~20 Å) to undergo further interactions.

Since the formation of free radicals is a chemical process, it can be modulated by biological processes. For example, peroxide-destroying enzymes such as catalase and peroxidase can reduce the amount of intracellular H_2O_2 and thus reduce the amount of biological damage that otherwise would have been produced. Furthermore, aminothiols (sulfhydryl-containing compounds) like cysteine and glutathione can act as radical scavengers. That is to say, these sulfhydryl compounds reduce the reaction between free hydroxy radicals and oxygen to form peroxyl radicals by competing with oxygen as a hydrogen donor

$$2\ OH\cdot + 2\ R\text{-}SH \rightarrow 2\ H_2O + RSSR$$

Due to the effects of these biological processes, approximately half the damage to critical molecules is induced by the indirect effects of ionizing radiation. The other half of the damage is the result of direct effects, i.e., direct ionization in critical biological molecules.

C. Radiation Effects on Macromolecules

The mechanism whereby cells lose their reproductive integrity as a result of irradiation is not fully understood, and the sensitive or critical target sites have not been identified. The deposition of energy (directly or indirectly) by ionizing radiation induces chemical changes in large molecules, which may then undergo a variety of structural changes. These structural changes include (1) hydrogen bond breakage, (2) molecular degradation or breakage, and (3) intramolecular and intermolecular cross-linking. It has been proposed that one or more of these structural changes leads to altered molecular function and eventually cell death.

Many macromolecules are normally held in rigid conformation by intramolecular cross-linking bonds; that is, specific chemical groups are linked together, frequently by hydrogen bonds, to form a three-dimensional structure. The hydrogen bonds are among the weakest in the molecule and thus are the most susceptible to disruption by radiation. Such structural changes can lead to severe alterations in the biochemical properties of the molecule. Loss of protein (enzyme) function by irradiation may be the result of a break in the hydrogen or disulfide bonds that maintain the secondary and tertiary structure of the molecule. Such a break can lead to a partial unfolding of the tightly coiled peptide chains which, in turn, can lead to a disorganization of the internal structure, resulting in a change or loss in chemical or enzymatic activity. Furthermore, molecular unfolding may lead to a decrease in protein solubility, an increase in protein aggregation, and finally protein precipitation. Radiation also may alter the viscosity, refractive index, optical rotation, and electrical conductivity of cellular proteins.

Rupture of the hydrogen bonds that link base pairs in DNA also may be an important effect of irradiation. It is not known whether irreversible damage is dependent on only one or a large number of H-bond ruptures. Finally, changes in the secondary and tertiary structure of proteins and DNA may affect the association of nuclear proteins with DNA. Such alterations could eventually alter genetic transcription and translation.

Molecular degradation or breakage into smaller units has been shown to occur when macromolecules are irradiated. Moreover, in molecules that contain a series of identical or similar repeating units, the breaks usually occur in the same bond, suggesting that energy absorbed anywhere in the molecule can be transmitted down the molecular chain to the weakest bond. The side chain groups of protein amino acids are the most radiosensitive portions of a protein molecule because they can act as traps for electrons. The specific changes that occur in these side chains are dependent on their chemical composition. The consequences of such a change will be similar to those observed for hydrogen bond breakage.

Molecular breakage also occurs in DNA. These breakages may occur as single-strand breaks, double-strand breaks (both chains of the double helix simultaneously), base loss (e.g., apurination), or base changes (e.g., deamination). Single-strand breakage between a sugar and a phosphate group often rejoin, provided there is no opportunity for the broken portions of the strand to separate. However, in the presence of oxygen, if at least one end of the broken strands is in a reactive state, this end may become peroxidized and thus unable to rejoin. A double-strand break will occur only if there is a break in each of the two strands less than about five nucleotides apart. This can occur when two single-strand breaks come into juxtaposition or when a single densely ionizing particle produces a break in both strands. The latter requires approximately 500 eV and could occur when a cluster of ionizations occurs within 20 Å (see Fig. 5-1).

Macromolecular cross-linking is another common macromolecular structural change. The molecule can undergo intramolecular cross-linking where the molecule becomes attached to itself, and intermolecular cross-linking where the molecule becomes attached to another molecule. In general, both proteins and nucleic acids can become cross-linked when a chemically active locus is produced and then comes into contact with another reactive area. Cross-linking of proteins will result in the inactivation of the molecule as well as decreased solubility and precipitation.

The formation of cross-links within the double helix of a DNA molecule, or between two DNA molecules, and/or between DNA and protein also can be attributed to the formation and rejoining of reactive sites. Cross-linking within the DNA double helix may occur between two bases, as has been demonstrated with ultraviolet radiation. Thymidine dimerization induced by ultraviolet radia-

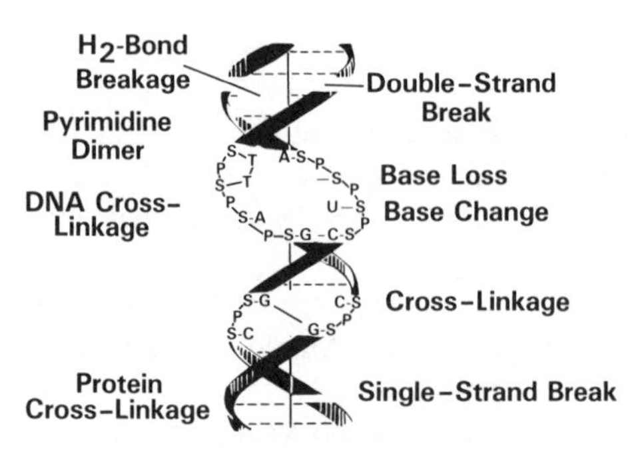

FIG. 5-1 Radiation damage to DNA.

tion forms a dimer linkage that is significantly more stable than a hydrogen bond. Similar base dimerization can be induced by ionizing radiation as well.

DNA complexed with protein is less sensitive to radiation-induced base cleavage or chain breakage than is noncomplexed DNA. Part of the protection afforded by nuclear proteins results from competition for and shielding from free radicals. However, due to the close association of DNA with nucleoproteins (histone and nonhistone proteins) and the nuclear membrane, intermolecular cross-linking among the elements of this complex can occur. These cross-links can impair both gene transcription and DNA replication.

Considerably less information is available regarding the radiation effects of RNA. However, it is likely that the changes produced are similar to those in DNA.

D. TARGET HYPOTHESIS

Although the critical lesion responsible for cell killing has not been identified, it has been established that the radiation-sensitive targets are located in the nucleus and not the cytoplasm of the cell. This was clearly demonstrated by Munro (1970) when he irradiated Chinese hamster cells with alpha particles emitted from a polonium source on the tip of a tungsten needle. These experiments demonstrated that a large dose of alpha radiation (50,000–100,000 cGy) to the cytoplasm does not kill the cell, whereas nuclear irradiation by several alpha particles (mean dose 26 cGy) can result in significant cell killing. Consequently, the main targets for lethal effects were postulated to be in or near (within 2 μm) the cell nucleus.

When cells are grown in the presence of precursor molecules labeled with the radioactive isotopes ^{3}H, ^{14}C, ^{32}P, and ^{125}I, the isotopes are incorporated into specific macromolecules. If the isotopes are concentrated in a specific site within the cell cytoplasm or nucleus, decay of the isotope will damage not only the labeled macromolecule but also the cellular site around the macromolecules. For example, the beta particle emitted from ^{3}H decay produces a densely ionized column with an average track length of 1 μm and an average of 160 ionizations along the track. Mouse leukemia cells labeled with [^{3}H]thymidine for one generation were killed three times more effectively than cells labeled with [^{3}H]uridine, and eight times more effectively than with [^{3}H]lysine or [^{3}H]histidine. This suggests that the effectiveness of ^{3}H decay depends on cell location and that macromolecular sensitivities follow the sequence DNA > mRNA > rRNA and tRNA > amino acids. Consequently, these studies implicate DNA as the most effective target molecule for cell killing.

If DNA were the target molecule, organisms with a high DNA content should have a greater chance of being damaged by radiation than organisms with a low DNA content. After reviewing the nucleic acid content and radiosensitivity of various organisms, Terzi (1961) suggested that organisms can be categorized into four classes, each with a constant ratio of target volume to nucleic acid content. Kaplan and Moses (1964) extended this concept to include a linear relationship between the logarithm of both the nucleic acid content and radiosensitivity. The ratio of the target volume to nucleic acid content is approximately 0.033 (one-thirtieth) for haploid organisms. This can be inferred to mean that (1) only 3% of DNA is vital to the cell, (2) 1 out of 30 hits produces one lethal event in DNA, (3) 97% of the damaged DNA is restored by an effective repair mechanism, (4) 3% of the DNA is vulnerable to the action of radiation while 97% is protected by proteins, and (5) DNA is not the target molecule. The last inference can be rejected based on the label incorporation data, while the second and third possibilities are the most likely to occur.

A direct correlation has been observed in hamster cells between the loss of the cells' reproductive integrity and the induction of radiation-induced chromosome aberrations. Chromosome aberrations are abnormal configurations or abnormal spatial arrangements of chromosomes. After chromosomal breakage, the broken end may recombine with the piece from which it was detached (restitution), or it may remain uncoupled and result in loss of genetic material during cell replication (deletion), or it may recombine with fragments of different chromosomes. Examples of chromosome aberrations include terminal deletions, interstitial deletions, inversions, translocations, and dicentrics. The seriousness of chromosomal aberrations depends on how many chromosomes are broken and how many breaks are induced per chromosome. Furthermore, the production of chromosomal aberrations is dependent on the radiation dose, the quality (photon, particle) of the radiation, the energy of the radiation, and irradiation in the presence or absence of oxygen.

The molecules responsible for the induction of chromosome aberrations most likely consist of DNA and protein. When structural analogs of thymidine (e.g., 5-bromodeoxyuridine, BUdR) are incorporated into DNA, this substitution dramatically increases the radiosensitivity of mammalian cells to a degree that is directly proportional to the amount of the analog incorporated. Furthermore, incorporation of BUdR results in an increase in radiation-induced chromosomal breaks, whereas substituted deoxyuridines that are not incorporated into DNA have no such effect on cellular radiosensitivity.

E. Summary of Radiation Chemistry

The chain of events from the incident X-ray photon or fast neutron to the observed biological change can be summarized (see Fig. 5-2). The described mechanisms apply to all types of ionizing radiations.

III. COMPARISON OF IONIZING RADIATIONS

The nature of the radiation exposure makes a considerable difference quantitatively in the biological response produced in the system, even when the same amount of energy has been deposited and the same number of ion pairs have been produced. In order to understand the reasons for this, it is necessary to have a means of comparing different radiations.

Directly ionizing radiations (fast electrons produced by X or gamma rays, recoil protons, beta particles, and fission neutrons) have very different charge-to-mass ratios and velocities, which result in markedly different spatial distributions of the ionizing events they produce. In general, the average separation of primary events (ionizations) along the track of an ionizing particle decreases with increasing charge and mass. For example, the tracks of fast electrons can be traced by ion pairs that are well separated in space; hence, X rays are described as sparsely ionizing radiation. Conversely, alpha particles (which carry two positive charges on a particle, are four times heavier than a proton and 8000 times heavier than an electron, but have shorter path lengths due to their increased mass and slower velocities) give rise to individual ionizing events that occur so closely together that they tend to overlap, giving rise to tracks that consist of well-delineated columns of ionization. Hence alpha particles and fission fragments are referred to as densely ionizing radiations. Neutrons, on the other hand, have intermediate velocities and mass, and the spatial distribution of ionizations fall between these two extremes. Consequently, neutrons are referred to as intermediate ionizing radiations.

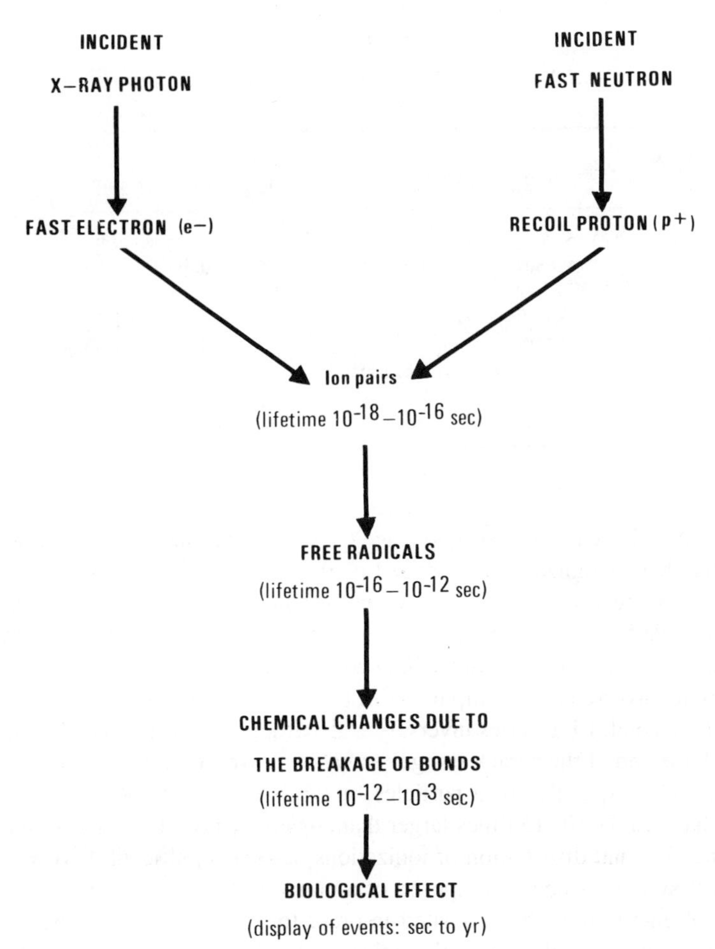

FIG. 5-2 Radiation chain of events from incident X-ray photon or fast neutron to observed biological change.

A. LINEAR ENERGY TRANSFER (LET)

The biological effects produced by radiation depend on the total energy deposited and the distribution of this energy. Zirkle (1954) introduced the term linear energy transfer (LET) to describe the average energy (*dE*) transferred locally to the biological material by a charged particle per unit length of the track (*dL*). The unit used to describe this quantity is kilo-electron-volt per micrometer of unit density material. Since charged particles possess a wide spectrum of kinetic

TABLE 5-1

LET VALUES IN WATER FOR VARIOUS RADIATIONS OF DIFFERENT SOURCES

Radiation	LET (keV/μm)	RBE
4 MeV X ray	0.3	0.6
^{60}Co gamma ray (1.2–1.3 MeV)	0.3	0.8
250 keV X ray	3	1.0
0.6 keV beta from ^{3}H	5.5	1.3
7 MeV protons	10	1.5
Recoil proton from fission neutron	45	1.8
5.3 MeV alpha from Po	190	3.5
Fission fragments from ^{239}Pu	4000–9000	0.7

energies, the LET can only be an average quantity and cannot imply a knowledge of the number of ionizations produced or the exact energy deposition of each ionization. Some values of LET in water for a number of types of radiation are given in Table 5-1. Several generalizations can be drawn from these LET values. First, LET is greater for particulate than nonparticulate radiation. This difference is due to the inverse relationship between the mass and the charge of the particle and LET. Second, LET varies inversely with particle kinetic energy. As particles approach the end of their track (range) and slow down, the density of ionizations increases. Consequently, beta particles, due to their lower energy and shorter ranges, have LETs 10–15 times larger than 3-MeV X rays or ^{60}Co gamma rays.

The number and distribution of ionizations become significant if we view the biological system as being composed of a number of targets in which a certain amount of energy must be deposited in order to produce an effect. Sometimes these targets are as small as a single critical molecule or as large as a cell or entire tissue. Consequently, a single ionization or multiple ionizations may be required to produce an effect. If the target is large and only one ionization is needed to produce the effect, low-LET radiation distributes ionizations widely and hits several targets (Fig. 5-3). However, for small targets that require multiple hits to produce an effect, high-LET radiations will be most effective.

B. RELATIVE BIOLOGICAL EFFECTIVENESS (RBE)

As a result of the LET difference in energy deposition between different types of radiation, it should not be surprising to find that equal doses of radiation do not produce equal biological effects. For example, 600 cGy of 250-kV X rays are

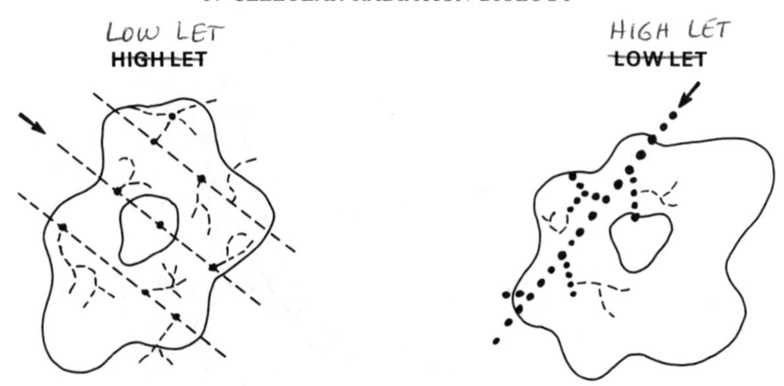

FIG. 5-3 Distribution of ionizations from particulate and electromagnetic radiations. Number of ionizations is same for each cell. Ionizations from neutrons are clustered. Ionizations from X and gamma rays are diffusely distributed.

required to kill 50% of a mouse colony (LD_{50} = 600 cGy), whereas only 300 cGy of fast neutrons are required for the same effect. Therefore, neutrons in this situation may be considered to be twice as effective as X rays.

When comparing the relative biological effectiveness (RBE) of two types of radiation, it has become customary to use 250-kV X rays as a standard reference radiation. In retrospect, this was probably a poor choice because gamma emitters like ^{60}Co have lower LETs and emit a more monoenergetic radiation, but X-ray effects were well documented and more widely used at the time of its selection. The formal definition of RBE is as follows:

$$\text{RBE} = \frac{\text{Biological efficiency of radiation under investigation}}{\text{Biological efficiency of 250-kV X rays}}$$

$$\text{RBE} = \frac{\text{Dose (rad to produce effect with 250-kV X rays)}}{\text{Dose (rad) to produce effect with radiation under investigation}}$$

RBE is initially proportional to LET. As the LET for ionizing radiation increases, so does the RBE (Table 5-1). This increase is attributed to a higher ionization density and therefore a more optimal energy deposition for high-LET radiations. However, beyond 100 keV/μm of tissue, the RBE decreases with increasing LET. As more ionizations are produced in the biological system, part of the energy deposited is wasted due to an overkill effect. The variation of the RBE as a function of LET is illustrated for damage to the clone-forming capacity of human kidney cells (Fig. 5-4).

The RBE of a particular radiation depends on not only the effect being studied but also the magnitude of effect, the dose, dose rate, dose fractionation, presence or absence of oxygen, and postirradiation conditions. These factors that affect RBE will be elaborated on in the next section.

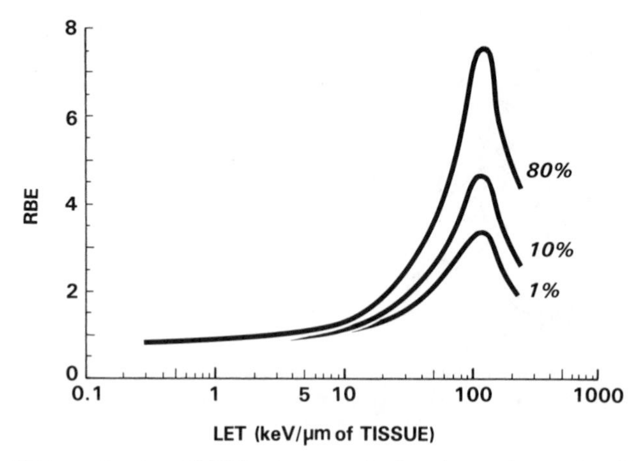

FIG. 5-4 RBE as a function of LET for damage to *in vitro* clonogenic capacity of human kidney cells.

IV. CELLULAR RADIATION BIOLOGY

A host of biological effects are induced by ionizing radiation. The magnitude to which each of these effects is induced depends on the amount of radiation absorbed and the amount of time required for the effect to be expressed. One particular biological effect that has been used to study the effects of radiation is cell death or the loss of reproductive integrity. The usefulness of measuring cell death is limited in that it is an effect that is applicable only to proliferating cell systems (e.g., hemopoietic and gut). For differentiated cells, which do not proliferate (e.g., nerve, muscle, or secretory), death can be defined as the loss of a specific metabolic function. Consequently, the concepts of cell death for these two cell systems are quite divergent. For example, the average lethal dose that induces reproductive death in proliferating cell systems is approximately 200 cGy. Conversely, tens of thousands of cGy are necessary to destroy cell function in nonproliferating cell systems.

A. CELL SURVIVAL CURVES

The most direct method for perceiving the ability of a single cell to proliferate is to wait until it gives rise to a colony. This technique, known as viable counting, dates back nearly 100 years to the microbiological methods developed by Robert Koch. It was not until 1955, when Puck and Marcus (1956) developed a method of culturing mammalian cells *in vitro,* that viable counting techniques

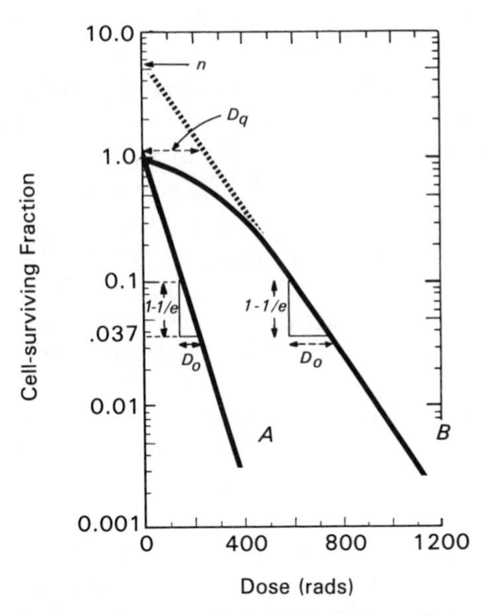

FIG. 5-5 Typical survival curve for mammalian cells exposed to radiation. A, Densely ionizing radiation; B, sparsely ionizing radiation. Reprinted with permission from Hall (1978). Copyright © 1978 by Lippincott/Harper & Row.

could be applied to mammalian cells in much the same manner as bacteria. In tissue culture, a radiation-killed cell may fail to show evidence of morphological change for long periods, but reproductive death eventually occurs. It is conventional to refer to a non-colony-forming cell as dead and to describe treated or irradiated cells that have retained their ability to give rise to a colony by successfully dividing five or more times as survivors.

The loss of the ability to form a colony as a function of radiation exposure can be described by the dose–survival curve. Survival curves are generally presented with the dose plotted on a linear scale and the fraction of treated surviving cells plotted on a logarithmic scale (Fig. 5-5). For densely ionizing radiations, such as alpha particles or fission fragments, the survival data closely approximate a straight line from the origin (curve A). The shape of the survival curve can be described by a single parameter, the slope of the line. The slope is expressed in terms of the dose required to reduce the number of clonogenic cells to $1/e$ (37%) of their former value, and this is designated as the D_0 value. Using classic target theory, the D_0 value also represents the dose required to induce an average of one lethal event per cell. Changes in cellular radiosensitivity are often described in terms of the D_0 value.

In the case of low-LET radiation, the survival curve is characterized by an initial shoulder followed by a portion that becomes linear on a semilogarithmic plot. These curves (illustrated in Fig. 5-5, curve B) are described by two parameters: the extrapolation number, *n*, and the slope, D_0. The extrapolation number is determined by extrapolating the linear portion of the survival curve until it intersects the surviving fraction axis. The extrapolation number is significant because it is proportional in size to the cells' capacity to accumulate and repair radiation damage. Typical values for the extrapolation number range from 1.5 to 12, and the slope of various mammalian-cell radiation survival curves ranges from 100 to 200 cGy.

One hypothesis that has been proposed to explain the shape of radiation survival curves, the single-hit/multitarget model, postulates that a mammalian cell can continue to reproduce until damage occurs in several targets. Damage to some of these targets occurs in the shoulder region of the curve, but reproductive capacity is preserved in essentially all the cells. With increasing dose, more targets are hit, until a point is reached where the cell can no longer accumulate damage without being killed. At this point the survival curve becomes similar to the exponential behavior of a single-hit/single-target system.

Ways have been devised to assess the viability of populations of irradiated mammalian cells *in vivo*. If colonies are to be counted, it is necessary that the colony formed from a single cell remain discrete from other colonies. McCulloch and Till (1962) reported that discrete nodules appeared on the spleens of mice that were exposed to a sublethal dose of radiation. The nodules, 10–20/spleen, could be seen and counted only because so few of the bone marrow cells irradiated retained the ability to proliferate. Kember (1967) used microscopic observation to count colonies arising from cells in irradiated cartilage, and Withers (1967) developed techniques to assay cell viability in the irradiated epidermis of the mouse. Finally, Withers and Elkind (1968) measured the radiosensitivity of mouse intestinal crypt cells by counting the number of microvilli per circumference that renewed after irradiation.

In each of these *in vivo* systems, the radiation survival curves appeared similar to those illustrated in Fig. 5-5; each possessed shoulders (*n* ranged from 1.5 to >12) and straight-line portions (D_0 ranged from 95 to 130 cGy). As a result, changes in the radiation survival curves, *in vitro* and *in vivo,* could be used to examine the effects of dose fractionation, low-dose-rate irradiation, oxygen, and LET, as well as the effect of repair on cellular radiosensitivity.

B. Factors That Affect Cellular Radiosensitivity

1. Recovery from Radiation Damage

Damage to mammalian cells produced by ionizing radiation can be divided into three categories: (1) lethal damage, which is nonrepairable and leads to cell

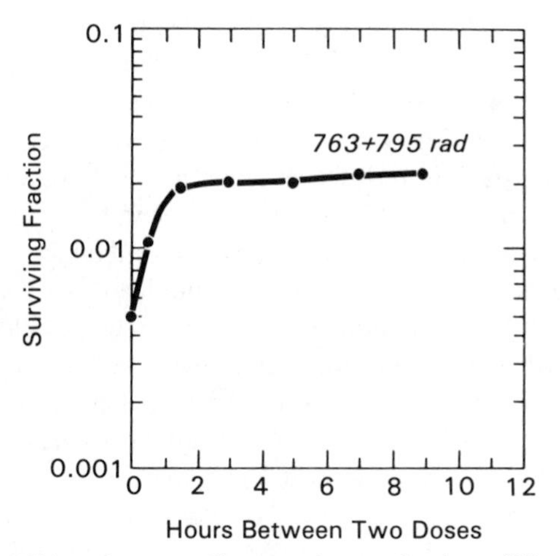

FIG. 5-6 Survival of Chinese hamster cells exposed to two fractions of X rays incubated at room Reprinted with permission from Sinclair and Morton (1966).

death; (2) sublethal damage (SLD), which is repairable; and (3) potentially lethal damage (PLD), which is that portion of SLD that may or may not be repairable, depending on the postirradiation environmental conditions.

The shoulder of the mammalian survival curve suggested to Elkind and coworkers that survival might be increased if a given radiation dose were delivered as two separate doses spaced at variable time intervals. For example, the surviving fraction was determined for a total dose of 1558 cGy delivered in approximately two equal doses separated by 0–6 hr (Fig. 5-6). As the interval between fractions was increased, the surviving fraction increased until a plateau was reached at 2 hr. This increase was said to be due to the repair of sublethal damage from the first exposure. Repair of SLD has been demonstrated in virtually every biological test system, including the response of intact normal skin, the tumor-control dose of experimental tumors in animals, and the lethal effects of whole-body irradiation in animals.

The amount of sublethal damage that is repaired depends on the magnitude of the shoulder of the radiation survival curve; thus it will vary with the LET of the radiation exposure. Consequently, less repair occurs between fractionated doses of high-LET radiation. For example, fractionating equal doses of alpha particles results in the same biological effect as the total dose in a single fraction. The repair of sublethal damage from fission neutrons, which have a LET intermediate between X rays and alpha particles, is intermediate between X rays and alpha particles.

Because of repair, a series of fractionated doses would require a greater total dose to produce the same biological effect of a single dose. This extra radiation, said to be wasted or lost because of dose fractionation, is quite consistent, ranging from 300 to 600 cGy under oxygenated conditions for normal and malignant tissues. For *in vitro* survival curves, repair is calculated to correct less than 200 cGy of cellular injury by radiation. The difference between *in vivo* and *in vitro* repair may be due to different environmental conditions.

Varying environmental conditions after exposure to X rays can influence the fraction of cells that survive a given dose. The damage is potentially lethal because under physiological conditions it would lead to cell death; but if postirradiation conditions are suboptimal (i.e., temperatures below 37°C, poor nutrients, overcrowded densities) for cell growth, cell killing is reduced. This reduction in cell killing, attributed to PLD repair, has been observed only in *in vitro* cultures and in tumors *in vivo,* not in normal tissues. The mechanism by which PLD repair occurs is not known, but it may be related to environmentally induced increases in cell cycle time. For example, conditions that prevent or retard cell growth, and hence cell division, also promote PLD repair. Consequently, a cell that is afforded more time to repair radiation damage before cell division may have a greater probability for recovery. However, regardless of how long irradiated cells are prevented from undergoing cell division, not all damage can be repaired. Liver was considered many years ago to be a radioresistant organ, since quite large doses of radiation did not noticeably affect its function. However, if hepatic cells were stimulated to divide, regardless of how long postirradiation, a great deal of morphological damage appeared; i.e., the latent lethal damage was expressed once the cells underwent mitotic division.

As was observed with SLD, no PLD repair occurs after irradiation with high-LET radiation. Although the inhibition of repair is not fully understood, it is believed to be a function of higher ionization density and the production of more lethal, nonrepairable damage.

2. Radiosensitivity and the Mitotic Cycle

Elkinds' split-dose recovery curves demonstrated an increase and plateau in cell survival when two X-ray doses were separated by 0–6 hr and the cells were maintained at room temperature to prevent cell growth. However, when the cells were maintained at physiological temperatures (37°C) between fractionated radiation doses, survival initially increased (recovery) but then decreased with increasing time between the dose fractions. This decrease was thought at the time to be associated with cell progression (redistribution) from radiation-resistant portions to radiation-sensitive portions of the cell cycle.

All mammalian cells, whether in culture or growing normally in tissue, have a DNA synthetic cycle that can be divided into four phases as first described by Howard and Pelc (1953): (1) cell division or mitosis, M, which lasts about 1 hr; (2) the first interval or gap between mitosis and DNA synthesis, G_1, which is

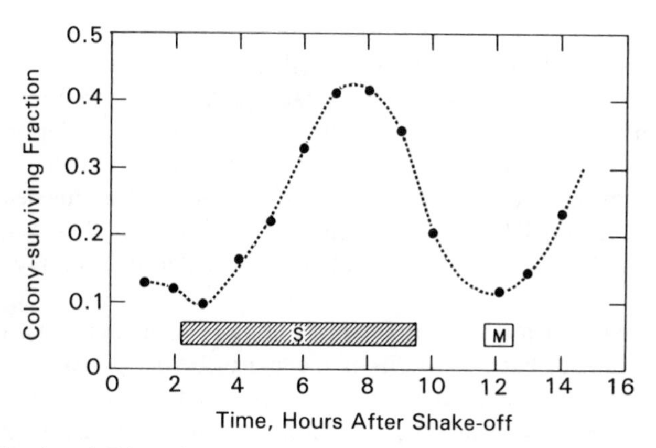

FIG. 5-7 Fraction of Chinese hamster cells surviving a dose of X rays as a function of time. Reprinted with permission from Sinclair and Morton (1966).

highly variable in length; (3) the period of DNA synthesis or replication, S, which, depending on the cell type, ranges between 6 and 15 hr; and (4) the second interval or gap between DNA synthesis and the next mitosis, G_2, which in general never varies more than 3–4 hr in length. A population of cells that is in exponential growth is made up of cells that are in each of these four phases. The proportion of cells in each phase of the cycle will be approximately the same as the fraction of the cell cycle time occupied by that phase. Therefore, in order to study the radiosensitivity of each of these four cell cycle phases, cells must be manipulated in order to acquire homogeneous populations. Two techniques that have been used to obtain synchronously dividing cell populations include chemical synchronization and mitotic selection. Chemical synchronization involves the addition of drugs (e.g., excess thymidine and hydroxyurea) that reversibly inhibit DNA synthesis and block the cells at the G_1–S border. When the drugs are removed, the surviving cells (some agents are toxic to S-phase cells) progress through the cell cycle fairly uniformly for several hours. Mitotic selection, first described by Terisima and Tolmach (1963), involves the gentle agitation of exponentially growing cells from which rounded, loosely attached mitotic cells are selectively detached and collected. If these mitotic cells are plated into flasks and incubated at 37°C before irradiation, characteristic radiation survival curves can be generated (Fig. 5-7).

A number of investigators have performed comparable experiments with different mammalian cells and obtained similar results. The following summarizes the main characteristics of the variation of radiosensitivity with cell age in the DNA synthetic cycle: (1) cells are most sensitive at or close to mitosis; (2) resistance is usually greatest in the latter part of the S-phase; (3) if the cell has a long G_1 phase, a resistant period is evident in early G_1 followed by a sensitive period at the G_1–S border; and (4) G_2 is usually sensitive, perhaps as sensitive as M.

These generalities are valid for proliferating cells *in vitro* and *in vivo,* and the variation in radiation response through the cell cycle is not dependent on the presence or absence of oxygen. Furthermore, as the LET of the radiation increases, the magnitude of the cell cycle radiosensitivity patterns decreases. Yet even with oxygen (LET $\simeq$ 350 keV/μm) and neon ions (LET $\simeq$ 500 keV/μm), slight radioresistance is still observable (probably due to delta rays) in late S-phase.

The mechanisms for the cell cycle variation in radiation resistance are not clearly understood, but since DNA is thought to be the primary target for cell killing, several hypotheses have been proposed. First, variations in cellular radiosensitivity have been related to DNA configurations or condensation. The more tightly condensed the genetic material, the less accessible to repair enzymes the damaged regions would become (i.e., mitosis), whereas during the DNA synthetic phase, the loosely coiled DNA is readily accessible and thus less radiation-sensitive. Second, variations in the level of naturally occurring sulfhydryl compounds like cysteamine correlate well with cell cycle radiosensitivity changes. Furthermore, sulfhydryl-binding agents, like *N*-ethyl maleimide, reduce the shoulder of survival curves and prevent the development of S-phase-specific radiation resistance. Finally, the activity of a number of DNA repair enzymes exhibits cell cycle-specific variations that also correlate well with radiation sensitivity. One or more of these mechanisms may be important and result in substantial changes in radiosensitivity through the cell cycle.

3. Dose Rate Effects

Dose rate is one of the principal factors that determines the biological consequences of a given dose. As the dose rate is lowered and the exposure time is increased, the biological effect of a given dose is decreased. Two processes are involved with this effect: repair of sublethal damage and repopulation by cell division (Fig. 5-8). As the dose rate during continuous exposure is decreased, more and more sublethal damage is repaired during the exposure. This is exemplified by a gradual reduction in the shoulder (n approaches unity) and a decrease in the slope (D_0 increases) of the radiation survival curve. Ultimately, the dose rate can be sufficiently low so that all sublethal damage is repaired during the exposure. The dose rate effect, observed *in vitro,* varies enormously for different cell types, and it is proportional to the size of the shoulder region of the single acute-exposure survival curve.

In both *in vitro* and *in vivo* systems, the dose rate can be reduced even further so that cell proliferation may occur during radiation exposure. In rapidly proliferating cell systems (gut and hemopoietic), a stem cell population can be

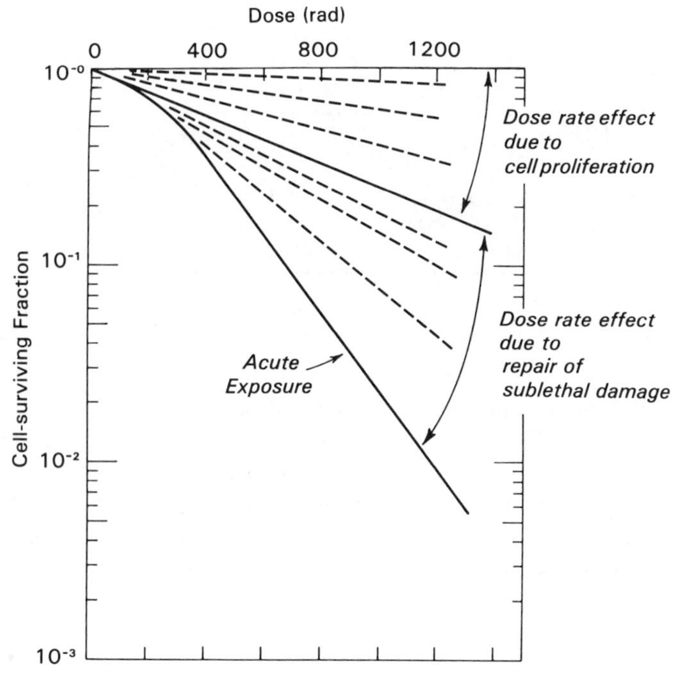

FIG. 5-8 Dose rate effect due to repair of SLD and cell proliferation. Reprinted with permission from Hall (1978). Copyright © 1978 by Lippincott/Harper & Row.

maintained in a steady state provided that cell death due to irradiation does not exceed the renewal rate. The critical dose rate is defined as the dose rate at which cell populations can continue to grow. For various cellular systems, the dose rate per cell cycle at which cell proliferation stops and depopulation occurs varies between 720 and 990 cGy/cell cycle.

4. Oxygen Effects

One of the most effective modifiers of radiation damage in biological systems is oxygen. For a given dose of X irradiation, as the concentration of oxygen is increased from 0 mm Hg (hypoxic) to 30 mm Hg (oxic) at 37°C, the amount of induced cell killing rapidly increases. Furthermore, the ratio of radiation doses under hypoxic and oxic conditions required to produce the same biological effect [referred to as the oxygen enhancement ratio (OER)] is constant; therefore, oxygen is said to be a dose-modifying agent. The effectiveness of oxygen as a dose modifier is LET-dependent (Table 5-2), and oxygen must be present during irradiation to enhance damage production.

TABLE 5-2

EFFECT OF LET ON OXYGEN ENHANCEMENT RATIO

Radiation	LET	OER
X Rays	0.3	2.5
15 MeV neutrons	45	1.6
Alpha particles	190	1.0

The mechanism by which oxygen acts as a dose-modifying agent is not fully understood, but it is thought to be related to its ability to form a reactive free radical. In review, oxygen interacts with free radicals to produce both hydrogen peroxides, $HO_2\cdot$, after the radiolysis of water, and organic peroxides from interactions with protein, lipids, and DNA. Consequently, oxygen enhances the indirect effect of radiation damage and prevents the repair of damaged molecules by fixing the radiation lesions.

5. Radioprotective Agents

These agents are sulfhydryl-containing compounds, the simplest of which include cystine and cysteamine. Animals injected with these compounds require higher doses of low-LET radiations to produce the same biological effect; hence the reduced effectiveness of the dose is called the dose reduction factor (DRF).

The mechanism of action for radioprotective agents appears to be either (1) a free radical scavenging process whereby the protective agent reacts with free radicals in competition with oxygen or (2) increased repair of radiation injury. Either mechanism results in reduced indirect effect or decreased fixation of damage in critical target molecules. Consequently, the protective effect of these agents parallels both the oxygen effect and the LET effect.

The usefulness of radioprotectors is limited at present because the agents available are quite toxic and are rapidly oxidized and removed from the body. Hence, they must be administered prior to irradiation in carefully controlled amounts.

WR-2721, a thiophosphate derivative of cysteamine, is a fairly new agent, and it is the best radioprotective agent to date. Its dose reduction factor is approximately two, it has reduced toxicity, and it can be administered up to 4 hr before irradiation. The potential application of radioprotectors will be addressed in a later chapter.

6. Radiation Sensitizers

Radiosensitizers are chemical agents, like oxygen, that have the capacity to increase the lethal effects of radiation. There are two types of sensitizers: apparent and true sensitizers. Apparent sensitizers include widely different compounds like antibiotics (actinomycin D and puromycin), alkylating agents (nitrogen and sulfur mustards), and antimetabolites (methotrexate and 5-fluorouracil), which potentiate the lethal effects of radiation. They act through diverse mechanisms (inhibition of protein or DNA synthesis) that tend to enhance radiation effects and interact with radiation additively, rather than synergistically like true sensitizers. The true sensitizer actually enhances the radiation effect. Halogenated pyrimidines (ClUdR, BUdR, and IUdR) have a halogen substituted in place of a methyl group, and they are incorporated into DNA in place of thymidine. These substitutions tend to weaken the DNA chain, thus rendering the cell more susceptible to DNA damage. The usefulness of these compounds as radiosensitizers is limited because the liver rapidly dehalogenates the compounds prior to cellular incorporation.

The best sensitizer of hypoxic cells at present is misonidazole (Ro-07-0582). This compound is very electron affinic and therefore sensitizes hypoxic cells in much the same manner as does oxygen. At a 10-m*M* concentration, misonidazole in culture increased the radiosensitivity of hypoxic cells to nearly that of oxygenated cells. Furthermore, misonidazole (1 m*M*) in mice reduced the tumor control dose (TCD_{50}) by a factor of almost two (2410 versus 4380 rads). Unlike oxygen, misonidazole is not rapidly metabolized and, consequently, it is able to diffuse further into hypoxic tissues or tumors.

V. TISSUE AND ORGAN SENSITIVITY TO IONIZING RADIATION

The radiosensitivity of tissues to X rays was first described by Bergonie and Tribondeau (1959). Their observations, which have formed the foundation of cancer radiotherapy, were that "The sensitivity of cells to irradiation is in direct proportion to their reproductive activity and inversely proportional to their degree of differentiation." The response of a tissue or organ to a dose of ionizing radiation depends primarily on two factors: the kinetics of the tissue as a whole and the radiosensitivity of the cells. In the case of fully differentiated tissue systems, radiation responses in terms of survival curves are irrelevant because the cells do not undergo a mitosis-linked death.

Cell populations in tissues may be compartmentalized according to their kinetics or reproductive capabilities. At least four basic pools or compartments can be recognized.

(1) *Stem type population.* This is a self-maintained system, whose function is to produce cells for another population. These represent the least differentiated cells and may be relatively few in number. Examples are stem cells of the hemopoietic system, intestinal crypt cells, and germinal cells of the epidermis. Based on radiation pathology observations, this pool, also referred to as vegetative intermitotic, is extremely sensitive to radiation.

(2) *Dividing transit population.* This population receives cells from the stem cell pool. The cells begin to differentiate into several cell types but also continue to divide several times during transit. Examples are the marrow reticulocyte pools, cells lining the walls of villi, and cells in the basal layer of the skin. This pool is also referred to as the differentiating intermitotic pool. The radiosensitivity of this pool is slightly greater than that of the stem type population.

(3) *Simple transit pool.* This population receives cells from the dividing transit pool and releases them into the mature or closed static pool. Further cell differentiation occurs in this pool. Examples include the immature erythroid and lymphoid elements and the cells lining the upper villi walls. This pool also has been referred to as the reverting postmitotic pool. This pool is relatively radioresistant and generally long-lived.

(4) *Closed static population.* This population, also referred to as fixed postmitotic, is composed of mature, fully differentiated cells with no mitotic activity. These cells serve their function, live a designated life span, and die. Adult nervous tissue is an excellent example of a closed static population; other examples are muscle cells and granulocytes. Because of its fixed postmitotic state, this population is extremely radioresistant.

As a result of cell population kinetics, tissues and organs that depend on an active stem cell pool will be more radiosensitive than those systems composed mainly of mature cellular pools with little or no stem cell activity. The expression of radiation injury at the tissue or organ level after irradiation will depend on the time required for functional cells to be renewed and the degree of cellular injury (cell killing–depletion) at the stem cell level (Fig. 5-9). In normal growth phase, cell populations increase in size following a log-normal pattern. After irradiation, surviving cells will replicate rapidly in an initial abortive attempt to repopulate the tissues. The abnormal, reproductively dead cells will die off quickly, and finally normal repopulation will ensue. The time interval between irradiation and repopulation of the tissues is a function of the number of stem cells remaining and thereby a function of dose. Survival depends on whether the tissue can adequately function with the reduced number of cells until repopulation occurs (see Chapter 9, this volume). Furthermore, the time interval between irradiation and expression of tissue or organ damage depends on the cell renewal rate or transit time from the stem pool to the functional pool. Since the time is quite

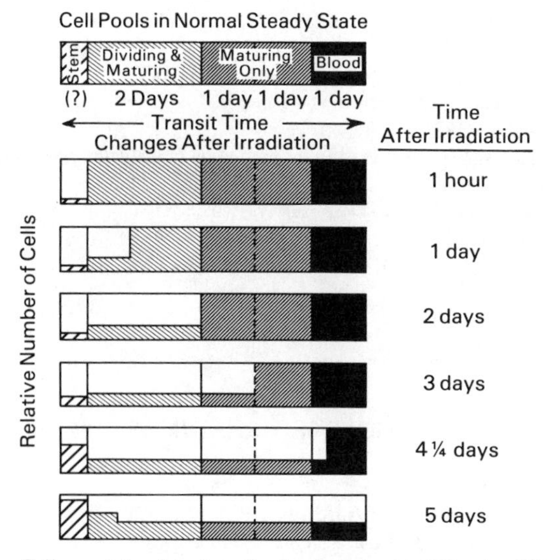

FIG. 5-9 Cell population kinetics after irradiation with 500 rads of X rays.

variable even for different tissues that exhibit similar radiosensitivities, the expression of tissue damage is highly variable, i.e., days to months.

REFERENCES

Bergonie, J., and Tribondeau, L. Interpretation of some results of radiotherapy and an attempt at determining a logical technique of treatment. *Radiat. Res.* **11,** 587 (1959).

Elkind, M. M., and Whitemore, G. F. *The Radiobiology of Cultured Mammalian Cells.* Gordon & Breach, New York, 1967.

Howard, A., and Pelc, S. R. Synthesis of deoxyribonucleic acid in normal and irradiated cells and its relation to chromosome breakage. *Heredity, Suppl.* **6,** 261–273 (1953).

Kaplan, H. S., and Moses, L. E. Biological complexity and radiosensitivity. *Science* **145,** 21–25 (1964).

Kember, N. F. Cell survival and radiation damage in growing cartilage. *Br. J. Radiol.* **40,** 496–505 (1967).

McCulloch, E. A., and Till, J. E. The sensitivity of cells from normal mouse bone marrow to gamma radiation *in vitro* and *in vivo*. *Radiat. Res.* **16,** 822–832 (1962).

Munro, T. R. The relative radiosensitivity of the nucleus and cytoplasm of Chinese hamster fibroblasts. *Radiat. Res.* **42,** 451–470 (1970).

Puck, T. T., and Markus, P. I. Action of X rays on mammalian cells. *J. Exp. Med.* **103,** 653–666 (1956).

Sinclair, W. K., and Morton, R. A. *Radiat. Res.* **29,** 450–474 (1966).
Terasima, T., and Tolmach, L. J. Growth and nucleic acid synthesis in synchronously dividing populations of HeLa cells. *Exp. Cell Res.* **30,** 344–362 (1963).
Terzi, M. Comparative analysis of inactivating efficiency of radiation on different organisms. *Nature (London)* **191,** 461–463 (1961).
Withers, H. R. The dose-survival relationship for irradiation of epithelial cells of mouse skin. *Br. J. Radiol.* **40,** 187–194 (1967).
Withers, H. R., and Elkind, M. M. Dose–survival characteristics of epithelial cells of mouse intestinal mucosa. *Radiology* **91,** 998–1000 (1968).
Zirkle, R. E. In *Radiation Biology* (A. Hollaender, ed.), Vol. I, pp. 315–350. McGraw-Hill, New York, 1954.

SUGGESTED READINGS

Alper, T. *Cellular Radiobiology.* Cambridge Univ. Press, London and New York, 1979.
Altman, K. I., Gerber, G. B., and Okada, S. *Radiation Biochemistry,* Vol. 1. Academic Press, New York, 1970.
Bacq, Z. M., and Alexander, P. *Fundamentals of Radiobiology,* 2nd ed. Pergamon, Oxford, 1961.
Elkind, M. M. *et al. Radiat. Res.* **25,** 359 (1965).
Hall, E. J. *Radiobiology for the Radiobiologist,* 2nd ed. Harper & Row, New York, 1978.

CHAPTER 6

Radiation Effects on the Lymphohematopoietic System: A Compromise in Immune Competency

RODNEY L. MONROY*

Armed Forces Radiobiology Research Institute
Bethesda, Maryland 20814-5145

The hematopoietic syndrome is anticipated when a dose of radiation greater than 100 cGy is received. In most cases, the syndrome is uncomplicated by the effects of gastrointestinal damage until the dose received is between 800 and 1000 cGy (1). This syndrome is characterized by a depression in the peripheral blood levels of the white blood cells, red blood cells, and platelets. The degree of that depression and its duration have been shown to be dose dependent (2). The primary effect of the radiation is on the rapidly proliferating hematopoietic elements of the bone marrow. This leads to the observed decline in the circulating mature cells: monocytes, granulocytes, lymphocytes, red blood cells, and platelets. The resulting clinical situation is potentially life threatening because of opportunistic infections. This enhanced susceptibility after irradiation is caused primarily by depression of the immune response, both specific and nonspecific.

The following sections describe the relationship of the pluripotent stem cell to the cellular elements of the immune response. The effects of radiation on this stem cell and its progeny are discussed. The level of immune competency after irradiation is addressed. Finally, the potential use of autologous and allogeneic bone marrow transplantation as a source of pluripotent stem cells for hematopoietic and immunological reconstitution is evaluated.

*Present address: Transplantation Research Program Center, Naval Medical Research Institute, Bethesda, Maryland 20814-5145.

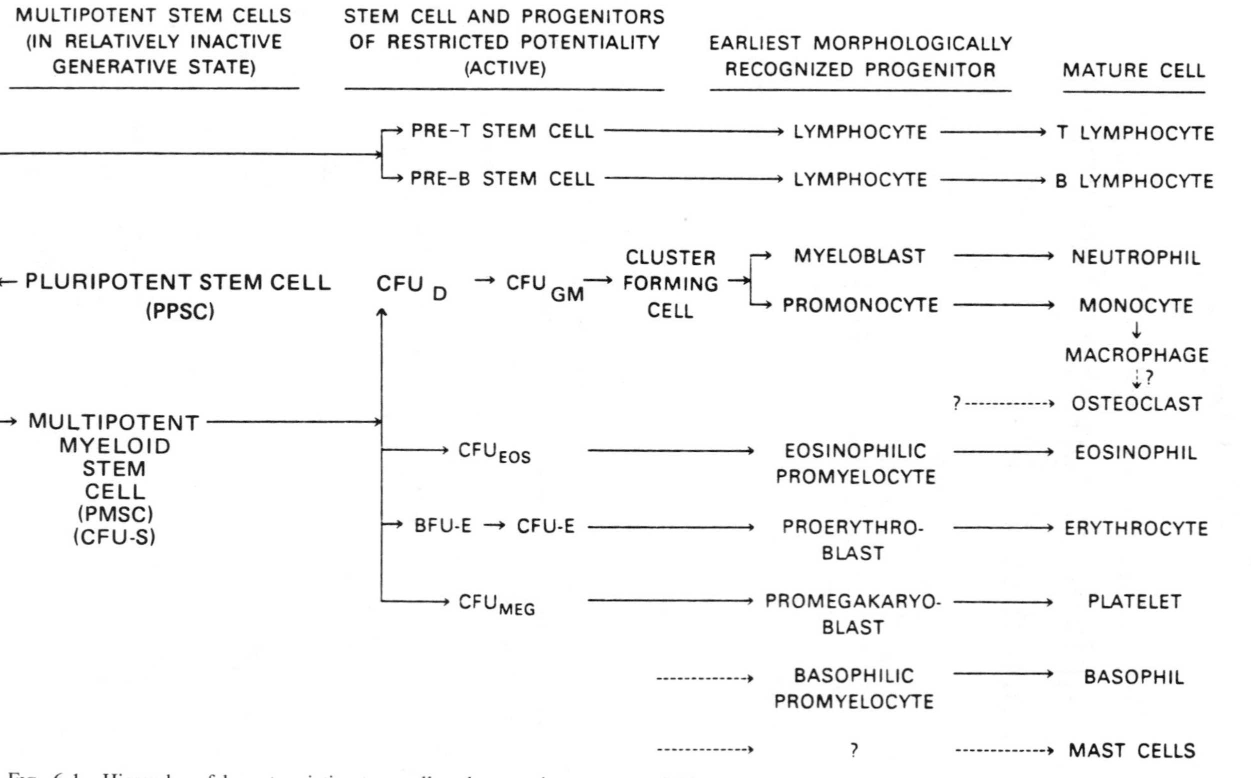

FIG. 6-1 Hierarchy of hematopoietic stem cell and progenitor systems. BFU, Burst-forming unit; CFU, colony-forming unit; CFU-S, colony-forming unit-spleen; D, diffusion chamber; E, erythroid; EOS, eosinophil; MEG, megakaryocyte; GM, granulocyte–monocyte; S, stem cell. (Reprinted with permission from Wintrobe *et al.*: *Clinical Hematology*. Lea & Febiger, Philadelphia, 1981.)

I. PLURIPOTENT STEM CELL

The mature cellular elements of the peripheral blood have limited life spans, and they are expended in their normal functions. This places a constant demand on the hematopoietic elements of the bone marrow for a high capacity of proliferation. Within the bone marrow, a pluripotent stem cell (PPSC) population meets the demand of the system. The PPSC is unique since it has the capability to self-renew and differentiate. The self-renewal characteristic is critical in order for the cell to perpetuate itself and not be used up over time. The ability to differentiate allows this cell to give rise to the various cell populations of the lymphohematopoietic system and the reticuloendothelial system.

Figure 6-1 (3) outlines the hierarchy of the lymphohematopoietic system, from the pluripotent stem cell down to the mature functional cells. These mature cells have specific functions within the body. Granulocytes are important for phagocytosis and microbiocidal activity against invasive bacteria, which are functions related to the nonspecific immune response. Lymphocytes are responsible for cell-mediated and humoral responses, which are related to the specific immune response. The monocyte enters the peripheral blood and migrates into specific tissues, where it differentiates and establishes itself as a macrophage (e.g., Kupffer's cells, alveolar macrophage) of the reticuloendothelial system. Thus, the hematopoietic, lymphoid, and reticuloendothelial systems are ultimately dependent on the PPSC for the replacement of their functional cell types. Therefore, if a specific agent such as radiation damages the PPSC, then all of these associated systems are compromised.

Effect of Radiation on the PPSC

In vitro culture techniques have been developed that identify morphologically undetectable progenitor cells (e.g., colony-forming unit-spleen, CFU-S; burst-forming unit-erythroid, BFU-E; colony-forming unit-granulocyte–macrophage, CFU-GM) (Fig. 6-1). These progenitor cells have proliferative capacity and give rise to a colony of specific cell types (e.g., erythroid and granulocyte). These progenitor cells do not measure the PPSC directly, but investigators using these techniques have been able to evaluate the physiology of the hematopoietic system and indirectly understand the PPSC.

An individual given a sublethal dose of radiation will not succumb. Instead, a small hematological change may occur. The effect of this exposure on the early hematopoietic progenitor cells is quite noticeable, and the changes in those cells can be determined using certain assays. The CFU-S assay in the murine system represents a multipotent stem cell of the hematopoietic system, and has been accepted as a close approximation of the PPSC. In the mouse model, a sublethal dose of only 95 cGy will kill 63% of the CFU-S population (a dose in which 37% of the population will survive, designated D_0), yet these irradiated mice will survive apparently uneventfully (4–6). The survival of the CFU-S population in the mouse at the $LD_{50/30}$ dose (520 cGy) is only 1.0–0.1% of normal values (7). Despite this 2–3 log kill of CFU-S, 50% of the animals survive. It is uncertain whether a 50% survival can be obtained in large animals and man with a 2–3 log kill in the PPSC, since the CFU-S assay or its equivalent is limited to the murine system. However, it is speculated, based on progenitor cell analysis in large-animal models (dogs and monkeys) (8), that such survival in humans is possible with this level of PPSC kill (2,7). At the $LD_{50/30}$ dose level, the cytological restoration of the bone marrow begins on the tenth day in the mouse and about the twenty-fifth day in man (9).

Radiation injury (at $LD_{50/30}$ doses) to the hematopoietic tissues has killed more than 99% of the critical hemopoietic cells. In this situation, the PPSC can (1) self-replicate and fill the niches that were left vacant by the radiation and/or (2) differentiate and give rise to progenitor cells, which will, in turn, give rise to the required mature cells. A theory is postulated in which the PPSC is "compartmentalized" (see Fig. 6-2) in the normal state in order to prevent complete utilization of the PPSC (3,10). In this model, the PPSC self-replicates before it differentiates. It is further postulated that, after irradiation, the PPSC must undergo self-replication in order to repopulate the "compartment" to a high enough level (it is uncertain what this level is) for differentiation to occur. Boggs *et al.* (11) have observed that the CFU-S do not allow cells to be directed down

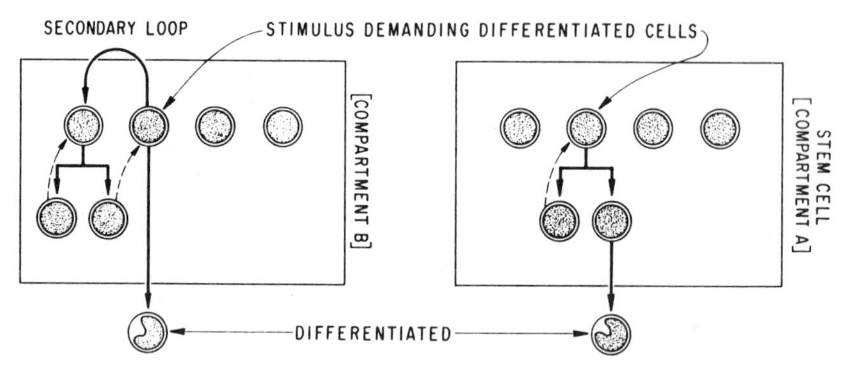

FIG. 6-2 Models of stem cell replication. It has been suggested that division in stem cell compartments may be asymmetric (compartment A) or symmetric (compartment B). Compartment size is maintained in compartment A by differentiating stimuli triggering cellular division in which one daughter matures and the other remains a stem cell. In compartment B the differentiating stimulus depletes the compartment by inducing cellular differentiation, but a secondary "compartment depletion recognition loop" induces stem cell division to maintain compartment size. (Reprinted with permission from Wintrobe *et al.*: *Clinical Hematology*. Lea & Febiger, Philadelphia, 1981.)

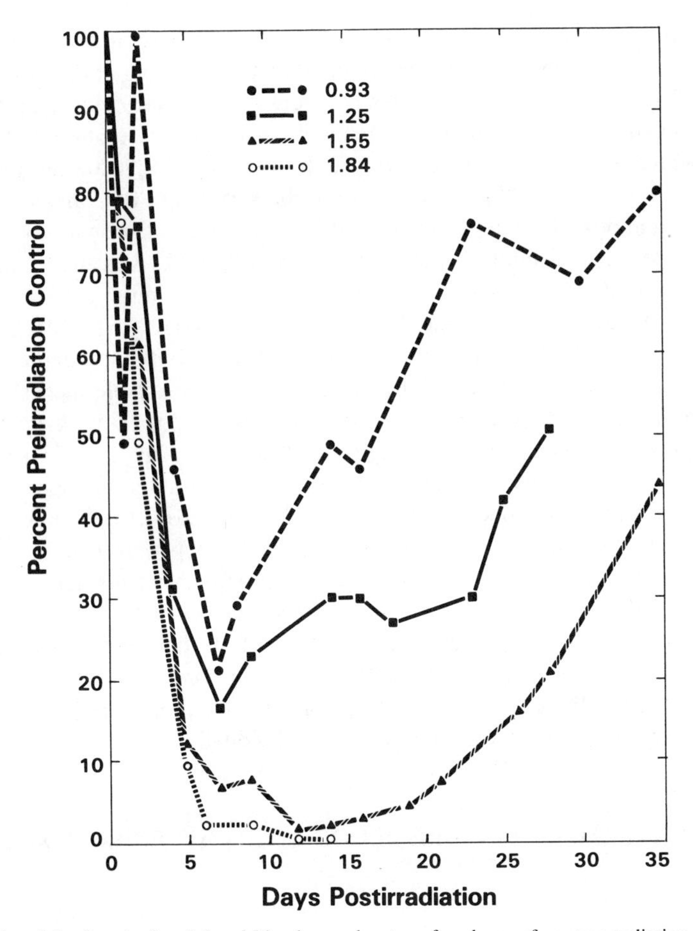

FIG. 6-3 Level of peripheral blood granulocytes after doses of neutron radiation.

identifiable cell lines until the CFU-S pool exceeds 10% of its steady-state size. However, the effect of radiation will create various stimuli (chemical and cellular) for those surviving cells to differentiate. Thus, the PPSC appear to ensure self-perpetuation before differentiating.

The time frame for PPSC repopulation, progenitor cell proliferation, and eventual maturation of the end cells (e.g., granulocytes) can be illustrated in a series of neutron-irradiated studies on dogs (12). Dogs irradiated with 155-cGy mixed neutron–gamma irradiation and not given therapeutic support (antibiotics and platelets) showed only a 33% survival, with the mean death occurring on day 15. The survivors did not show an increase in their circulating granulocytes until day 21 (Fig. 6-3). If therapeutic support was administered after the 155-cGy dose, then the survival increased to 100%. But the increase in circulating granulocytes was no earlier for these survivors than for survivors of the unsupported group. The point is that it took about 3 weeks for the different phases of the hematopoietic hierarchy to recover enough to contribute mature functioning cells, which eventually ensured survival. It is proposed that, during the initial period of 7–14 days in this 3-week period, the surviving PPSC underwent self-replication in order to refill the niches left vacant by the radiation death of PPSC. Vos (13) described a similar rapid proliferation of the CFU-S in mice during days 3–14 after a sublethal dose of irradiation (460 cGy). This initial phase is believed to take the longest period of time. Bone marrow analysis of the earliest assayable progenitor (CFU-GM) in the dog model showed no activity until about day 12–14. This suggests that the PPSC filled enough of the available niches and was differentiating into the various progenitor cells. In the Vos study, the cellularity of the bone marrow became strikingly identifiable by day 15–17 in both the immature myeloid cells and the erythroid cells. Maturation and appearance of granulocytes in the peripheral blood occurred by day 20–21. The nonsurvivors (mean time of death 15 days) of the unsupported group were unable to defend themselves against opportunistic bacterial infection because no significant level of functional granulocytes was present. This was when the PPSC was self-replicating. Thus, the percentage survival of the PPSC dictates the survival of the individual by the level of mature cells that it can contribute to the peripheral blood. Mature cells (e.g., granulocytes, platelets, and monocytes) are critical in maintaining the immunocompetency of an individual.

II. SURVIVAL OF MATURE BLOOD CELLS AFTER IRRADIATION

The mature blood cells (monocytes, platelets, red blood cells, and granulocytes) are functionally radioresistant to doses of radiation that produce the bone marrow syndrome. In fact, it appears that these cells circulate in the peripheral blood and function normally after irradiation. The primary problem is that the production of new cells will halt as a result of the irradiation. The length of time of this stoppage, as mentioned, will depend on the dose of radiation received. Thus, as old cells of the peripheral blood are used up, they will not be replaced, and the levels of those cells in the blood will fall. The rate of fall is closely related to the life span of each cell type. The maintenance of these circulating cells is essential to the survival of an individual.

A. Granulocytes

One of the principal causes of death after total-body irradiation is infection. The specific cause is neutropenia, wherein the decrease in production of anti-

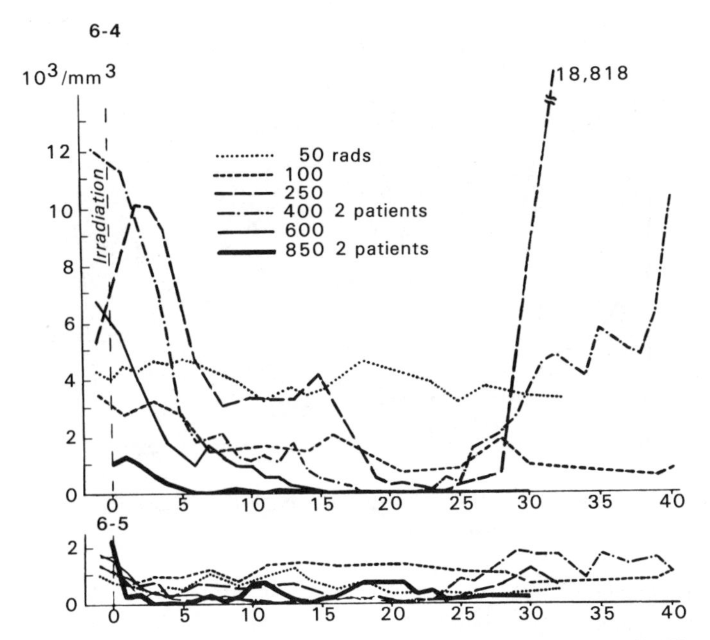

FIGS. 6-4 AND 6-5 Alterations in counts of various leukocytes [polynuclear cells (Fig. 6-4) and lymphocytes (Fig. 6-5)] in man caused by total-body irradiations with different doses of gamma rays. (Reprinted with permission from Ref. 14.)

bodies and the failure of protective barriers (e.g., skin, intestine, and lungs) are considered additive factors. Mathé (9) reported that neutropenia in man (and in animals) plays a much more important role in the pathogenesis of these infections than does any other mechanism. In fact, in patients who were neutropenic with a high fever, a fever drop has been observed 1–3 days after the beginning of the return of the granulocytes. Thus, production of the granulocytes appears to be the key cellular element in survival after irradiation.

The granulocyte levels reach an initial nadir 5–10 days postirradiation, and the level of depression corresponds to the radiation dose received (see Figs. 6-4 and 6-5) (9). The length of time before recovery is also dose dependent, so that, at the higher yet potentially survivable doses, a prolonged period of neutropenia occurs. Lethal infections develop during this period. Production of granulocytes is most critical for the recovering bone marrow, reflected in the fact that granulocytes are the first to reappear after irradiation.

B. Platelets

The platelet has a life span of 9–12 days, and it is a critical cellular element in the hemostasis of an individual. In the clinical setting, pathological conditions

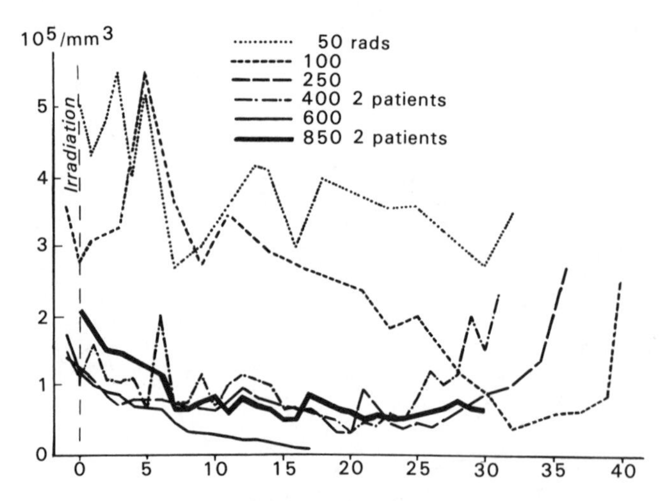

FIG. 6-6 Evolution of thrombocyte count in patients deliberately irradiated with variable doses of cobalt-60. (Reprinted with permission from Ref. 14.)

such as aplastic anemia are frequently characterized by thrombocytopenia capillary fragility, abnormal bleeding, and purpura. In these conditions, platelet transfusions are required to keep the patient's platelet count above 20,000/mm^3. The irradiated individual faces a clinical problem similar to that of the patient with aplastic anemia. The platelet levels fall off after irradiation to an initial nadir on about day 10, and the falloff is dose dependent, as with the granulocyte (see Fig. 6-6) (9,14). As observed in the treatment of the patient with aplastic anemia, an irradiated individual must be transfused with platelets to prevent bleeding episodes. Thus, a potentially lethal synergism exists when neutropenia is combined with thrombocytopenia, as experienced in the irradiated individual and the patient with aplastic anemia. The platelet recovery profile is similar to that of the granulocytes. Platelets, as has been inferred, are the most critical "transfusable" blood product for the irradiated individual. They may be required for at least 2–3 weeks, depending on the level of radiation exposure.

C. Red Blood Cells (RBC)

The red blood cell has an estimated life span in the human of 120 days. Although the production of bone marrow red blood cells is reduced or lost after exposure to radiation, the red blood cell does not become a limiting cell to ensure survival. The clinical RBC parameters (e.g., hematocrit and hemoglobin) after irradiation are observed to fall off at a gradual rate consistent with normal cell turnover. However, this takes place only in the uncomplicated case, where no

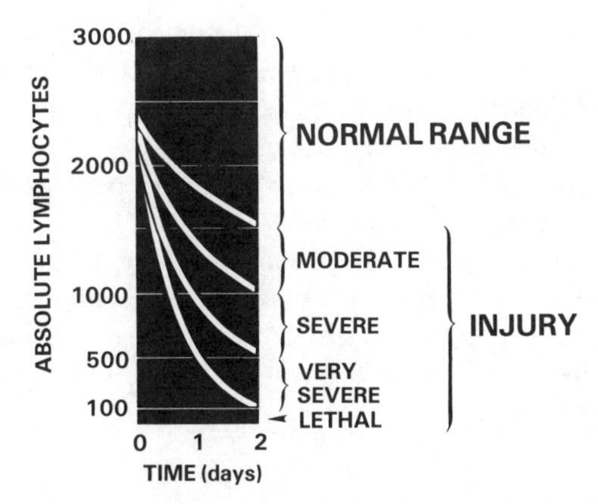

FIG. 6-7 Relationship between decrease in early lymphocyte counts and amount of radiation exposure.

blood loss occurs with thrombocytopenia or infection. In these cases, packed RBC transfusions may be required. RBC recovery after irradiation occurs after the granulocytes and platelets have begun to recover.

D. LYMPHOCYTES

In contrast to the cells of peripheral blood described above, the circulating lymphocytes are extremely radiosensitive and have been suggested as biological dosimeters (15,16). Figure 6-7 illustrates this dose–response decline in the circulating lymphocytes. In addition to the loss from the circulation, morphological damage rapidly appears, and the lymphoid tissues (spleen, thymus, and lymph nodes) quickly decrease in size (17).

However, the lymphocytes are heterogeneous with respect to function (cellular response of T cells and humoral response of B cells), life span (a few hours to years), tissue of origin, proliferation behavior, maturation state of migrational response, and radiosensitivity. Thus, when the circulating lymphocytes are reduced by 90% after exposure to radiation, each unique population needs to be evaluated separately in order to understand its radiosensitivity. Lymphocytes are subdivided into T cells and B cells, based on the tissue of origin. T cells are from the thymus, and B cells are from the bursa equivalent. T lymphocytes are further subdivided by function into helper, killer, suppressor, amplifier, and possibly even more populations that are yet to be identified.

The effect of radiation on lymphocytes is dramatic. Doses as small as 15–20 cGy are able to kill a small fraction of circulating lymphocytes (17). Lymphocytes show rapid morphological changes after irradiation, and are rapidly cleared from circulation. Some lymphocytes survive this early phenomenon but later undergo functional proliferation and experience an interphase death. Many investigators have calculated general lymphocyte sensitivity to be similar to that of proliferating cells of the hematopoietic tissue, with a D_0 of 70–100 cGy, depending on the investigator (17–19). More specifically, the peripheral T lymphocytes (human) have been reported to have a D_0 of about 50 cGy (17–20). A more radioresistant T lymphocyte population does exist with a D_0 of about 550 cGy. The circulating B lymphocytes (human) have been reported to have a D_0 of 50 cGy (17,19). With respect to the phase of antigenic activation, the various lymphocyte populations will have different radiosensitivities. This effect is discussed further in the section on the specific immune response.

III. EFFECT OF RADIATION ON IMMUNE COMPETENCY

A. Nonspecific Immune Response: Radiation Effects

Cells that undergo phagocytosis and kill foreign microorganisms are the key elements of the nonspecific immune response. These cells are primarily the granulocytes and the macrophages. Radiation depresses the nonspecific immune response by depressing the levels of circulating monocytes and granulocytes.

1. Macrophage

The macrophage or macrophage cell type is proposed to have its origin from the circulating monocyte (bone marrow origin), which enters the specific tissues and differentiates into a macrophage. However, the tissue macrophages are a heterogeneous population with apparent widespread distribution in the body (liver, Kupffer cell; lung, alveolar macrophages; peritoneal cavity, peritoneal macrophages). These cells have the capability to phagocytize and catabolize foreign substances (e.g., microorganisms and toxins). Their role is accentuated further by their involvement in the recognition phase of most new antigens. It is proposed that the macrophage catabolizes the foreign substance and presents it as an antigen or a series of antigens for the continuation of the specific immune response (21). In general, the macrophage is an end cell without further proliferation, although in some cases a low level of proliferation has been observed. Thus, it is not surprising to find that this cell is reported as being quite radioresistant. The effects of radiation on the various essential macrophage functions are listed in Table 6-1 (17). The macrophage therefore seems to retain normal function even after doses higher than associated with the bone marrow syndrome. It appears from this that radiation will affect only the monocyte pool as a source for macrophage replacement. The macrophages (e.g., Kupffer cell, liver)

TABLE 6-1

EFFECTS OF RADIATION ON MACROPHAGE ACTIVITIES

Macrophage function or activity	Effect of radiation
Morphology	No change
Migratory activity	No change
Phagocytosis of antigens, particles, etc.	No change
Catabolism of ingested antigens	Variables reduced, particularly in kilorad range
Intracellular biochemical activities (lysosomal enzyme levels, etc.)	Variable increase
Membrane-bound antigen	Increased levels
Immunogenicity of macrophage-associated antigen	No change in most protein and erythrocyte antigens; marked reduction in Shigella antigen

appear to have a long life span, and this monocyte shortage may not be significant in the overall individual survival after irradiation.

2. Granulocyte

The granulocyte is another cell type associated with the nonspecific response. This cell has the ability to migrate to the sites of infection, and to ingest and kill large quantities of the infectious organisms. It represents a first line of bactericidal defense at a wound. This cell type, as previously discussed, is rapidly used up after irradiation, and it appears to be the limiting mature cell in survival after irradiation. So, without this cell, the tissue macrophages must clear the blood of the additional infectious organisms that are normally cleared by the granulocytes.

B. Effects of Radiation on the Specific Immune Response

The specific immune response is a complex interrelationship of cells, soluble factors, and antibodies. There are two functional arms of this type of response: the humoral and the cell mediated. Sharp and Watkins (19) summarized the immunological response as the maintenance of immunological homeostasis provided by an intercommunicating differentiation network of morphologically similar, but functionally heterogeneous, sets of cells that provide all the necessary interactions and terminal effects. They state, "Perturbations caused by the selective expansion of sets of cells responding to an antigen are heavily regulated, and a successful response is accompanied by a return to a new steady state with a set

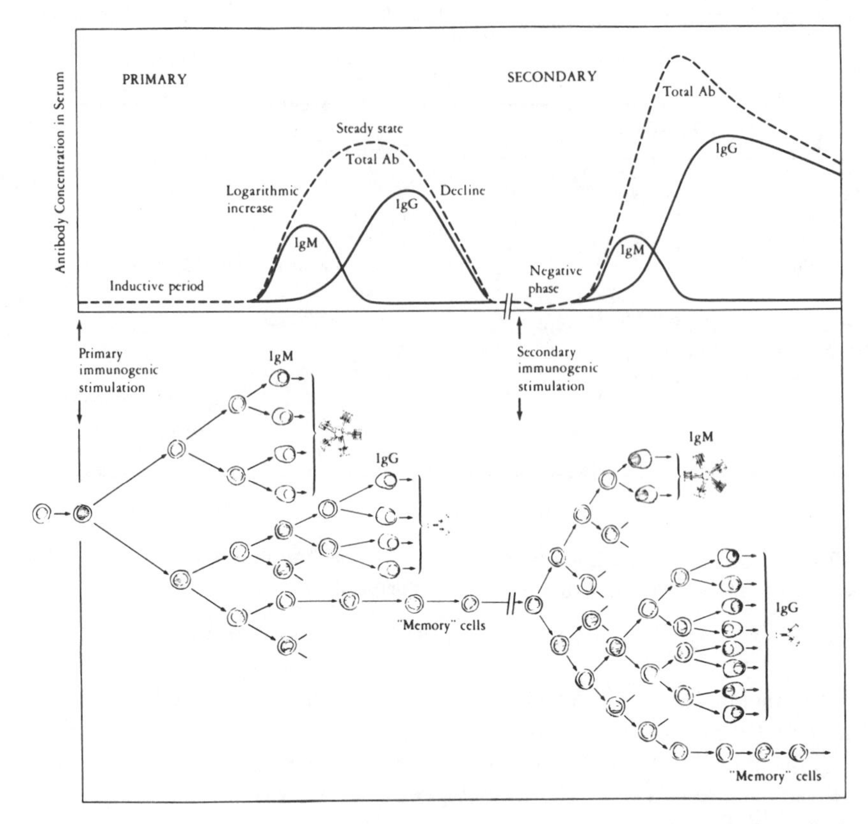

FIG. 6-8 Scheme of humoral and cellular events in primary and secondary (anamnestic) antibody responses. Ab, Antibody; IgG, immunoglobulin G; IgM, immunoglobulin M. (Reprinted with permission, Herscowitz, H. B. Immunophysiology: Cell function and cellular interactions. In *Immunology II,* Chapter 7, p. 155. Saunders, Philadelphia, Pennsylvania, 1978.)

of memory cells incorporated into the network.'' In evaluating the immune response after irradiation, there will be perturbations at different phases of the response and with different cell populations, thus affecting the homeostasis of this complex system. This section discusses the effect of radiation on the immune responses, humoral and cell mediated, and on the cells that make up each system.

1. Humoral Response

The humoral response is the production of antibodies in response to some foreign antigen. Antibody production is divided into either a primary or a secondary (anamnestic) response (Fig. 6-8). The primary response is associated with antibody production to a new antigen. During the primary response, an initial time period exists in which no serum antibody titers can be measured. This is the characteristic lag phase associated with antigen processing (a macrophage function), activation, and proliferation of the cells responsible for antibody production. In the secondary response, the serum antibody titers increase more rapidly and reach higher serum levels than in the primary response. The production of memory cells (both T and B lymphocytes) during the primary response are the reason for this rapid response to the reinfused antigen.

Numerous investigations have evaluated the radiosensitivity of each of those humoral responses. The complexity of the cellular systems involved in the responses results in data that are not black or white upon first inspection. In fact, the primary humoral response may be depressed, or enhanced, or not altered by exposure to total-body irradiation. It was observed that if the antigenic challenge was given within 24 hr after irradiation, there was a depression of the primary humoral response to doses as low as 50 cGy (22). A sublethal dose in mice of 400 cGy was shown to suppress the primary immune response 100% (Fig. 6-9). In contrast, investigators changed the experimental protocols to vary the time of antigen injection, the antigen dose, the form of antigen, and the dose of irradiation. The results of these types of experiments showed frequent enhancement of the humoral response (Fig. 6-10). From these results, it was concluded that (1) the degree of optimum stimulation varied from one antigen to another, (2) the time interval between antigen challenge and irradiation differed in terms of optimal stimulation for different antigens, and (3) the dose of irradiation for optimal stimulation also varied for different antigens (22). However, the enhancement of the immune response was limited to doses less than 300 cGy, and it occurred when the antigenic stimulation had been delayed long enough to allow the steady-state population of precursor cells (and possibly the antigen-reactive cells) to recover. The recovery of a steady-state population is characterized by an overshoot before returning to the steady state, which results in the observed enhancement of antibody production. As the radiation dose is increased, a prolonged period of precursor population depression is observed, with potentially insufficient time for recovery of the population before presentation of the antigen. This results in a depressed antibody response. If the antigen is presented at a later time, then the antibody response is increased only if the antigen presentation coincides with the precursor-cell overshoot. Therefore, the primary humoral antibody response after irradiation shows a dependency on the survival of precursor T and B lymphocytes, which in turn depend on the lymphoid stem cell, and eventually back to the pluripotent stem cells of the bone marrow. The D_0 of the primary humoral response has been shown to be 80 cGy (22), which is similar to the D_0 of the PPSC.

Stoner *et al.* (23) demonstrated that the secondary response was about three times more radioresistant than the primary response (Fig. 6-11). However, the

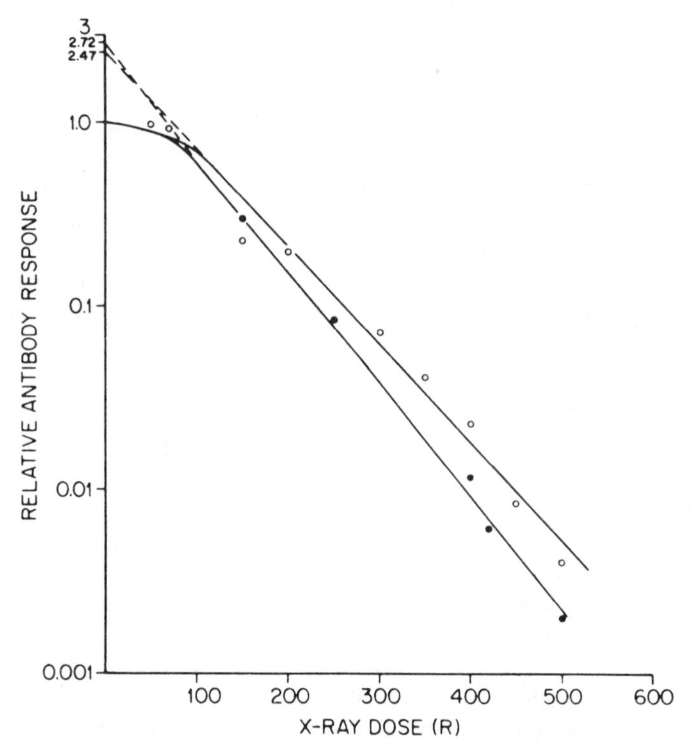

FIG. 6-9 Inactivation of immune response. (○) Rat, (●) mouse. (Reprinted with permission, Najarian, J. S., and Simmons, R. L., eds. *Transplantation*. Lea & Febiger, Philadelphia, Pennsylvania, 1972.) [After Simić, M. M., *et al.*, *Bull. Boris Kidric Inst. Nucl. Sci.* **16** (Suppl. 1), 1 (1965); and after Makinodan, T., *et al.*, *J. Immunol.* **88,** 31 (1962).]

radiosensitivity of the secondary response has also been found to depend on the radiation dose, the antigen dose used for sensitization (primary response), and the interval between sensitization and irradiation. In addition, it has been difficult to separate out a primary type response from the secondary response itself. Interpretation of the secondary response is not clear-cut; it has been proposed that some cell types are responsible for mediation in both responses (19,22). The secondary response is believed to represent specific differentiation products of a memory cell pool (T and B lymphocytes) induced by the primary antigenic stimulus. The size of the memory cell pool decreases as the level of the primary antigenic stimulus subsides and time passes. Thus, it was proposed that radiation could substantially damage the memory cell pool, and that, without further antigenic stimulus, the secondary response to the antigen would be lost (23,24).

a. Humoral Response: Radiation Effects at the Cellular Level. The interaction of T and B lymphocytes is illustrated in Fig. 6-12. It has been established

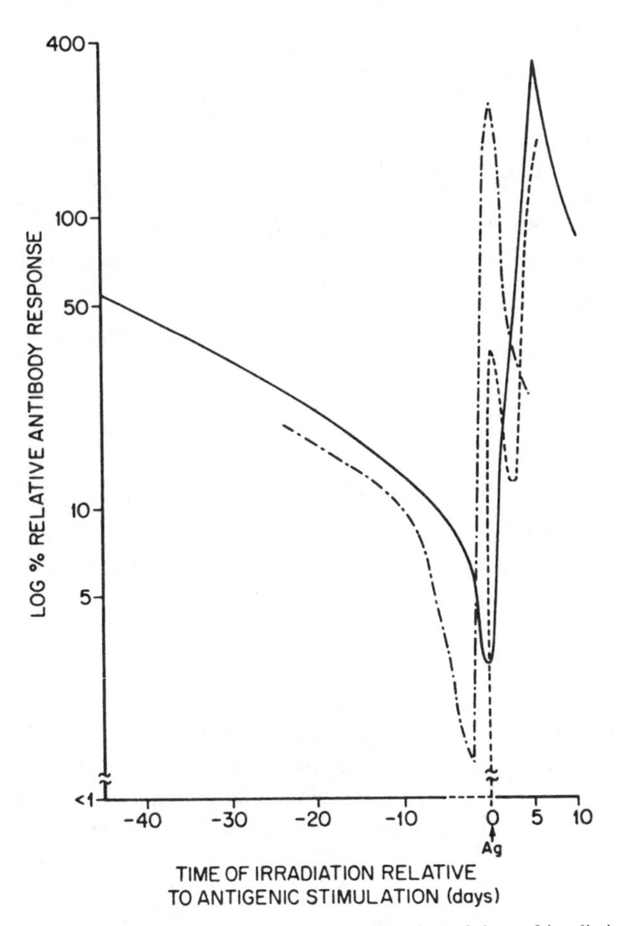

FIG. 6-10 Radiosensitivity of immune response as a function of time of irradiation relative to antigenic stimulation. ——, Mouse, 710 R; -··-, rabbit, 500 R; - - -, rat, 500 R. (Reprinted with permission, Najarian, J. S., and Simmons, R. L., eds. *Transplantation*. Lea & Febiger, Philadelphia, Pennsylvania, 1972.)

that most humoral antibody responses follow this pathway. Two functional subgroups of T lymphocytes are involved in the regulation of the humoral response: helper T lymphocytes and suppressor T lymphocytes. The radiosensitivity of each of these cell types must be considered when studying the effect of radiation on this immune response. The helper T and suppressor T lymphocytes are in a homeostatic balance, which regulates the immune response. So a perturbation such as radiation changes this balance. It is the degree of effect on each functional cell type that determines how the immune system responds. Cells that have been stimulated by the presentation of the antigen become activated, and their

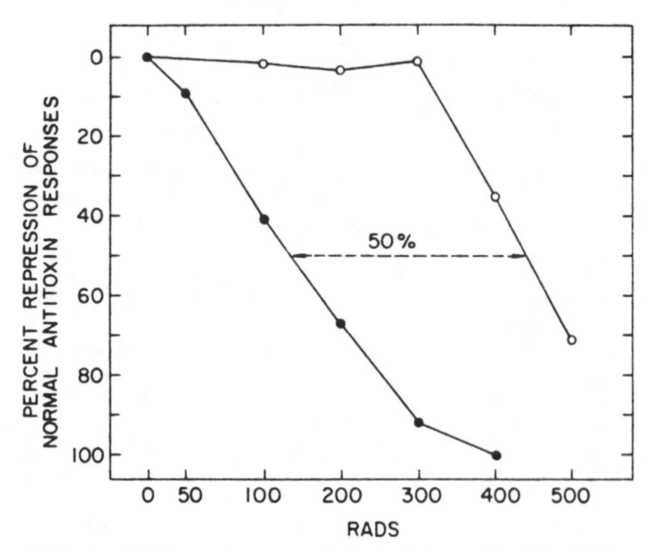

FIG. 6-11 Effect of graded doses of gamma radiation on primary (●) and secondary (○) antitoxin responses.

functional properties change accordingly. Activated helper T lymphocytes produce a soluble factor that stimulates (or helps to activate) B lymphocytes to transform, proliferate, and produce immunoglobulins. Stewart and Perez (25) reported the relative radiosensitivities of the lymphocyte subpopulations (see Fig. 6-13). Other investigators have shown that the antigen-activated helper T

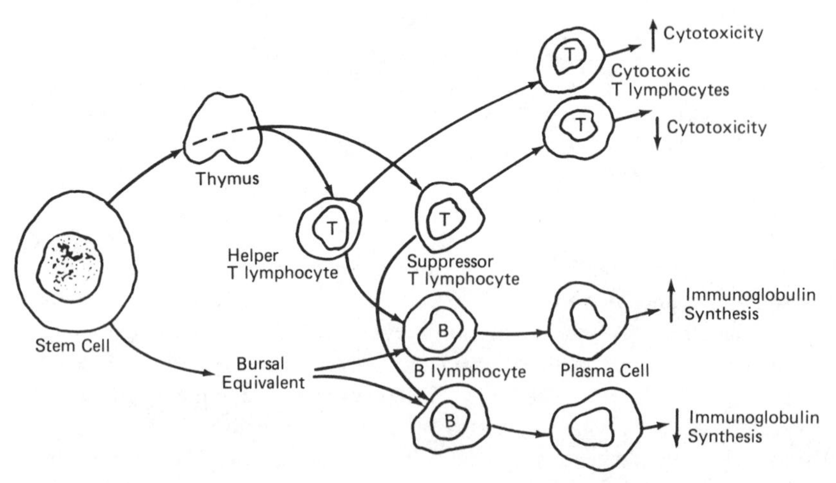

FIG. 6-12 Immunoregulatory function of T lymphocytes in T–B lymphocyte and T–T lymphocyte interactions. (Reprinted with permission from Ref. 21.)

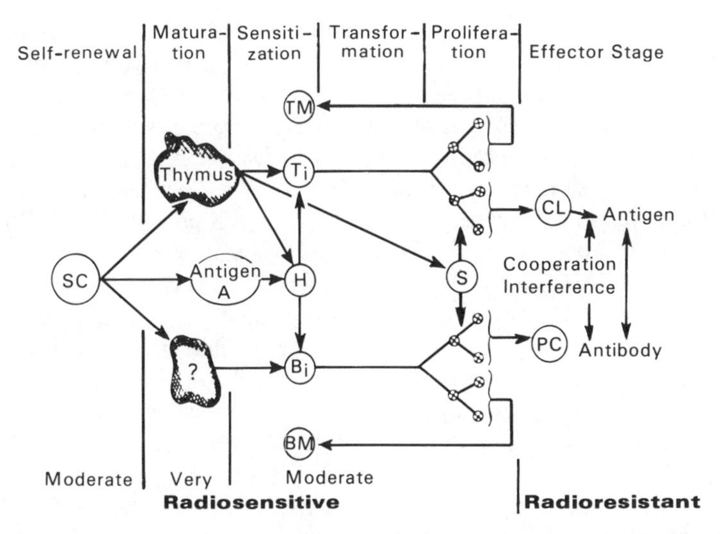

FIG. 6-13 Interaction and radiosensitivity of lymphocyte subpopulations. Each functional compartment is noted at top of figure, and radiation sensitivity is noted at bottom. In this model, bone marrow stem cells (SC) maturate in thymus and become T lymphocytes. These lymphocytes can be divided into stem precursors (T_i) which respond to specific antigens, helper cells (H) which are involved in initiation of antibody- and cell-mediated immune responses, and suppressor cells (S) which regulate response. Progeny of T cell proliferation are memory cells (TM) and effector lymphocytes (CL) which interact directly with antigen. Although control lymphoid organ is not known, bone marrow stem cells also mature into B lymphocytes. Through cooperation with thymus-derived helper (H) and suppressor (S) cells, these stem precursors (B_i) respond to specific antigens, and progeny differentiate into antibody-producing plasma cells (PC) and long-lived memory cells (BM). [Reprinted with permission, Stewart, C. C., and Perez, C. A. *Radiology* **118,** 201–210 (1976).]

lymphocytes are relatively radioresistant, with a D_0 value of 230 cGy. The unactivated cell is more radiosensitive, with reported D_0 values similar to those of other proliferative cell types (23). The activated suppressor T lymphocyte regulates the immune response (shown in Fig. 6-12). It limits the production of immunoglobulins, via another soluble factor, to the activated B lymphocytes. Various studies have shown that the suppressor T lymphocytes are more radiosensitive than the helper T lymphocytes (25). Again, the activated form was found to be more radioresistant than the unactivated form. The B lymphocyte is the ultimate responding cell in the humoral cascade. This cell, via a series of stimuli, activates, proliferates, and transforms into the cell of immunoglobulin production, the plasma cell. The unactivated B lymphocyte is radiosensitive, with D_0 values in the range of 50–70 cGy. The activated B lymphocyte, like the activated T lymphocyte, is suspected of being more radioresistant. The plasma cell population, which has both short- and long-lived cells, is an extremely radioresistant cell (22,25). Plasma cells have been shown to be capable of producing antibodies even after radiation doses as high as 10,000 cGy (22).

b. Lymphocyte Subpopulations. The radiosensitivities of these various populations have been established, but also important is the rate of recovery of these populations in reestablishing an effective immune response. From investigative studies with total lymphoid irradiation (TLI), it has been shown that all lymphoid subpopulations were effectively eliminated in laboratory animals and in man (26). But the rate of recovery for the suppressor T cell population was found to differ from the rate of recovery for the helper T cell population (27–29). The suppressor T cell population was found to recover to near normal levels by 4–6 weeks post-TLI, whereas the helper T cell population was found to recover at a much slower rate. The helper T cell recovery may take as long as 12 months. These observations are further reflected in the immunological responses post-TLI. The ratio of helper T cells to suppressor T cells in the peripheral blood of man was shown to be approximately two before irradiation. As a result of the differing recovery patterns of the lymphocyte subpopulations, the ratio of helper T cells to suppressor T cells was shown to be 0.65 for as long as 6–8 months post-TLI (28). Thus, a cellular imbalance exists in an individual post-TLI. This cellular imbalance of T cell subpopulations results in a diminished antibody response. Kotzin *et al.* (30) showed that peripheral blood mononuclear cells from TLI patients produced tenfold less immunoglobulin (both IgM and IgG) in response to the *in vitro* stimulus (pokeweed mitogen) than before radiotherapy. The observed decrease in immunoglobulin production was associated with the presence of suppressor T cells. Experiments designed to evaluate the differing primary antibody responses to either proteins (T cell dependent) or polysaccharides (T cell independent) have further separated the humoral immune response after TLI. The humoral response to foreign proteins has been shown to be suppressed for an extended time postirradiation (31,32). These investigators again attribute the suppressed response to the cellular imbalance between the helper and suppressor T cells, wherein the suppressor cell population controls the immune response. This is illustrated further in the antibody response to polysaccharides in which the response is T cell independent. Tanay and co-workers studied this response in mice (31) and humans (32) after TLI, and showed that the response was lost 3 days after treatment. But when it was reevaluated at 30 days after treatment, the antibody response was significantly higher than for controls. Thus, the cellular imbalance impairs the T cell-dependent antibody response, but it does not affect the T cell-independent response. This is consistent with the observations that the B cells recover after TLI and their function is not impaired (33).

The humoral response, an intimate balance of various cell types, is clearly affected by radiation. The nonstimulated cells (B, helper T, and suppressor T lymphocytes) are radiosensitive (D_0 values 50–100 cGy). Thus, these cells are not able to fully respond to a new antigen (primary response) immediately after irradiation. The time of recovery of these unstimulated cell populations dictates the time when a primary humoral response can be mounted. In general, at doses less than the $LD_{50/30}$ value, the primary response may be suppressed as much as 1–2 weeks. At doses greater than the $LD_{50/30}$ value (assuming survival), the suppression would extend from weeks to possibly months. The secondary humoral response is less radiosensitive, as are the activated cell types and plasma cells. Therefore, a humoral response is possible to a previously sensitized antigen. But the extent of this response may be suppressed by the fact that some unactivated cells may be required to maximize the response.

2. Cell-Mediated Immunity

Induction of cell-mediated immunity (CMI) is initiated when T lymphocytes come into contact with a foreign antigen (e.g., microorganism) either at the site of infection or in the lymph nodes draining the infected tissue. T lymphocyte transformation and proliferation follow the antigen stimulation, and from this point, the lymphocytes perform the functions of CMI (Fig. 6-14). Similar to the B lymphocytes of the humoral response, some of these activated T lymphocytes become long-lived memory cells, which are able to respond quickly to a previously sensitized antigen.

The T lymphocytes involved in CMI protect the host by a variety of mechanisms. One mechanism is the secretion of lymphokines, which are soluble factors that stimulate or inhibit various levels of the immune response (see Table 6-2). Two examples are the migration inhibitory factor (MIF) and the lymphocyte-derived chemotactic factor (LDCF). These lymphokines act to recruit cells to the site of infection; the cells may be monocytes, eosinophils, or neutrophils, which are associated with the nonspecific immune response. These recruited cells destroy the infectious agent. Another mechanism involves the direct killing of mammalian cells that are infected with certain organisms. These cells are called cytotoxic T lymphocytes.

The T lymphocytes of CMI are also under the control of a sophisticated genetic recognition system that is associated with the major histocompatibility complex (MHC). The gene products of the MHC give rise to a set of specifically defined surface markers. It has been proposed that these surface markers are important for the presentation of antigen and that they may be essential in identifying an allogeneic mammalian cell as foreign.

The radiation sensitivity of this cellular arm of the immune response is variable, depending on the specific phase involved. In general, it has been shown that the induction of CMI is more radioresistant than that of the humoral response. In experiments with guinea pigs, a 300-cGy whole-body dose given before sensitization with diphtheria toxoid suppressed the development of circulating antitoxin but failed to inhibit the induction of the CMI response (17,34). Production of the antitoxin decreased proportionally as the radiation dose was

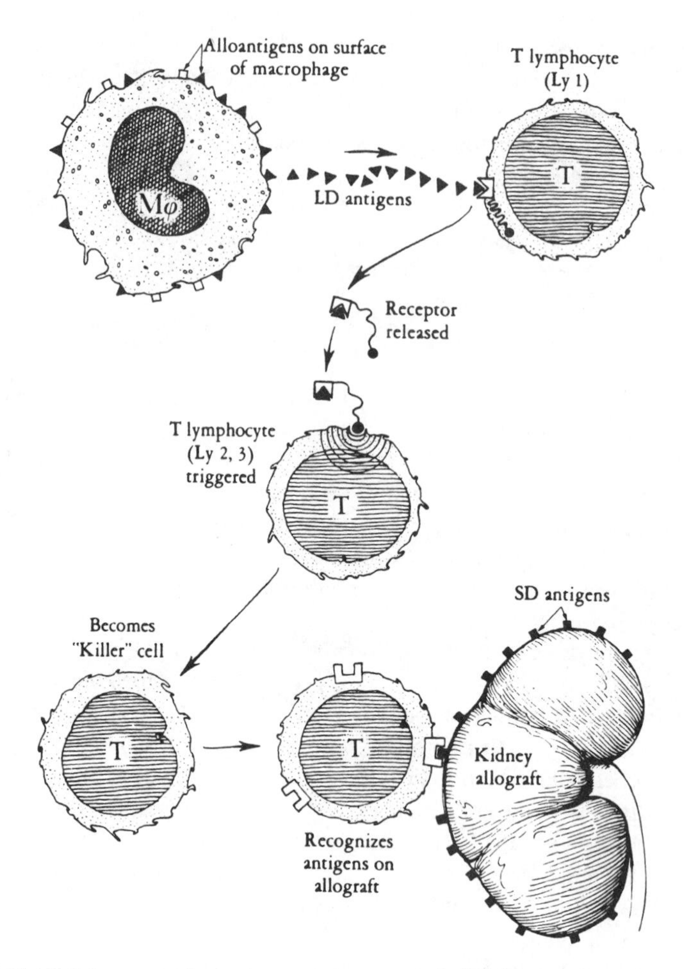

FIG. 6-14 Cellular events that make up tertiary stage of CMI in a murine system, soliciting interaction of antigen with T lymphocytes. (Reprinted with permission, Bellanti, J. A., and Rocklin, R. E. *Immunology II*. Saunders, Philadelphia, Pennsylvania, 1978.)

increased from 50 to 300 cGy. The CMI response was not affected in this dose range. However, when the hematopoietic syndrome occurs after higher dose levels (800 cGy), Uhr and Schraff (35) showed that both the humoral and cell-mediated responses of the rabbit were lost. Thus, the cell-mediated response and its associated cells are more radioresistant than are the humoral response and its associated cells. The literature suggests that, at doses below the $LD_{50/30}$ values in all animal models (even in man), the humoral response is suppressed and the amount of suppression is dose dependent. However, the CMI response is not

TABLE 6-2

PRODUCTS OF ACTIVATED LYMPHOCYTES

Mediators affecting macrophages
Migration inhibitory factor (MIF)
Macrophage-activating factor (indistinguishable from MIF)
Macrophage aggregation factor (MAF) (? same as MIF)
Factor causing disappearance of macrophage from peritoneum (? same as MIF)
Chemotactic factor for macrophages
Antigen-dependent MIF
Mediators affecting neutrophil leukocytes
Chemotactic factor
Leukocyte inhibitory factor (LIF)
Mediators affecting lymphocytes
Mitogenic factors
Antibody-enhancing factors
Antibody-suppressing factors
? Chemotactic factor
Mediators affecting eosinophils
Chemotactic factor
Migration stimulation factor
Mediators affecting basophils
Chemotactic augmentation factor
Other cells
Cytotoxic factors—lymphotoxin
Growth inhibitory factors
Clonal inhibitory factor
Proliferation inhibitory factor
Osteoclast-activating factor (OAF)
Skin reactive factor
Interferon
Immunoglobulin

appreciably affected. At doses greater than the $LD_{50/30}$ values, both arms of the specific immune response are severely depressed.

In conclusion, the immune competency of an individual after irradiation exposure is altered. An interdependence of the cellular functions in the humoral and cell-mediated responses exists, and it must be considered in discussions of immune competency. Several points need to be made about the radiosensitivity of the various cellular phases of the specific immune response. First, cells that have been activated by an antigen near the time of irradiation appear to be more radioresistant than the unactivated cells. So in this case, an immune response to the antigen is likely to occur. The converse to this observation is that unsensitized cells, either T or B lymphocytes, are more radiosensitive, and that a humoral immune response to a new antigenic stimulus immediately after moderate doses of radiation (200–400 cGy) is suppressed. On further interpretation of the data, a relationship is seen between the radiation dose received, the time of antigenic challenge, the type of antigen, and whether a prior sensitization had occurred. This all leads back to understanding the cells involved in each response and how they survive and recover after exposure to radiation. It is the return of that homeostatic balance that is critical. It should be emphasized that the ultimate recovery of the functional cells is still dependent on the recovery of the PPSC in the bone marrow. Finally, although the CMI response is more radioresistant than the humoral response, the CMI response rarely operates alone. In fact, in a host infected with organisms against which CMI is important, the presence of antibodies and components of the complement system is frequently critical to the host's ultimate survival. The cells of the nonspecific immune responses (monocytes, macrophages, and neutrophils) are solicited to the site of infection by lymphokines produced by T lymphocytes of the CMI response. However, after irradiation, the loss of these cells from circulation will limit the ability of an individual to mount a nonspecific immune response even if the T lymphocytes survive and produce the lymphokines. Therefore, the immune competency of an individual is changed by radiation. It is how much the immune-responsive cells are affected and how soon they will return to normal that are critical to the survival of an individual.

REFERENCES

1. Jones, T. D. Hematologic syndrome in man modeled from mammalian lethality. *Health Phys.* **41,** 83–103 (1981).
2. Robinson, C. V. Relationship between animal and stem cell dose survival curves. *Radiat. Res.* **35,** 318–344 (1968).
3. Wintrobe, M. M., Lee, G. R., Boggs, D. R., Bithell, T. C., Forester, J., Athens, J. W., and Lukens, J. N. *Clinical Hematology,* 8th ed. Lea & Febiger, Philadelphia, Pennsylvania, 1981.
4. Till, J. E., and McCulloch, E. A. Direct measurement of the radiation sensitivity of normal mouse BM cells. *Radiat. Res.* **14,** 213–222 (1961).
5. McCulloch, E. A., and Till, J. E. The sensitivity of cells from normal mouse bone marrow to gamma irradiation *in vitro* and *in vivo*. *Radiat. Res.* **16,** 822–832 (1962).
6. Till, J. E., and McCullouch, E. A. Repair processes in irradiated mouse hemopoietic tissue. *Ann. N.Y. Acad. Sci.* **114,** 115–125 (1964).
7. Bond, V. P., and Robinson, C. V. A mortality determinant in nonuniform exposures of the mammal. *Radiat. Res., Suppl.* **7,** 265–275 (1967).
8. MacVittie, T. J., Monroy, R. L., Patchen, M. L., and Darden, J. H. Relative biologic effect: The canine hemopoietic response to sublethal total body irradiation with cobalt-60 gamma or mixed neutron:gamma radiation. In *Proceedings of the Seventh International Congress of Radiation Research* (J. J. Broerse, G. W. Barendsen, H. B. Kal, and A. J. van der Kogel, eds.), Abstr. No. C1-21. Martinus Nijhoff, The Hague, 1983.
9. Mathé, G. Total body irradiation injury: A review of the disorders of the blood and hematopoietic tissues and their therapy. In *Nuclear Hematology* (E. Szirmai, ed.), pp. 275–338. Academic Press, New York, 1965.

10. Cronkite, E. P., Carsten, A. L., and Brecher, G. Hemopoietic stem cell niches, recovery from radiation and bone marrow transfusions. In *Proceedings of the Sixth International Congress of Radiation Research* (S. Okada, M. Imamura, T. Terashima, and H. Yamaguchi, eds.), pp. 649–656. Topan Printing Co., Ltd., Tokyo, 1979.
11. Boggs, S. S., Chervenich, P. A., and Boggs, D. R. The effect of post-irradiation bleeding or endotoxin on proliferation and differentiation of hematopoietic stem cells. *Blood* **40,** 375–389 (1972).
12. Monroy, R. L., Zeman, G. H., Jemionek, J. F., Contreras, T. J., Darden, J. H., and MacVittie, T. J. A re-evaluation of fission neutron effectiveness for bone marrow failure in a large animal model. (In preparation.)
13. Vos, O. Repopulation of the stem cell compartment in haemopoietic and lymphatic tissues of mice after X-irradiation. In *Effects of Radiation on Cellular Proliferation and Differentiation,* pp. 149–162. IAEA, Vienna, 1968.
14. Mathé, G., Amiel, J. L., and Schwarzenberg, L. Treatment of acute total body irradiation injury in man. *Ann. N.Y. Acad. Sci.* **114,** 368–392 (1964).
15. Dolphin, G. W. *Handling of Radiation Accidents.* IAEA, Vienna, 1969.
16. Conklin, J. J., Kelleher, D. L., and Walker, R. I. Evaluation and treatment of nuclear casualties. Part I. Acute radiation syndrome and triage. *Med. Bull.* **40,** 9–16 (1983).
17. Anderson, R. E., and Warner, N. L. Ionizing radiation and the immune response. *Adv. Immunol.* **24,** 215–335 (1976).
18. Cronkite, E. P., Chanana, A. D., Joel, D. D., and Laissue, J. Lymphocyte repopulation and restoration of cell mediated immunity following radiation: Whole-body and localized. In *Conference on Interaction of Radiation and Host Immune Defense Mechanisms in Malignancy,* pp. 181–206. Brookhaven Nat. Lab., Upton, New York, 1974.
19. Sharp, J. G., and Watkins, E. B. Cellular and immunological consequences of thymic irradiation. In *Immunopharmacologic Effects of Radiation Therapy* (J. B. Dubois, B. Serron, and C. Rosenfeld, eds.), pp. 137–179. Raven Press, New York, 1981.
20. Kwan, D. K., and Norman, A. Radiosensitivity of human lymphocytes and thymocytes. *Radiat. Res.* **69,** 143–151 (1977).
21. Sharma, S. D., and Remington, J. S. The role of cell-mediated immunity in resistance to infection in the immunocompromised host. In *Infections in the Immunocompromised Host: Pathogenesis, Prevention and Therapy* (J. Verhoef, P. K. Peterson, and P. G. Quie, eds.), pp. 59–76. Elsevier/North-Holland Biomedical Press, New York, 1980.
22. Mankinodan, T., and Price, G. B. Circumventing graft rejection. In *Transplantation* (J. S. Najarian and R. L. Simmons, eds.), pp. 251–271. Lea & Febiger, Philadelphia, Pennsylvania, 1972.
23. Stoner, R. D., Hess, M. W., and Terres, G. Primary and secondary antibody responses related to radiation exposures. In *Conference of Interaction on Radiation and Host Immune Defense Mechanisms in Malignancy,* pp. 152–166. Brookhaven National Laboratory, Upton, New York, 1974.
24. Nettersheim, P., Williams, M. L., and Hammons, A. S. Regenerative potential of immunocompetent cells. III. Recovery of primary antibody-forming potential for X-irradiation: The role of the thymus. *J. Immunol.* **103,** 505–518 (1969).
25. Stewart, C. C., and Perez, C. A. Effects of irradiation on immune responses. *Radiology* **118,** 201–210 (1976).
26. Slavin, S., Yatziv, S., Zanibar, I., Fuks, Z., Kaplan, H. S., and Strober, S. Nonspecific and specific immunosuppression by total lymphoid irradiation (TLI). In *Immunology 80, Progress in Immunology IV* (M. Fougereau and J. Dausset, eds.), p. 1160. Academic Press, New York, 1980.
27. Strober, S. Overview: Effect of total lymphoid irradiation on autoimmune disease and transplantation immunity. *J. Immunol.* **132,** 968–984 (1984).

28. Haas, G. S., Halperin, E., Doseret, D., Lingood, R., Russell, P. S., Colvin, R., Barrett, L., and Cosimi, A. B. Differential recovery of circulating T-cell subsets after nodal irradiation for Hodgkins' disease. *J. Immunol.* **132,** 1026–1030 (1984).
29. Field, E. H., Engleman, E. G., Terrell, C. P., and Strober, S. Reduced *in vitro* immune responses of purified human Leu-3 (helper/inducer phenotype) cells after total lymphoid irradiation. *J. Immunol.* **132,** 1031–1034 (1984).
30. Kotzin, B. L., Strober, S., Kansas, G. S., Terrell, C. P., and Engleman, E. G. Suppression of pokeweed mitogen stimulated immunoglobulin production in patients with rheumatoid arthritis after treatment with total lymphoid irradiation. *J. Immunol.* **132,** 1049–1055 (1984).
31. Tanay, A., Strober, S., Logue, G. L., and Schiffman, G. Use of the total lymphoid irradiation (TLI) in studies of the T-cell dependence of autoantibody production in rheumatoid arthritis. *J. Immunol.* **132,** 1036–1040 (1984).
32. Tanay, A., and Strober, S. Opposite effects of total lymphoid irradiation on T-cell dependent and T-cell independent antibody responses. *J. Immunol.* **132,** 979–984 (1984).
33. May, R. D., Slavin, S., and Vitetta, E. S. A partial characterization of suppressor cells in the spleens of mice conditioned with fractionated total lymphoid irradiation (TLI). *J. Immunol.* **131,** 1108–1114 (1983).
34. Slavin, S. B., and Smith, R. F. Delayed hypersensitivity in the development of circulatory antibody. The effect of X-irradiation. *J. Exp. Med.* **109,** 325–338 (1959).
35. Uhr, J. W., and Schraff, M. Delayed hypersensitivity. The effect of X-irradiation on the development of delayed hypersensitivity and antibody formation. *J. Exp. Med.* **112,** 65–76 (1960).

CHAPTER 7

Effect of Ionizing Radiation on Gastrointestinal Physiology

PAMELA J. GUNTER-SMITH

Armed Forces Radiobiology Research Institute
Bethesda, Maryland 20814-5145

The effects of ionizing radiation on gastrointestinal physiology can be grouped into two main categories: prodromal effects, which are acute effects occurring at low doses of exposure (1 Gy), and the gastrointestinal syndrome, which occurs at higher doses of radiation (10 Gy). The prodromal effects consist of nausea and vomiting at early times postexposure. While not life threatening, they can result in severe incapacitation of the exposed individual. The gastrointestinal syndrome is marked by progressive deterioration of the mucosal cell lining of the gastrointestinal system, which leads to loss of fluids and electrolytes and also bacteremia. Without intervention, death from the gastrointestinal syndrome will follow within a well-defined period of time, 4–5 days postirradiation.

In spite of these pronounced effects of radiation exposure on gastrointestinal physiology, our current understanding of the mechanisms underlying these effects is incomplete. As a result, methods to prevent and treat radiation-induced gastrointestinal dysfunction are uncertain. Also, it has not been determined yet if the prodromal effects and the gastrointestinal syndrome have the same etiology. However, for ease of presentation, in this chapter they will be treated as separate processes. The symptomatology of the prodromal effects and the gastrointestinal syndrome will be described. The normal physiologic processes that are involved will be discussed, followed by current concepts of the interaction of radiation with these processes, which results in the observed dysfunction.

I. PRODROMAL EFFECTS

A. Nausea and Vomiting

The prodromal effects following exposure to ionizing radiation are nausea, vomiting, and diarrhea. The time of onset, duration, and severity of these symptoms depend on the magnitude of the dose received. At doses of 1 Gy, emesis associated with the prodromal effects has been observed as early as 30 min postexposure and lasting as long as 2 hr (1). At higher doses, the onset of prodromal effects may be earlier. Indeed, in a well-documented case of human exposure, nausea and vomiting occurred almost immediately after exposure, accompanied by bloody diarrhea (2). The precise relation between the time of onset of emesis and its severity and the dose of exposure has been of interest, but the relationship has not been adequately defined. It may be speculated that such information is useful in determining the dose of an exposure. In the canine model, the 50% emesis dose is 1.7 Gy (3) to 5.4 Gy (4), with the 90% level observed with 8-Gy gamma photons (5).

Vomiting is a complex multifaceted event that requires the coordinated response of neural, respiratory, and gastrointestinal centers. It occurs in response to a variety of stimuli, including local irritation and distention of the gut, introduction of various drugs and hormones into the systemic circulation, psychogenic stimuli, and exposure to ionizing radiation. The sequence of events involved in vomiting are as follows (6). Indications that vomiting is imminent are those of widespread autonomic discharge: hypersalivation, tachypnea, and dilation of the pupils. Events leading to the expulsion of gastric contents are initiated by slow and deep inspiration against a closed glottis. This reduces the intrathoracic pressure below atmospheric pressure. The abdominal muscles contract strongly, raising the intra-abdominal pressure; the resulting pressure gradient forces the gastric contents into the esophagus. If the hypopharyngeal sphincter is closed, the contents do not enter the mouth but return to the stomach. These events produce the retching that generally precedes vomiting (7). But if the sphincter is opened (by drawing the larynx and the hyoid bone forward), the gastric contents are expelled into the mouth. Thus, expulsion of the gastric contents is a passive process that does not involve active contractions of the body of the stomach.

As indicated above, the vomiting process is faithfully reproduced in response to a variety of what appear to be unrelated stimuli. Indeed, nausea and vomiting can be major complicating factors of various therapeutic regimens, especially radiotherapy and chemotherapy. But the identification of a common mechanism by which these various stimuli elicit these events has been elusive. In 1952, Wang and Borison (8) demonstrated that emesis stimulated by a class of pharmacological agents such as apomorphine arose from neural activity in a small area of the medulla under the floor of the fourth ventricle called the area postrema. This area, which they called the chemoreceptor trigger zone, lies outside the blood–brain barrier (9), where it continuously samples the constituents of the systemic circulation. Although ablation of the area postrema in experimental animal models was found to abolish the emetic response to pharmacologic agents (4,10,11), it was ineffective in preventing emesis that arose from other stimuli, such as local irritation of the gastrointestinal tract produced by ingestion of copper sulfate (8). Thus Wang and co-workers suggested that neurons from the area postrema projected to some other area, the vomiting center, which also received afferent input directly from the gastrointestinal tract. This area is located in the medulla near the tractus solitarius. The sensation of nausea is thought to be related to subconscious recognition of the stimulation of this or closely adjacent centers (6). Since nausea does not always result in vomiting, the sensory center is probably a distinct center.

Ablation of the area postrema has also been observed to abolish radiation-induced emesis in the dog (10,11), monkey (12), and man (13). As a result, it was concluded that stimulation of the chemosensory trigger zone by some radiation-released humoral agent was responsible for radioemesis. The notion that these agents may be released locally from mast cells is gaining attention. While speculative, the proposal is given some credence by the facts that the area postrema is itself heavily populated by mast cells (14), these cells contain emetic substances (15), and mast cells appear to be degranulated following radiation exposure (16).

However, not all of the available data are consistent with the view that radioemesis results from local effects of humoral agents on the area postrema. For example, Brizzee (12) observed that supradiaphragmatic vagotomy also prevented emesis in monkeys. Borison (17) reported that ablation of the area postrema did not consistently prevent radioemesis in cats although other maneuvers that decrease afferent input from the abdomen did confer protection. While some of the disparity between these results can be attributed to differences of species, these studies suggest that the vomiting motor center receives input from other areas in the abdomen that may be important in radioemesis. Thus, although the major neural centers involved in the responses have been determined (Fig. 7-1), the precise etiology of radiation-induced emesis remains unknown.

During the past decade, much effort has been invested in the search for an antiradioemetic that confers protection but has few side effects and does not affect performance. Clinically used antiemetics include antihistamines, anticholinergics, phenothiazines, butyrophenones, cannabinoid, benzamidazoles, procainamide derivatives, and corticosteroids (18). The fact that so many different classes of drugs are effective implies that the neuropharmacology of vomiting is extremely complex. The situation is further complicated by the observations that not all of these agents are effective against all of the emetic stimuli and that efficacy is species dependent. For example, some of the antihistamines that were found to be effective antiradioemetics in dogs (3,19) were found to be ineffective clinically (20). The phenothiazines (e.g., chlorpromazine) are especially useful clinically (18), but many of these agents have undesirable neurologic side effects that decrease their suitability for use by the Armed Forces.

Currently, the dopamine antagonists metaclopramide and domperidone are

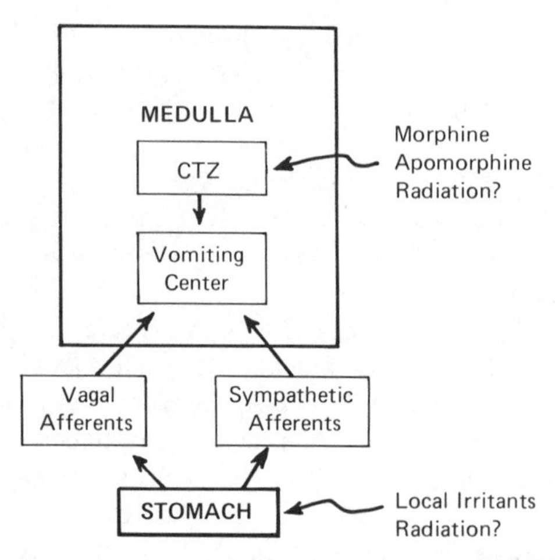

FIG. 7-1 Inputs to the vomiting center. (Adapted from Ref. 6.)

being considered for use as antiradioemetics. Metoclopramide has both central and peripheral actions (21,22). Domperidone does not cross the blood–brain barrier (23) and so is thought to have only peripheral actions. Both have been shown to be effective in preventing radioemesis (18,24) in some instances. Dubois *et al.* (5) observed that domperidone prevented radioemesis in 90% of dogs exposed to 8 Gy of radiation. Extension of these results to subhuman primates was unsuccessful. Metoclopramide, on the other hand, has been shown to be effective in humans (18).

In summary, several classes of compounds have been found to be clinically useful antiradioemetics. The species dependency of efficacy has complicated the search for an effective agent that does not compromise performance. A number of compounds are being considered for their antiemetic properties, including the dopamine antagonists and several of the newer antihistamines. The variety of emetic agents that may be released, which could include agents released secondary to radiation-induced degranulation of mast cells, makes it likely that a combination of agents may have to be given to prevent radioemesis.

B. Gastric Motility

In addition to eliciting nausea and vomiting, radiation exposure produces concomitant changes in motility of the gastrointestinal tract at earlier times postexposure. Following irradiation with doses as low as 1 Gy, motility of the small intestine of rodents increases (25). Intestinal propulsion increases immediately but returns to normal, and then may be suppressed for 3 days (25). Borowska *et al.* (26) reported an increase in small intestinal motility in mice by 24 hr postexposure after an initial decrease. In this study, the increase was correlated with an increase in prostaglandin levels. Both could be inhibited by pretreatment with indomethacin (a prostaglandin synthesis inhibitor). Radiation also affects gastric emptying. Gastric emptying is suppressed at 20 min postexposure and may remain so for 3 hr (5). So the effects on gastrointestinal motility are seen to be varied and time dependent.

The relation between these effects on motility and radiation-induced vomiting is unclear, and some evidence exists that changes in motility are not essential for vomiting. For example, some emetic agents such as apomorphine suppress motility initially (27,28), whereas radiation exposure is generally associated with increased motility. Gastric emptying is delayed in species that vomit (5,29) and those that do not vomit (25,30,31). Further, the antiradioemetic domperidone is effective in preventing emesis in the canine but not in preventing the concomitant delay in gastric emptying (5). Thus vomiting and delayed gastric emptying may be independent and may occur by separate pathways.

The etiology of the diarrhea associated with the prodromal effects of ionizing radiation is also unknown. It may be secondary to changes in motility of the small intestine, or it may be stimulated by some neurohumoral agent released very early after exposure. Histamine, shown to be released within minutes postirradiation (32), has been demonstrated to stimulate secretion in the small intestine (33).

II. THE GASTROINTESTINAL SYNDROME

In 1956 Quástler (34) described a mode of acute radiation death whose time course was well defined and invariant of dose in the range of 10–100 Gy gamma or X irradiation. Death of the irradiated animal occurred 4–5 days postexposure, and could be correlated with changes in the morphology and physiology of the small bowel. This mode of death and its accompanying pathology were referred to as the gastrointestinal syndrome. The intestinal epithelial stem cell is the target of radiation damage, and the resulting decrease in mitotic activity leads to denudation of the intestinal mucosa, fluid and electrolyte imbalance, and bacteremia (Fig. 7-2).

Evidence to date suggests that the pathophysiology arising from the gastrointestinal syndrome is perhaps of greater consequence in neutron exposure than is damage to the hematopoietic system. Otto and Pfeiffer (35) demonstrated that the mean survival time decreases with increasing doses of X irradiation. The effect of increasing dose on survival was more pronounced in studies with neutron exposure. At a neutron dose of 3.6 Gy, the mean survival decreased to 5 days,

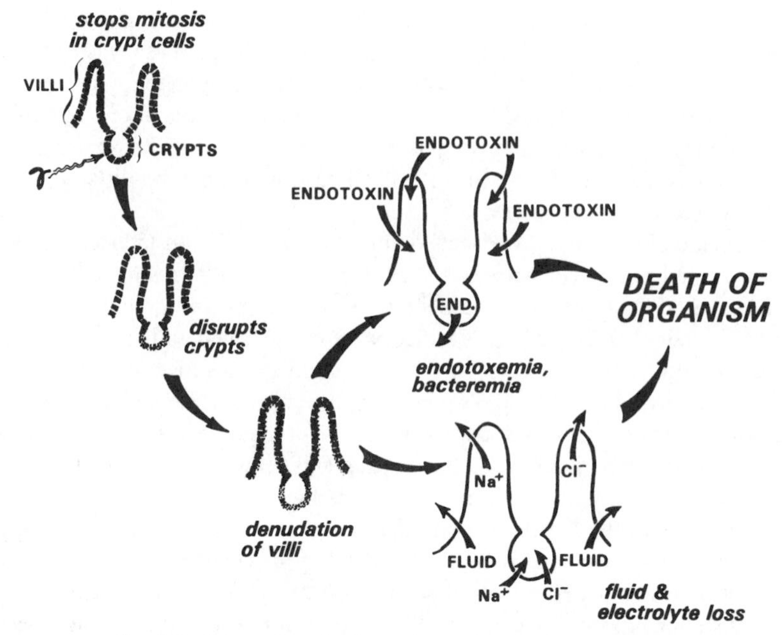

FIG. 7-2 Proposed mechanisms of gastrointestinal syndrome.

whereas at 6.3 Gy X irradiation, mean survival was 12 days. Other studies have supported this shift in survival time after neutron exposure (36–38). The relative biological effectiveness for neutron to X irradiation for the gastrointestinal syndrome has been estimated at 2.4 (37). Thus, the prevention or treatment of radiation-induced gastrointestinal dysfunction is of particular importance after exposure to neutrons.

While the exact cause of lethality remains uncertain, the events after denudation (fluid and electrolyte imbalance and also bacteremia) severely compromise the irradiated animal and contribute to the lethality. Despite the pronounced effect of radiation on gastrointestinal physiology, very little is known concerning the mechanisms associated with radiation's induced alterations in electrolyte transport and bacteremia. The loss of fluids and electrolytes may be attributed to the direct effects on cellular transport processes and the secondary effects due to the action of radiation-released hormones, the action of bile salts on the intestinal mucosa, changes in intestinal blood flow, or loss of absorptive cells secondary to denudation of the intestinal villi. Bacteremia has been related to a change in the intestinal flora and to a decrease in integrity of the intestinal mucosa. The following section focuses on (1) the effect of radiation on the physiology of the small bowel and (2) current concepts of how these changes lead to the dysfunction associated with the gastrointestinal syndrome.

A. Morphology and Physiology of the Small Intestine

The mammalian small intestine is morphologically and functionally heterogeneous. Morphologically, the intestinal epithelial cell lining can be subdivided into villus epithelium and crypt epithelium. Functionally it can be divided into proliferative and absorptive areas. The villus epithelium is composed of cells that line the fingerlike projections called villi, which protrude into the intestinal lumen. Cells lining the glandular structures that lie in the submucosal space comprise the crypt epithelium. Cells in the crypts have access to the surface via openings called crypt ostia, which have been compared to pits in a forest of trees. At least five distinct cell types can be identified along the intestinal mucosa. The villus epithelium is made up of two cell types: columnar epithelial cells (which carry out the absorptive functions of the mucosa) and goblet cells (which secrete mucus). Less differentiated forms of these two types of cells are also found in the crypt region, along with three additional cell types. They are as follows: the Paneth cells, which secrete some as-of-yet-undefined substance; Argentaffin cells, which secrete serotonin; and a population of rapidly cycling columnar stem cells (39,40).

This morphological heterogeneity of the small intestine gives rise to the functional differences observed. Cells occupying the crypt area play a major role in proliferation of the intestinal mucosa, whereas villus cells perform the absorptive functions of the gut. Germinative properties of the small intestine are carried out by the rapidly cycling stem cells in the crypt, which are thought to correspond to the undifferentiated columnar cells (41). These cells actively undergo mitosis, and as they divide, new cells are pushed up toward the villus region of the mucosa. Ultimately, because of continual mitosis at the base of the crypt, these cells reach the top of the villus, where they are eventually sloughed off into the lumen. The elegant studies of Messier and Leblond (42) demonstrated that this process requires approximately 2–4 days in the rat and mouse. The turnover time for human small intestine appears to be similar (43). As the cells reach the villus, they differentiate and lose their ability to divide. Although they no longer play a role in proliferation, these cells acquire new proteins that are crucial for absorption.

Under normal conditions, the intestinal mucosa is an absorptive organ that is responsible for the absorption of nutrients, such as sugars, amino acids, and electrolytes, from the intestinal lumen into the blood. Absorption of fluid is simply a consequence of this solute transport. The transport of solute is accomplished by means of specialized transport proteins located within the cell membranes of mature villus epithelial cells. The cellular processes underlying these transport phenomena for many of the commonly absorbed solutes have been elucidated within the last two decades (44).

In some cases, the normally absorptive organ is converted into a secretory organ that transports fluid and electrolytes from the blood to the intestinal lumen. It is thought that a small amount of secretion occurs normally, but in the stimulated state, active secretion from the gut can result in the loss of several liters of fluid each hour (45). The intestinal secretagogues are numerous, including a variety of neurohumoral agents, such as serotonin, vasoactive intestinal peptide, acetylcholine, enkephalins, and prostaglandins, as well as a variety of bacterial toxins such as those released by cholera and some strains of *Escherichia coli* (46,47). Studies of the mechanisms underlying the secretory responses to these agents have revealed a commonality of mechanism. With very few exceptions, all of the secretagogues are thought to act by raising intracellular cyclic adenosine monophosphate and/or calcium. The latter presumably serves as a second messenger that initiates changes in a constituent of the cellular membrane, so that absorption of ions is inhibited and a secretory process is either unmasked or induced (48). Since the current consensus is that absorptive processes are associated with mature villus cells, the inhibition of absorption by secretagogues is thought to occur in the villus cells, while the stimulated active secretory processes are localized to the less differentiated crypt cells (48).

B. Effect of Radiation on Proliferation of Intestinal Epithelial Cells

The events leading to cellular reproduction have been incorporated into what is called the cell cycle. The cell cycle consists of two main phases: mitosis (during which the cell is actively dividing) and interphase (during which mitotic figures are not observed) (49). During interphase, the cell is not idle; it is either performing specialized tasks or preparing to divide. Immediately after mitosis, the cell enters the G_1 portion of interphase or the preduplication period. The amount of time spent in G_1 depends on the cell type. Indeed, some cells (differentiated cells) appear to be locked into G_1 for the remainder of their lifetimes. Proliferative cells, however, move from G_1 to S, during which time duplication of their genetic code occurs. This may require several hours, depending on the particular cell type. From the S phase, cells move into the G_2 part of interphase (postduplication), which precedes mitosis. The exact time required for completion of the cell cycle by an intestinal stem cell is species-dependent. In the rat it has been shown that the cell cycle requires approximately 10 hr (50).

The effects of ionizing radiation on the intestinal stem cell have been studied extensively. Lescher and co-workers, in a series of studies (51–54), defined the pattern of radiation-related changes in the intestinal stem cell cycle and its recovery. They demonstrated that while cells irradiated during mitosis were extremely radiosensitive and unable to divide, cells in interphase were also affected. The movement of cells from the preduplication G_1 phase into the DNA duplication S phase was slowed, and irradiated S phase cells and G_2 cells were blocked there. The severity of these alterations was dose dependent. At doses <10 Gy, some cells survive and are capable of resuming mitosis. So with sublethal doses, these cells can effect recovery of the mucosa if given sufficient time (55). It also appears that proliferation in the surviving population may be accelerated. The mechanism by which the apparent acceleration occurs is not understood. It has been suggested that more cells are recruited into the rapidly cycling population (56,57) or that the turnover rate of surviving cells is increased (52).

While mitosis in the surviving cells can ultimately repopulate the villus, studies of the long-term effects of radiation occurring months after exposure suggest that not all of their progeny are normal and capable of performing all the functions observed in nonirradiated animals. Clearly, this results from radiation-related alterations in the genetic code that is duplicated and conferred to daughter cells.

Many of the symptoms of the gastrointestinal syndrome are a consequence of damage to the intestinal stem cell by radiation. The mucosa loses its epithelial cell lining with the loss of proliferative function of the intestinal crypt and also continual migration of the villus cells toward the villus crest. The villus shortens, and in approximately 4 days it becomes devoid of its epithelial cell lining (34). The time required for denudation can be correlated with the time normally required for the cells to migrate up the villus (58). Recognition of the importance of maintaining the integrity of the intestinal mucosa has generated considerable interest in finding an effective radioprotectant for the intestinal epithelial cell. Most of the advances in the area of cytoprotection result from research on the gastric ulcer. Recent evidence has revealed that prostaglandins confer protection to gastrointestinal epithelial cells following exposure to a variety of noxious stimuli. The mechanism for this protection is unknown, but it is not related to the inhibition of gastric acid secretion (59). The protective effect of prostaglandins may arise from the stimulation of mucous secretion (60). Recently, Hanson and Thomas (61) extended this work to the injury of crypt cells resulting from radiation. They observed that prostaglandins are also effective in protecting (presumably) intestinal stem cells in irradiated rats. Unfortunately, these agents are not without undesirable side effects (62), so the potential benefits must be weighed against the injurious effects of these agents. Additional work in this area is crucial to preventing the gastrointestinal syndrome.

C. Radiation-Induced Loss of Fluids and Electrolytes

The decrease in absorption of fluids and electrolytes postirradiation has been of interest for some time. Curran *et al.* (63) demonstrated the inhibition of sodium and water absorption 13 hr after X irradiation. At longer intervals postirradiation, net secretion was observed in rats. Curran *et al.* further speculated that the early appearance of the net secretion did not correspond to denudation of the villi, and so must be due to some physiologic effect on the intestinal epithelium. These observations have since been substantiated by numerous investigators. It was assumed that the secretion arose from inhibition of the normal absorptive properties of the gut rather than stimulation of active secretion. However, recent studies conducted at the Armed Forces Radiobiology Research Institute (P. J. Gunter-Smith, unpublished observations) are consistent with stimulation of an active transport process.

The mechanisms by which radiation exposure inhibits absorption and/or stimulates secretion are not known. They may be related to the direct effects of radiation on cell membrane properties and also secondary effects. To date no clear evidence exists that the fluid and electrolyte secretions result from the direct effect of ionizing radiation. Although it has been observed that radiation of the exteriorized small bowel is sufficient to initiate the gastrointestinal syndrome (64), we cannot exclude the possible effects of locally released agents such as histamine and other hormones. Direct effects of radiation on cell membranes have been reported, which include increased membrane permeability, inactivation of enzymes, and peroxidation of membranes (65). Some of the work most suggestive of direct effects on epithelial transport is that of Porvaznik (66) and Moran *et al.* (67). Moran *et al.* (67) observed that radiation caused decreased sodium-dependent sugar transport by epithelial cells in culture. Porvaznik (66) demonstrated a decrease in integrity of the tight junctional complexes associated with the goblet cells in irradiated mice. These breaks may provide a low-resistance pathway between the cells for the secretion of substances from the blood to the lumen. Clearly, in view of the techniques of cell and organ culture now available, the question of direct versus indirect effects should be resolved in the near future.

A number of agents released in response to radiation exposure can secondarily elicit the secretion of fluids and electrolytes. They are too numerous to list here, but several deserve mention. Histamine has been shown to stimulate not only gastric secretion but also secretion by the small bowel (33). Cockerham *et al.* (68) have shown that histamine is increased postirradiation in the small intestine of irradiated canines. Given the large concentration of mast cells in the gut, the possibility that histamine plays a role in radiation diarrhea is certainly plausible. Other agents that may increase in response to radiation exposure include the prostaglandins, which are potent secretagogues (46). Increases in prostaglandin-like activity in irradiated rat gut were reported by Borowska *et al.* (26). Further, Eisen and Walker (69) observed decreased prostaglandin 15-OH dehydrogenase in mouse small intestine that would inhibit prostaglandin catabolism. In view of the number of potential secretagogues released, secondary effects due to these agents are likely. Defining the role of each of these agents in the radiation response is difficult.

The role of bile and bile salts in radiation-induced diarrhea has received considerable attention. A series of studies by Jackson and co-workers (70,71) demonstrated that radiation-induced diarrhea and sodium loss in the rat could be prevented by prior bile duct ligation. Their interpretation was that loss of sodium and fluids resulted from the inhibition of sodium reabsorption from bile. These observations and others prompted speculation that bile salts have some direct effect on the intestinal mucosa, which produces postirradiation diarrhea. Other data supporting this concept include those indicating that (1) inclusion of high concentrations of bile salts in the diet of rats decreased the proliferation of mucosa (72), (2) introduction of these agents into the intestinal lumen resulted in diarrhea (73), and (3) bile salts had a mucous-depleting action on the mucosa (74). However, very recent studies by K. L. Jackson and J. P. Geraci (personal communication) further support the view that bile and bile salts do not directly stimulate radiation-induced diarrhea.

Changes in intestinal vascular blood flow have also been implicated as causative factors in the gastrointestinal syndrome. Kabal and co-workers (75) reported a decreased intestinal blood flow 72 hr postirradiation in the canine. They suggested that the gastrointestinal syndrome was perhaps similar to intestinal ischemia. Eddy and Casarett (76), using microangioradiographic and histological techniques, also demonstrated radiation-related vasculature damage in rats at later times postexposure. In contrast, recent studies by Cockerham *et al.* (77) demonstrated an increase in intestinal blood flow at very early times beginning 10 min postirradiation, and the increase could be prevented by prior treatment with antihistamines (68). These experiments were not carried past 100 min, so they cannot be directly compared with those of Kabal (75). Eddy and Casarett (76) and Cockerham *et al.* (77) demonstrated radiation changes in intestinal morphology, especially in the crypt region. Although in some cases intestinal ischemia can disrupt the mucosal lining and thus be responsible for fluid loss, there has yet to be a clear demonstration of a relationship between the two in radiation-induced diarrhea. In addition, a correlation between vasoactivity and malabsorption has not been determined.

Studies comparing the gastrointestinal syndrome in irradiated germfree and conventional mice suggest a possible role for enteric bacteria in acute radiation death. Porvaznik *et al.* (78) demonstrated a change in the composition of the bacterial population in the intestine postirradiation. Such changes may result in an increased number of opportunistic pathogenic bacteria. As indicated earlier, a variety of toxins of bacterial origin can elicit fluid and electrolyte loss due to their interaction with receptors on epithelial cell membranes (46). Evidence that the intestinal flora may play a role in the gastrointestinal syndrome includes the facts that (1) treatment of irradiated animals with electrolyte solutions is ineffective unless combined with antibiotics (79), and (2) the onsets of diarrhea and death are delayed in germ-free animals, compared to conventional animals (80). Thus, while the role of bacteria in the loss of fluids and electrolytes is not totally understood, evidence to date suggests that bacteria are of major consequence. Changes in the intestinal flora postirradiation and subsequent bacteremia will be discussed in detail in another chapter.

In summary, the etiology of fluid and electrolyte loss following exposure to ionizing radiation is uncertain. Evidence to date suggests that it is complex and most likely results from a combination of any of the factors discussed above.

Is the Loss of Fluids and Electrolytes Responsible for Lethality?

The role of fluid and electrolyte loss in the lethality of the gastrointestinal syndrome has been controversial. Jackson *et al.* (81) proposed that the increased loss of sodium and water via the gut is sufficient to cause death. This proposal was supported by observations indicating a decrease in the exchangeable sodium concentration and extracellular fluid space and also a decrease in plasma volume (82,83). These observations further suggested that fluid and electrolyte loss is of major consequence. In contrast, Lushbaugh (84,85) maintained that loss of sodium from the gastrointestinal tract is not the main contributor to radiation lethality and that appreciable amounts of total-body sodium are not lost via this route. His view was supported by studies indicating that electrolyte replacement is not sufficient to prevent death and that survival is increased by the inclusion of antibiotics (80). Subsequent studies by Gits and Gerber (86) indicated no significant decrease in plasma volume. Although considerable time has passed since these observations were made, the role of gastrointestinal fluid and electrolyte loss in death due to intestinal irradiation remains unclear.

Studies indicate that antibiotic treatment in combination with fluid and electrolyte replacement is more effective than the latter alone, and they point to the role of bacteremia in radiation lethality. Porvaznik (66) demonstrated that radiation exposure decreases the integrity of the tight junctional complexes between some cells lining the intestinal mucosa. While the role of the junctional integrity in fluid and electrolyte transport has been alluded to above, Porvaznik speculated that the disrupted region may provide an easy route of entry for agents that are normally confined to the intestinal lumen (such as bacterial toxins) and ultimately lead to sepsis. Thus, potentially pathogenic substances may have access to the systemic circulation long before the intestinal mucosa becomes denuded. Obviously, challenge of an organism with an infectious agent will have severe consequences in an animal whose immunologic system is suppressed due to radiation effects on the hematopoietic system. Very recent studies, however, suggest that bacteremia may not directly contribute to lethality of the gastrointestinal syndrome (J. P. Geraci, personal communication). The role of intestinal bacteria in postirradiation infection and its consequences are discussed in detail in a later chapter.

D. Other Effects of Radiation on Gastrointestinal Physiology

Prodromal effects and the changes associated with the gastrointestinal syndrome have been emphasized. But radiation affects gastrointestinal function in other ways that may pose health problems for the irradiated individual and thus play a role in determining survival postexposure. In addition to the fact that radiation exposure changes the electrolyte transport characteristics of the small intestine, changes in the secretory pattern of the gastric mucosa have been described. Suppression of basal and stimulated gastric acid secretion has been observed after exposure to ionizing radiation (29). The role of these changes in overall gastrointestinal dysfunction is not clear. But since the secretion of gastric acid is affected as early as 40 min after exposure, the prodromal effects may be involved.

Although this chapter has focused on the changes in gastrointestinal physiology that occur within 4–5 days postirradiation, some changes in gastrointestinal physiology do not become apparent until weeks or months postexposure. Thus the individual exposed to ionizing radiation who ''survives'' the gastrointestinal and the hematopoietic syndromes may still be faced with serious pathology of the gastrointestinal tract in the future. For example, decreases in the absorptive capacity of the intestinal mucosa for various nutrients have been observed at weeks postirradiation (87,88). Furthermore, decreases in glucose and proline absorption were observed in sublethally irradiated rats at 180 days postexposure, when morphology of the small intestine appeared to be normal (87). In addition, chronic intestinal complications after radiotherapy (such as ''spontaneous necrosis'' and fibrosis) are of considerable concern to the clinical community (89,90).

In conclusion, the effects of ionizing radiation on gastrointestinal physiology are extensive. In spite of the magnitude of these changes, little is known about the mechanisms underlying the observed dysfunction. While the dysfunctions associated with prodromal effects and the gastrointestinal syndrome have been treated as separate entities, the reader should be aware that there can be considerable overlap of the two, especially at sublethal doses. Clearly, successful treatment will require a combination of approaches to prevent the prodromal effects and to mitigate the dysfunction associated with the gastrointestinal syndrome.

REFERENCES

1. Gerstner, H. B. Acute clinical effects of penetrating nuclear radiation. *J. Am. Med. Assoc.* **168,** 381–388 (1958).
2. Franger, H., and Lushbaugh, C. C. Radiation death from cardiovascular shock following a criticality accident. *Arch. Pathol.* **83,** 446–460 (1967).

3. Cooper, J. R., and Mattsson, J. L. Control of radiation-induced emesis with promethazine, cimetidine, thiethylperazine, or naloxone. *Am. J. Vet. Res.* **40,** 1057–1061 (1979).
4. Chinn, H. J., and Wang, S. C. Locus of emetic action following irradiation. *Proc. Soc. Exp. Biol. Med.* **85,** 472–474 (1954).
5. Dubois, A., Jacobus, J., Grissom, M., Eng, R., and Conklin, J. J. Altered gastric emptying during radiation-induced vomiting. *J. Nucl. Med.* **24,** 56A (1983).
6. Guyton, A. C. *Textbook of Medical Physiology,* 6th ed., pp. 832–833. Saunders, Philadelphia, Pennsylvania, 1981.
7. Davenport, H. W. *A Digest of Digestion,* pp. 15–18. Year Book Publishers, Chicago, Illinois, 1978.
8. Wang, S. C., and Borison, H. L. A new concept of organization of the central emetic mechanism: Recent studies on the sites of action of apomorphine, copper sulfate and cardiac glycosides. *Gastroenterology* **22,** 1–12 (1952).
9. Wislocki, G. B., and Putnam, T. J. Note on the anatomy of the area postrema. *Anat. Rec.* **19,** 281–287 (1970).
10. Brizzee, K. R., Calton, F. M., and Vitale, D. E. Effects of selective placement of lesions in lower brain stem structures on X-irradiation emesis in the dog. *Anat. Rec.* **130,** 533–541 (1954).
11. Wang, S. C., Renzi, A. A., and Chinn, H. I. Mechanism of emesis following X-irradiation. *Am. J. Physiol.* **193,** 335–339 (1958).
12. Brizzee, K. R. Effect of localized brain stem lesions and supradiaphragmatic vagotomy of X-irradiated emesis in the monkey. *Am. J. Physiol.* **187,** 567–570 (1956).
13. Lindstrom, P. A., and Brizzee, K. R. Relief of intractable vomiting from surgical lesion in the area postrema. *J. Neurosurg.* **19,** 228–236 (1962).
14. Cammermeyer, J. Mast cells in the mammalian area postrema. *Z. Anat. Entwicklungsgesch.* **139,** 71–92 (1972).
15. Cutz, E., Chan, W., Track, N. S., Goth, A., and Said, S. I. Release of vasoactive intestinal polypeptide in mast cells by histamine liberator. *Nature (London)* **275,** 661–662 (1978).
16. Conte, F. P., Melville, G. S., and Upton, A. C. Effect of graded doses of whole-body X-irradiation on mast cells in the rat mesentery. *Am. J. Physiol.* **187,** 160–167 (1956).
17. Borison, H. L. Abdominal receptor site for emetic action of X-irradiation. *Fed. Proc., Fed. Am. Soc. Exp. Biol.* **15,** 21–22 (1956).
18. Editorial: Drugs used in vomiting. *Br. Med. J.* **1,** 481–483 (1970).
19. Gralla, E. J., Sabo, J. P., Hayden, D. W., Yochmowitz, M. G., and Mattsson, J. L. The effect of selected drugs in first stage radioemesis in beagle dogs. *Radiat. Res.* **75,** 286–295 (1979).
20. Frytak, S., and Moertel, C. G. Management of nausea and vomiting in the cancer patient. *J. Am. Med. Assoc.* **245,** 393–396 (1981).
21. Lind, B., and Breivik, H. Metoclopramide and perphenazine in the prevention of postoperative nausea and vomiting. *Br. J. Anaesth.* **42,** 614–617 (1970).
22. Seigel, L. J., and Longo, P. L. The control of chemotherapy-induced emesis. *Ann. Intern. Med.* **95,** 352–359 (1981).
23. Laduron, P. M., and Leysen, J. E. Domperidone, a specific *in vitro* dopamine antagonist devoid of in vivo central dopaminergic activity. *Biochem. Pharmacol.* **28,** 2161–2165 (1979).
24. Bernier, J., and Huy, J. Domperidone in the symptomatic treatment of radiotherapy-induced nausea and vomiting. *Postgrad. Med. J.* **55,** Suppl. 1, 50–54 (1979).
25. Conard, R. A. Side effects of ionizing radiation on the physiology of the gastrointestinal tract: A review. *Radiat. Res.* **5,** 1267–1288 (1956).
26. Borowska, A., Sierakowski, S., Mackowaik, J., and Wisniewski, K. A prostaglandin-like activity in the small intestine and postirradiation gastrointestinal syndrome. *Experientia* **35,** 1368–1370 (1979).

27. Stewart, J. J., Burks, T. F., and Weisbrodt, N. W. Intestinal myoelectric activity after activation of central emetic mechanism. *Am. J. Physiol.* **233,** E131–137 (1977).
28. Weisbrodt, N., and Christensen, J. Electrical activity of the cat duodenum in fasting and vomiting. *Gastroenterology* **63,** 1004–1010 (1972).
29. Dorval, D. E., Eng, R., Colomboraro, P., Durakovic, A., Conklin, J. J., and Dubois, A. Effects of ionizing radiation on rhesus monkey. *J. Nucl. Med.* **24,** 98A (1983).
30. Smith, W. W., Ackerman, I. B., and Smith, F. Body weight, fasting, and forced feeding after whole body X-irradiation. *Am. J. Physiol.* **168,** 382–390 (1952).
31. Husle, E. V. Gastric emptying in rats after part-body irradiation. *Int. J. Radiat. Biol.* **10,** 521–532 (1966).
32. Cockerham, L. G., Doyle, T. F., Donlon, M. A., and Helgeson, E. A. Canine postradiation histamine levels and subsequent response to compound 48/80. *Aviat. Space Environ. Med.* **55,** 1041–1045 (1984).
33. Lee, J. S., and Silverberg, J. W. Effect of histamine secretion in the dog. *Am. J. Physiol.* **231,** 793–798 (1976).
34. Quástler, H. The nature of intestinal radiation death. *Radiat. Res.* **4,** 303–320 (1956).
35. Otto, F., and Pfeiffer, U. Mortality response of mice after whole-body exposure to 1.7-5 MeV neutrons and X-rays. *Radiat. Res.* **50,** 125–135 (1972).
36. Bond, V. P. Comparison of the mortality response of different species to x-rays and fast neutrons. In *Biological Effects of Neutrons and Proton Irradiations,* Vol. II, pp. 365–377. IAEA, Vienna, 1965.
37. Geraci, J. P., Jackson, K. L., Christensen, G. M., Thower, P. D., and Fox, M. S. Cyclotron fast neutrons RBE for various normal tissues. *Radiology* **115,** 459–463 (1975).
38. Vogel, H. H., and Jordan, D. L. Comparison of damage to the small intestine by fission neutrons or gamma rays from 60-Co. In *Gastrointestinal Radiation Injury* (M. F. Sullivan, ed.), pp. 327–329. Excerpta Med. Found., Amsterdam, 1966.
39. Cheng, H., and Leblond, C. P. Origin, differentiation and renewal of the four main epithelial cell types in the mouse small intestine. I. Columnar cell. *Am. J. Anat.* **141,** 461–479 (1975).
40. Trier, J. S. Morphology of the epithelium of the small intestine. In *Handbook of Physiology* (C. F. Code, ed.), Sect. 6, Vol. III, pp. 1125–1175. Am. Physiol. Soc., Washington, D.C., 1968.
41. Cheng, H., and Leblond, C. P. Origin, differentiation and renewal of the four main epithelial cell types in the mouse small intestine. V. Unitarian theory of the origin of the four epithelial cell types. *Am. J. Anat.* **141,** 537–561 (1975).
42. Messier, B., and Leblond, C. P. Cell proliferation and migration as revealed by autoradiography after injection of thymidine-H3 into rats and mice. *Am. J. Anat.* **106,** 247–294 (1960).
43. Bertalanfly, F. D., and Nagy, K. P. Mitotic activity and renewal rate of epithelial cells of human duodenum. *Acta Anat.* **45,** 362–370 (1961).
44. Frizzell, R. A., and Schultz, S. G. Models of electrolyte absorption and secretion by gastrointestinal epithelia. *Int. Rev. Physiol.* **19,** 205–225 (1975).
45. Carpenter, C. C. J. Cholera. In *Principles of Internal Medicine* (R. G. Peterdorf, R. D. Adams, E. Braunwald, K. J. Isselbacher, J. B. Martin, and J. D. Wilson, eds.), 10th ed., pp. 864–865. McGraw-Hill, New York, 1983.
46. Binder, H. J. Net fluid and electrolyte secretion. In *Mechanisms of Intestinal Secretion* (H. J. Binder, ed.), pp. 1–16. Alan R. Liss, Inc., New York, 1979.
47. Miller, R. J., Kachur, J. F., Field, M., and Rivier, R. Neurohumoral control of ileal electrolyte transport. *Ann. N.Y. Acad. Sci.* **372,** 571–593 (1981).
48. Field, M. Intracellular mediators of secretion in the small intestine. In *Mechanisms of Intestinal Secretion* (H. J. Binder, ed.), pp. 83–93. Alan R. Liss Inc., New York, 1979.
49. Ham, A. W. *Histology,* pp. 96–102. Lippincott, Philadelphia, Pennsylvania, 1969.

50. Lipkin, M. Cell turnover in gastrointestinal mucosa. In *Basic Mechanisms of Gastrointestinal Mucosal Cell Injury and Protection* (W. Harmon, ed.), pp. 31–47. Williams & Wilkins, Baltimore, 1981.
51. Lesher, S. Compensatory reaction in the intestinal crypt cells after 300 R 60-Co gamma irradiation. *Radiat. Res.* **32,** 510–519 (1967).
52. Lesher, S., and Bauman, J. Recovery of reproductive activity and the maintenance of structural integrity in the intestinal epithelium of the mouse after single dose whole-body 60-Co gamma ray exposure. In *Effects of Radiation on Cellular Proliferation and Differentiation,* Proceedings of a Symposium on the Effects of Radiation on Cellular Proliferation and Differentiation, pp. 507–513. IAEA, Vienna, 1968.
53. Lesher, S., Sallese, A., and Jones, M. Effects of 1000 R whole-body X-irradiation on DNA synthesis and mitosis in the duodenal crypts of the BCF1 Mouse. *Z. Zellforsch. Mikrosk. Anat.* **77,** 144–148 (1967).
54. Hagemann, R. F., and Lesher, S. Irradiation of the G. I. tract: Compensatory response of stomach, jejunum, and colon. *Br. J. Radiol.* **44,** 599–602 (1971).
55. Potten, C. S., and Hendry, J. H. Differential regeneration of intestinal proliferative cells and cryogenic cells after irradiation. *Int. J. Radiat. Biol.* **27,** 413–424 (1975).
56. Hanson, W. R., Henninger, D. L., Fry, R. J. M., and Sallese, A. R. The response of small intestinal stem cell in the mouse to drugs and irradiation treatment. In *Cell Proliferation of the Gastrointestinal Tract* (D. R. Appleton, J. P. Sunter, and A. J. Watson, eds.), pp. 198–212. Pitman Medical, Marshfield, Massachusetts, 1980.
57. Yau, H. C., and Cairnie, A. B. Cell-survival characteristics of intestinal stem cells and crypts of Co-irradiated mice. *Radiat. Res.* **80,** 92–107 (1979).
58. Maisin, J., Maisin, J. R., and Dunjii, A. The gastrointestinal tract. In *Pathology of Irradiation* (C. C. Berdjis, ed.), pp. 296–344. Williams & Wilkins, Baltimore, Maryland, 1971.
59. Robert, A. Gastric and intestinal cytoprotection by prostaglandins. In *Basic Mechanisms of Gastrointestinal Mucosal Cell Injury and Protection* (W. Harmon, ed.), pp. 191–196. Williams & Wilkins, Baltimore, Maryland, 1981.
60. Kauffman, G. L., Jr. Does gastric gel mucus contribute protection? In *Basic Mechanisms of Gastrointestinal Mucosal Cell Injury and Protection* (J. W. Harman, ed.), pp. 369–372. Williams & Wilkins, Baltimore, Maryland, 1981.
61. Hanson, W. R., and Thomas, C. 16,16-dimethyl prostaglandin E-2 increases survival of murine intestinal stem cells when given before photon radiation. *Radiat. Res.* **96,** 393–398 (1983).
62. Northway, M., and Castell, D. O. Evidence in support of an injurious effect of prostaglandins on gastrointestinal mucosa. In *Basic Mechanisms of Gastrointestinal Mucosal Cell Injury and Protection.* (J. W. Harmon, ed.), pp. 237–248. Williams & Wilkins, Baltimore, Maryland, 1981.
63. Curran, P. F., Webster, E. W., and Housepain, J. A. The effect of x-irradiation on sodium and water transport in rat ileum. *Radiat. Res.* **13,** 360–380 (1960).
64. Sullivan, M. F., Marks, S., Hackett, P. L., and Thompson, R. C. X-irradiation of the exteriorized or *in situ* intestine of the rat. *Radiat. Res.* **11,** 653–666 (1959).
65. Edwards, J. C., Chapman, D., Cramp, W. A., and Yatvin, M. D. The effects of ionizing radiation of biomembrane structure and function. *Prog. Biophys. Mol. Biol.* **43,** 71–93 (1984).
66. Porvaznik, M. Tight junction disruption and recovery after sublethal γ irradiation. *Radiat. Res.* **17,** 233–250 (1979).
67. Moran, A., Handler, J., and Hagan, M. Regulation of sodium couple hexose transport in cultured kidney epithelia. The importance of cell proliferation. *Kidney Int.* **25,** 310 (1984).
68. Cockerham, L. G., Doyle, T. F., Donlon, M. A., and Gossett-Hagerman, C. J. Antihistamine block radiation-induced increased intestinal blood flow in canines. *Fundam. Appl. Toxicol.* **3,** 597–604 (1985).

69. Eisen, V., and Walker, R. I. Effect of ionizing radiation on prostaglandin 15-OH-dehydrogenase (PGDH). *Br. J. Pharmacol.* **62,** 461P (1978).
70. Forkey, D. J., Jackson, K. L., and Christensen, G. M. Contribution of bile to radiation-induced sodium loss via the rat small intestine. *Int. J. Radiat. Biol.* **14,** 49–57 (1967).
71. Jackson, K. L., and Entenman, C. The role of bile secretion in the gastrointestinal radiation syndrome. *Radiat. Res.* **10,** 67–79 (1959).
72. Fry, R. J. M., and Staffeldt, E. Effect of diet containing sodium deoxycholate on the intestinal mucousa of the mouse. *Nature (London)* **203,** 1396–1398 (1964).
73. Sullivan, M. F. Dependence of radiation diarrhea on the presence of bile in the intestine. *Nature (London)* **195,** 1217–1218 (1962).
74. Sullivan, M. F., Hulse, E. V., and Mole, R. H. The mucous-depleting action of bile in the small intestine of the irradiated rat. *Br. J. Exp. Pathol.* **46,** 235–244 (1965).
75. Kabal, J., Baum, S. J., and Parkhurst, L. J. Canine intestinal vasoactivity during development of the gastrointestinal syndrome. *Radiat. Res.* **50,** 528–538 (1972).
76. Eddy, H. A., and Casarett, G. W. Intestinal vascular changes in the acute radiation intestinal syndrome. In *Gastrointestinal Radiation Injury* (M. F. Sullivan, ed.), pp. 385–392. Exerta Med. Found., Amsterdam, 1966.
77. Cockerham, L. G., Doyle, T. F., Trumbo, R. B., and Nold, J. B. Acute postirradiation canine intestinal flow. *Int. J. Radiat. Biol.* **45,** 65–72 (1984).
78. Porvaznik, M., Walker, R. I., and Gillmore, J. D. Reduction of the indigenous filamentous microorganisms in the rat ileium following X-irradiation. *Scanning Electron Microsc.* **3,** 15–22 (1979).
79. Taketa, S. T. Water-electrolyte and antibiotic therapy against acute (3- to 5-day) intestinal radiation death in the rat. *Radiat. Res.* **16,** 312–326 (1965).
80. Matsuzawa, T. Survival time of germfree mice after lethal whole body X-irradiation. *Tohoku J. Exp. Med.* **85,** 257–263 (1965).
81. Jackson, K. L., Rhodes, R., and Entenman, C. Electrolyte excretion in the rat after severe intestinal damage by X-irradiation. *Radiat. Res.* **8,** 361–373 (1958).
82. Swift, M. N., and Taketa, S. T. Effect of circulating blood volume of partial shielding of rat intestine during X-irradiation. *Radiat. Res.* **8,** 516–525 (1958).
83. Zsebok, Z., and Petranyi, G. Changes in electrolyte balance in the gastrointestinal syndrome. *Acta Radiol. Ther. Phys. Biol.* **2,** 377–383 (1964).
84. Lushbaugh, C. C., Sutton, J., and Richmond, C. R. The question of electrolyte loss in the intestinal death syndrome of radiation damage. *Radiat. Res.* **13,** 814–824 (1960).
85. Bond, V. P., Osborne, J. W., Lesher, S., Lushbaugh, C. C., and Horsey, S. Mechanism of intestinal radiation death: A panel discussion. In *Gastrointestinal Radiation Injury* (M. F. Sullivan, ed.), pp. 351–364. Excerpta Med. Found., Amsterdam, 1966.
86. Gits, J., and Gerber, G. B. Electrolyte loss, the main cause of death from the gastrointestinal syndrome? *Radiat. Res.* **55,** 18–28 (1973).
87. Muhiuddin, M., Tamura, K., and DeMarie, P. Changes in absorption of glucose and protein following irradiation to the exteriorized ileum. *Radiat. Res.* **74,** 186–190 (1978).
88. Thomson, A. B. R., Chesseman, C. I., and Walker, K. Effect of abdominal irradiation on the kinetic parameters of intestinal uptake of glucose, galactose, leucine, and gly-leucine in the rat. *J. Lab. Clin. Med.* **102,** 813–827 (1983).
89. DeCrosse, J. J., Rhodes, R. S., Wentz, W. B., Reagan, J. W., Dworken, H. J., and Holden, W. D. The natural history and management of radiation induced injury of the gastrointestinal tract. *Ann. Surg.* **170,** 369–384 (1969).
90. Wellwood, J. M., and Jackson, B. T. Intestinal complications of radiotherapy. *Br. J. Surg.* **60,** 814–818 (1973).

CHAPTER 8

Postirradiation Cardiovascular Dysfunction

ROBERT N. HAWKINS* AND LORRIS G. COCKERHAM**

Armed Forces Radiobiology Research Institute
Bethesda, Maryland 20814-5145

I. INTRODUCTION

Cardiovascular dysfunction may be defined as the inability of any element of the cardiovascular system to perform adequately upon demand, leading to inadequate performance and nutritive insufficiency of various parts of the body. Cardiovascular dysfunction may be seen in many pathological conditions, such as myocardial infarction and its associated decreases in cardiac output and cardiac reserve, or varicose veins, which result in a decrease of venous return. Cardiovascular dysfunction also may result from extracorporal factors such as positive G forces, which cause blood to pool in the lower parts of the body. The severity of cardiovascular dysfunction may range from the inability of the circulation to support physical work to the complete collapse of the circulation, i.e., shock.

Exposure to supralethal doses of radiation (accidental and therapeutic) has been shown to induce significant alterations in cardiovascular function in man. These findings indicate that, after irradiation, cardiovascular function is a major determinant of continued performance and even survival. For the two persons who received massive radiation doses (45 and 88 Gy, respectively) in criticality accidents, the inability to maintain systemic arterial blood pressure (AP) was the

*Present address: Naval Medical Research and Development Command, Bethesda, Maryland 20814-5145.

**Present address: Department of Life Sciences, Air Force Office of Scientific Research, Bolling Air Force Base, Washington, D.C. 20332-6448.

immediate cause of death (1,2). In a study of cancer patients given partial-body irradiation, two acute lethalities were attributed to myocardial infarction after an acute hypotensive episode during the first few hours postexposure (3). The occurrence of hypotension in the entire group of patients in the same study was 80%. Similar cardiovascular alterations postirradiation have been observed in several but not all experimental animal models.

Although radiation-induced cardiovascular dysfunction has been observed in many species, its severity, duration, and even etiology may vary with the species, level of exposure, and dose rate. For this reason, our consideration of the effects of radiation on cardiovascular performance is limited to the circulatory derangements that (1) occur in rat, dog, and monkey after supralethal doses and (2) lead to radiation-induced cardiovascular dysfunction in these experimental models. We then consider other recent data as they pertain to the etiology of cardiovascular dysfunction in irradiated animals.

Experimental data concerning cardiovascular dysfunction in the rat are often conflicting and confusing. Conscious rats have been reported to experience a decrease in tail pressure during the first 30 min after receiving 100 Gy of 14.5-MeV electrons (4), in contrast to earlier studies that reported no acute postirradiation hypotensive episode in this species (5). But changes in tail pressure, which may be related to thermoregulatory function (4), have been shown to decrease in the irradiated rat when aortic pressure is unchanged (6).

More recent studies (7), on the other hand, have reported a transient but significant (within 1 min postirradiation) radiation-induced cardiovascular dysfunction in the rat anesthetized with pentobarbital. In these experiments, irradiation resulted in a prompt disruption in physiological function, including a marked drop in mean AP and a small but consistent decrease in heart rate. Both parameters usually returned to baseline values within 2 min postexposure. Although pentobarbital is known to depress autonomic reflexes, similar results have been shown in chloralose-anesthetized rats also exposed to 100 Gy of 14.5-MeV electrons (8).

Experimental data concerning postirradiation cardiovascular dysfunction in the dog is no less confusing than in the rat. Dogs irradiated with 190 Gy mixed gamma–neutron radiation presented an initial hypertension immediately (1–2 min) after exposure. This hypertension was accompanied by a significant increase (+60%) in heart rate. By 30 min postirradiation, AP and heart rate had returned to preirradiation values, and remained at this level throughout the remainder of the 60-min observation period (9). In support of these findings, other investigators (10) found no significant differences in the AP of irradiated dogs before 48 hr after mixed neutron–gamma irradiation (25 Gy). In contrast, dogs that had been irradiated with 100 Gy cobalt-60 experienced a decline in pressure that began within minutes postexposure and reached −31% within 60–90 min postexposure (11). The latter study also provided evidence of increased intestinal blood flow during the postirradiation period, suggesting splanchnic involvement in the hypotensive response.

As early as 1968, early transient incapacitation (ETI, discussed later in this chapter) was shown to have a temporal relationship with hypotension, and a causal relationship between these two phenomena was suggested (12). Indeed, extensive research into the etiology of ETI focused much of the early primate postirradiation cardiovascular studies on the cardiovascular dysfunction that occurs immediately after exposure. Many of these studies revealed cardiovascular dysfunction to some extent, but they consisted mostly of blood pressure measurements of animals trained to perform a cognitive task. However, no studies were performed to address decrements in physical performance. The maintenance of cardiovascular integrity during stress is determined by the changes in (1) the pumping action of the heart, (2) compliance of the vascular beds, (3) resistance of the peripheral circulation, (4) quantity of blood in the vascular system, and (5) viscosity of the blood. Failure of any of the mechanisms to respond properly during severe stress may compromise the integrity of the entire cardiovascular system, and more comprehensive studies are required to elucidate the extent of postirradiation cardiovascular dysfunction. Bruner (13), in a study of cardiovascular dysfunction in rhesus monkeys after 10 Gy cobalt-60 irradiation, observed an immediate decrease in total peripheral resistance and AP accompanied by increased cardiac output, which was primarily due to increased heart rate.

In the Bruner study, the hypotensive episode lasted approximately 15 min. By the end of the observation period (30 min postirradiation), total peripheral resistance and cardiac output had returned to near control levels and AP had returned to 90% of the preirradiation value. The data suggest that the immediate hypotension was due to arterial vasodilation. These observations have been confirmed in recent studies (14) reporting similar changes in rhesus monkeys after 40 Gy cobalt-60. In the latter studies, the severity of hypotension was greater, but AP had recovered to 75% of the control value by 15–20 min postirradiation. Arterial pressure then began to exhibit a downward trend, and by 90 min postirradiation had once again dropped to approximately 50% of the preirradiation blood pressure. Evidence provided by measurements of central venous pressure, cardiac output (CO), and total peripheral resistance (TPR) determinations (TPR = AP/CO) in these studies suggests a triphasic and differential response to radiation by these animals. Immediate hypotension was hypothesized to result from arterial vasodilation plus venous pooling. This conclusion is supported by the observations of decreased cardiac output and AP in the face of an increase in heart rate. Further, the data showed that the subsequent hypotensive response that began approximately 30 min postirradiation resulted from venous pooling and decreased cardiac preload while total peripheral resistance was at or near preexposure values. The site(s) of blood pooling could not be determined from these studies. But other studies using this model show that blood flow to the ileum increased to above the preirradiation level during the recovery period, even though AP was still depressed (15). These data suggest an intestinal involvement in the subsequent vascular pooling.

II. ETIOLOGY OF POSTIRRADIATION CVD

It may be observed that the two persons who received massive radiation doses in criticality accidents were unable to maintain systemic AP in the face of fluid and vasopressor therapy (1,2). This observation, along with later experimental animal data (14,16), indicates that postirradiation cardiovascular dysfunction may be similar to circulatory shock, which has given rise to the term radiation-induced shock. Moreover, other evidence indicates that even sublethal doses of radiation may induce a functional cardiovascular deficiency that manifests itself as early hypotension, but because of cardiovascular reserve, the lesion is masked during a period of circulatory deterioration (17). It has been hypothesized that if the initial insult is sufficient, a functional lesion may exist even though masked by cardiovascular reserve. This damage may then be seen later as radiation-induced shock when the cardiovascular reserve is no longer capable of maintaining homeostasis (14). On the other hand, if the initial radiation insult is large enough to overcome cardiovascular reserve, then irreversible deterioration of the circulation occurs and radiation-induced shock may ensue. So efforts to delve into the etiology of radiation-induced cardiovascular dysfunction must address those things that may induce functional but transient changes, and those that may overwhelm cardiovascular reserve and lead to radiation-induced shock.

The principal responsibility for stabilizing AP is delegated to the autonomic nervous system. In the normal animal, total cardiovascular function is modulated by these pathways. In response to an abrupt decrease in blood pressure, cardiopulmonary reflexes are activated, which results in increased cardiac function and regional vasoconstriction; these are usually sufficient to maintain adequate perfusion of critical organs (18). Failure of this system to perform will exacerbate any radiation-induced cardiovascular dysfunction, and evidence suggests that the baroreceptor reflex is impaired during the time of initial radiation-induced hypotension, at least in the monkey (19). Despite the change in reflex sensitivity, both monkey (16) and rat (7) experience a fivefold increase in the arterial plasma concentration of epinephrine after the initial hypotension (2 min postirradiation). Recovery of the autonomic nervous system of the rat after exposure to 100 Gy of 14.5-MeV electrons may be responsible for complete cardiovascular recovery (7), which may last up to 60 min after radiation (8); this is in contrast to the monkey, which experiences cardiovascular collapse during this time (14,20,21). Postirradiation cardiovascular response in the rat is characterized by cardioacceleration and an increase in venous return (8), while the response in the monkey involves peripheral vascular pooling or extravasation of blood into the periphery (14,22). The associated decrease in venous return and cardiac output in the monkey may proceed to radiation-induced shock.

When one consults the literature concerning shock, a vast array of biochemical processes can be implicated in postirradiation cardiovascular dysfunction. Among the candidates for a possible role are histamine (9), opiate peptides (23), bradykinin (24), and prostaglandins (25). The influence of most of these agents on the cardiovascular system after irradiation has been either completely neglected (in the case of bradykinin and prostaglandins) or only briefly considered (as in the case of opiate peptides) (7).

On the other hand, many investigators, beginning with the early work of Lewis (26), have proposed that histamine is the primary cause for several of the acute physiological responses. Histamine is a potent vasoactive agent, found within tissue mast cells and in circulating basophils. The actions of histamine in the circulatory system include increased capillary permeability, venoconstriction, and arteriolar dilation (27).

A great deal of evidence has accumulated that implicates histamine in postirradiation cardiovascular dysfunction, including the finding that plasma histamine increases in monkeys after exposure to radiation (16,28). Data also indicate that the postirradiation histamine release can be prevented by pretreating the animals with the mast cell content-depleting drug 48/80 (28). Following this lead, studies were performed to investigate the involvement of histamine in a radiation-induced shock model (14,16). In these studies, rhesus monkeys exposed to 40 Gy cobalt-60 experienced a 17-fold increase in plasma histamine levels at 2 min postirradiation, compared to preirradiation values. Plasma histamine levels and AP returned toward preirradiation levels after this initial change, and the immediate postirradiation hypotension correlated significantly with the increase in plasma histamine levels. Moreover, in another group of animals, these investigators prevented the immediate postirradiation hypotension by pretreating the animals with the H1 and H2 histamine receptor antagonists diphenhydramine and cimetidine. Pretreatment with antihistamines did not prevent either the initial increase in histamine or the secondary decline in blood pressure produced in this radiation-induced shock model. These data are supported in a more recent effort (21) in which monkeys were pretreated with the H1 and H2 antagonists mepyramine and cimetidine and the mast cell stabilizer disodium cromoglycate. In these experiments, immediate postirradiation hypotension was prevented by a gradual postirradiation decrease in AP (21). In other studies, however, these same investigators were unable to prevent the immediate postirradiation hypotension with mast cell stabilizers alone (20).

The efficacy of antihistamines in treating postirradiation hypotension in a dog model has also been investigated (29). Dogs exposed to 100 Gy cobalt-60 exhibited blood histamine levels 20% higher than preirradiated control values, and reached a peak within 5 min postirradiation. Administration of mepyramine and cimetidine before irradiation prevented this small postirradiation decrease in AP seen in the dog model. Compared to a nontreated control group, the rate of decrease of blood pressure usually seen in dogs was reduced by half. The postirradiation increase in intestinal blood flow reported previously by this group (11) and again seen in the nontreated control group of this study was ameliorated by pretreatment with antihistamines. These data suggest a possible role of splanchnic blood flow and histamine in the postirradiation cardiovascular response of the dog.

However, substantial support for a "histamine hypothesis" in the rat model does not exist. Measurement of plasma histamine using the rat model indicated no detectable changes after irradiation (7). Moreover, it was apparent that all tissue stores of histamine were not depleted in these experiments, because the administration of 48/80, within 30 min postirradiation, produced a marked increase in plasma histamine and a decrease in AP. Nevertheless, it remains possible that local release of histamine also occurs in the rat while cardiovascular dysfunction is masked by an extensive ability of the circulatory system to compensate. Yet levels of radiation up to 450 Gy have not been sufficient to overcome the cardiovascular reserve in the rat (7).

Although considerable evidence implicates histamine in the initial postirradiation hypotension in the monkey and dog, it cannot explain the subsequent cardiovascular deterioration in the monkey or the postirradiation response of the rat. In the monkey, it has been hypothesized that the cardiovascular responses to radiation and endotoxin may be mediated by similar mechanisms, because pretreatment with sublethal radiation has been shown to protect against the acute responses to endotoxin (30). On the other hand, although the monkey and the rat both undergo severe and prolonged hypotensive responses after the administration of endotoxin (23,31), the single transient hypotensive response of the rat (7,8) (as opposed to the prolonged and triphasic hypotensive response of the monkey) after irradiation (14) suggests a different causal mechanism.

In summary, radiation-induced cardiovascular dysfunction has been observed in different species, including man, but the extent of the CVD and its etiology may vary with the species, level of exposure, and potentially the dose rate. Evidence suggests that changes in the autonomic nervous system and histamine play major roles in the cardiovascular response of the monkey after irradiation, and these factors result in ultimate circulatory collapse during a shock-like sequela. In the rat, however, evidence suggests that histamine probably plays only a minor role. Interference with the autonomic pathways that modulate cardiovascular function may account for the postirradiation circulatory response of the rat. Moreover, the rat appears to possess enough cardiovascular compensatory ability to maintain cardiovascular performance after levels of radiation that lead to radiation-induced shock in the monkey. The etiology of postirradiation hypotension in the dog remains clouded. The fact that the dog undergoes a transient postirradiation hypertension suggests some interference with autonomic regulation. But, unlike the rat, the dog apparently undergoes a period of progressive hypotension that is slower and less extensive than that recorded in the monkey. Although histamine probably plays a role, the extent remains to be established.

III. NEUROVASCULAR DYSFUNCTION

Even though radiation-induced shock and the ultimate collapse of the circulation or other radiation effects lead to the demise of the irradiated animal, the early transient hypotension remains important because of the impact it may have on cognitive performance and, even more important, physical performance.

Early transient incapacitation is the complete cessation of performance of a trained task, occurring transiently and within the first 30 min after exposure to supralethal doses of ionizing radiation. Performance decrement is a reduction in some measured animal function at the same time. One obvious possible explanation for these neurologic dysfunctions is an inadequate supply of oxygen to the brain. A reduction in systemic blood pressure to below local autoregulatory capabilities will reduce the driving force required to maintain cerebral blood flow, leading to cerebral hypoxia. This in turn could result in decreased performance and eventually incapacitation.

Since exposure to supralethal levels of ionizing radiation (such as gamma photons) results in postirradiation hypotension (14,20,32,33) in which the AP often decreases to less than 50% of normal, such exposure is a possible explanation for postirradiation performance decrement. One investigation (33) was able to closely correlate performance decrement with postirradiation hypotension, usually with the decrement following within a few minutes of the initial fall in blood pressure.

Postirradiation hypotension may produce a decreased cerebral blood flow even though the central nervous system (CNS), through the mechanism of autoregulation, can often maintain cerebral blood flow under conditions of severe hypotension. One study (12) demonstrated a dramatic fall in cerebral blood flow immediately following a single 25-Gy dose of cobalt-60 exposure. Blood pressure fell 59% by 5 min postirradiation, with the cerebral blood flow falling to only 30% of its preirradiation value. A critical postirradiation mean AP of 50–60% of normal must be maintained for adequate autoregulation of cerebral circulation (34).

The initial precipitous decline in mean AP in rhesus monkeys (to 75% below preirradiation levels) may be associated with the similar immediate decrease in regional cerebral blood flow in the hypothalamus, hippocampus, visual cortex, and postcentral gyrus of the cerebral cortex (15,21,35). A similar decrease in cerebral blood flow accompanied by symptoms and signs of cerebral ischemia has been reported in man (36). Klatzo *et al.* (37) maintain that the initial stages of cerebral ischemia are dominated by a cytotoxic edema that appears to develop when the regional cerebral blood flow is reduced below 20 ml/100 g/min. Although the reduced regional cerebral blood flow reported by Chapman and Young (12) and Cockerham *et al.* (21,35) resulting from cobalt-60 irradiation did not fall below this threshold, some neurologic dysfunction may have been produced. This is supported by Suzuki *et al.* (38), who reported that spontaneous activity in both cortical and hippocampal neurons ceased within 60 sec after the onset of ischemia.

On the basis of the reported diminished regional cerebral blood flow after irradiation (15,21,35), one might predict functional impairment of the CNS in monkeys following irradiation. In fact, postirradiation ETI has been reported in monkeys starting as early as 2 min after irradiation, lasting 10–30 min. It is often accompanied by severe systemic hypotension, when AP is decreased to less than 50% of normal (38,39).

Another complication in cerebral ischemia is the opening of the blood–brain barrier, as reported by Suzuki *et al.* (40). This may be associated with the development of cerebral vasogenic edema (37). Gross *et al.* (41) reported increased permeability of the blood–brain barrier after infusion of histamine into the internal carotid artery. The postirradiation release of histamine may be involved with the production of cerebral edema and its associated neurologic dysfunctions. Studies have indicated the use of an anticholinesterase drug as a probe to measure damage to the blood–brain barrier (42–44). The utility of this method and the accuracy of the results have been confirmed independently (45). Using this technique, damage to the blood–brain barrier in rats exposed to 100 Gy of 6-MeV bremsstrahlung photons was detected 48 hr postirradiation (46). Recent studies, using a more sensitive anticholinesterase probe (42), showed damage to the blood–brain barrier of rats exposed to high-neutron or high-gamma radiation doses ranging from 2 to 10 Gy (47).

In summary, both the rat and the primate have been observed to experience postirradiation performance decrements. Yet the postirradiation hypotensive responses of these two species (and the dog) may be of different etiology, severity, and time course. Based on the analysis of data from radiation accidents involving humans, it appears that the monkey best models the cardiovascular response of the human after radiation exposure. A causal relationship between postirradiation performance decrement and hypotension is an attractive hypothesis, but remains to be proved. Nevertheless, it is intuitively obvious that postirradiation hypotension will exacerbate any performance decrement when combined with other cardiovascular demands, such as heat stress, physical effort, or positive G forces. Further evidence suggests that paralethal doses of radiation may cause cardiovascular dysfunction, although frank hypotension may be prevented by the reserve capacity of the circulatory system.

The effects of radiation on task performance and cardiovascular function, under conditions of added cardiovascular stress, remain a fruitful area of research. The relationships between blood–brain barrier damage and histamine release, blood–brain barrier damage and radiation, as well as blood–brain barrier damage and ischemia seem to be on solid ground. Nevertheless, before a temporal relationship, and definitely a causal relationship, can be firmly established between radiation, hypotension, histamine release, and ETI, the postirradiation measurements of blood chemistry, blood pressure, regional cerebral blood flow, and behavioral effects must be accomplished on the same animal subject. Moreover, it is important to determine if the high levels of circulating histamine that have been observed in the primate after irradiation do indeed cross the blood–brain barrier. Finally, the histamine hypothesis would be strengthened by any evidence showing that the hypotension and/or ETI response can be changed by blocking the postirradiation release of histamine.

REFERENCES

1. Fanger, H., and Lushbaugh, C. C. Radiation from cardiovascular shock following a criticality accident. *Arch. Pathol.* **83,** 446–460 (1967).
2. Shipman, T. L. A radiation fatality resulting from massive overexposure to neutrons and gamma rays. In *Diagnosis and Treatment of Acute Radiation Injury,* Proceedings of a Scientific Meeting Sponsored by International Energy Agency and World Health Organization, Geneva, 1961, pp. 113–131. Int. Doc. Serv., Columbia Univ. Press, New York, 1961.
3. Salazar, O. M., *et al.* Systemic (half-body) radiation therapy: response and toxicity. *Int. J. Radiat. Oncol. Biol. Phys.* **4,** 937–950 (1978).
4. Mickley, G. A. Behavioral and physiological changes produced by a supralethal dose of ionizing radiation: Evidence for hormone-influenced sex differences in the rat. *Radiat. Res.* **81,** 48–75 (1980).
5. Montgomery, P. O'B., and Warren, S. Mechanism of acute hypotension following total body irradiation. *Proc. Soc. Exp. Biol. Med.* **77,** 803–807 (1951).
6. Phillips, R. D., and Kineldorf, D. J. The effect of whole-body X-radiation on blood pressure in the rat. *Radiat. Res.* **18,** 86–95 (1963).
7. Alter, W. A., III, Catravas, G. N., Hawkins, R. N., and Lake, C. R. Effect of ionizing radiation on physiological function in the anesthetized rat. *Radiat. Res.* **99,** 394–409 (1984).
8. Hawkins, R. N. Effects of high energy electron radiation on venous return in the anesthetized rat. *Physiologist* **26**(4), A-76 (1983).
9. Turbyfill, C. L., Thorp, J. W., and Wise, D. *Cardiovascular Response of Beagles to a Supralethal Dose of Mixed Gamma-neutron Radiation,* Sci. Rep. SR70-4. Armed Forces Radiobiol. Res. Inst., Bethesda, Maryland, 1970.

10. Kabal, J., Baum, S. J., and Parkhurst, L. J. Canine intestinal vasoactivity during the development of the gastrointestinal radiation syndrome. *Radiat. Res.* **50,** 528–538 (1972).
11. Cockerham, L. G., Doyle, T. F., Trumbo, R. B., and Nold, J. B. Acute postradiation canine intestinal blood flow. *Int. J. Radiat. Biol.* **45,** 65–72 (1984).
12. Chapman, P. H., and Young, R. J. Effect of cobalt-60 gamma irradiation on blood pressure and cerebral blood flow in the *Macaca mulatta*. *Radiat. Res.* **35,** 78–85 (1968).
13. Bruner, A. Immediate changes in estimated cardiac output and vascular resistance after ^{60}Co exposure in monkeys. Implication for performance decrement. *Radiat. Res.* **70,** 391–405 (1977).
14. Hawkins, R. N., Alter, W. A., III, Doyle, T. F., and Catravas, G. N. Radiation-induced cardiovascular dysfunction in the rhesus monkey. *Radiat. Res.* **94**(3), 654 (1983).
15. Meeks, F. A., Alter, W. A., III, Hawkins, R. N., Eng, R. R., O'Neill, J. T., and Catravas, G. N. Radiation-induced changes in regional blood flow in rhesus. *Physiologist* **27,** 274 (1984).
16. Alter, W. A., III, Hawkins, R. N., Catravas, G. N., Doyle, T. F., and Takenaga, J. K. Possible role of histamine in radiation-induced hypotension in the rhesus monkey. *Radiat. Res.* **94**(3), 654 (1983).
17. Myers, J. H., Blackwell, L. H., and Overman, R. R. Early functional hemodynamic impairment in baboons after 1000 R or less of gamma radiation as revealed by hemorrhagic stress. *Radiat. Res.* **52,** 564–578 (1972).
18. Guyton, A. C., Jones, C. E., and Coleman, T. G. *Cardiac Output and Its Regulation*. Saunders, Philadelphia, Pennsylvania, 1973.
19. Nathan, M. A., and Craig, D. J. Effect of high-energy x-ray and pulsed gamma-neutron radiation on brain blood flow, vascular resistance, blood pressure, and heart rate in monkeys. *Radiat. Res.* **50,** 543–555 (1972).
20. Cockerham, L. G., Doyle, T. F., and Hampton, J. D. Postradiation cerebral blood flow in primates pretreated with disodium cromoglycate. *Fed. Proc., Fed. Am. Soc. Exp. Biol.* **43,** 957 (1984).
21. Cockerham, L. G., Pautler, E. L., and Hampton, J. D. Radiation-induced hippocampal ischemia altered with antihistamines and disodium cromoglycate. *Fed. Proc., Fed. Am. Soc. Exp. Biol.* **44,** 1357 (1985).
22. Kundel, H. L. The effect of gamma irradiation on the cardiovascular system of the rhesus monkey. *Radiat. Res.* **27,** 406–418 (1966).
23. Holaday, J. W., and Faden, A. I. Naloxone reversal of endotoxin hypotension suggest role of endorphins in shock. *Nature (London)* **175,** 450–451 (1978).
24. Nies, A. S., Forsyth, R. P., Williams, H. E., and Melmon, K. L. Contribution of kinins to endotoxin shock in unanesthetized rhesus monkeys. *Circ. Res.* **22,** 155–164 (1968).
25. Revenas, B., and Smedegard, G. Aggregate anaphylaxis in the monkey: Attenuation of the pulmonary response by pretreatment with indomethacin. *Circ. Shock* **8,** 21–29 (1981).
26. Lewis, T. *The Blood Vessels of the Human Skin and Their Responses*. Shaw & Sons, London, 1927.
27. Haddy, F. J. Effect of histamine on small artery and large vessel pressures in the dog foreleg. *Am. J. Physiol.* **198,** 161–168 (1969).
28. Doyle, T. F., and Strike, T. A. *Radiation-released Histamine in the Rhesus Monkey as Modified by Mast Cell Depletion and Antihistamine,* Sci. Rep. SR75-18. Armed Forces Radiobiol. Res. Inst., Bethesda, Maryland, 1975.
29. Cockerham, L. G., Doyle, T. F., Donlon, M. A., and Gossett-Hagerman, C. J. Antihistamines block radiation-induced increased intestinal blood flow in canines. *Fundam. Appl. Toxicol.* **5**(3), 597–604 (1985).
30. Liu, D. T., Griffin, M. J., and Hilmas, D. E. Effect of staphylococcal enterotoxin B on cardiorenal function and survival in X-irradiated rhesus macaques. *Am. J. Vet. Res.* **39,** 1213–1217 (1978).

31. Nies, A. S., Forsyth, R. P., Williams, H. E., and Melmon, K. L. Contribution of kinins to endotoxin shock in unanesthetized rhesus monkeys. *Circ. Res.* **22,** 155–164 (1968).
32. Doyle, T. F., Curran, C. R., and Turns, J. E. The prevention of radiation-induced, early, transient incapacitation of monkeys by an antihistamine. *Proc. Soc. Exp. Biol. Med.* **145,** 1018–1024 (1974).
33. Bruner, A. Immediate dose-rate effects of ^{60}Co on performance and blood pressure in monkeys. *Radiat. Res.* **70,** 378–390 (1977).
34. Farrar, J. K., Gamache, F. W., Jr., Ferguson, G. G., Barker, J., Varkey, G. P., and Drake, C. G. Effects of profound hypotension on cerebral blood flow during surgery for intracranial aneurysms. *J. Neurosurg.* **55,** 857–864 (1981).
35. Cockerham, L. G., Doyle, T. F., Pautler, E. L., and Hampton, J. D. Disodium cromoglycate, a mast cell stabilizer, alters postradiation regional cerebral blood flow in primates. *J. Toxicol. Environ. Health* **18,** 91–101 (1986).
36. Finnerty, F. A., Jr., Guillaudeu, R. L., and Fazekas, J. F. Cardiac and cerebral hemodynamics in drug induced postural collapse. *Circ. Res.* **5,** 34–39 (1957).
37. Klatzo, I., Suzuki, R., Orzi, F., Schuier, F., and Nitsch, C. Pathomechanisms of ischemic brain edema. In *Recent Progress in the Study and Therapy of Brain Edema* (K. G. Co and A. Baathmann, eds.), pp. 1–17. Plenum, New York, 1984.
38. Suzuki, R., Yamaguchi, T., Choh-Luh Li, and Klatzo, I. The effects of 5-minute ischemia in mongolian gerbils. II. Changes of spontaneous neuronal activity in cerebral cortex and CA1 sector of hippocampus. *Acta Neuropathol.* **60,** 217–222 (1983).
39. Doyle, T. F., Turns, J. E., and Strike, T. A. Effect of an antihistamine on early transient incapacitation of monkeys subjected to 4000 rads of mixed gamma radiation. *Aerosp. Med.* **42,** 400–403 (1971).
40. Suzuki, R., Yamaguchi, T., Kirno, T., Orzi, F., and Klatzo, I. The effects of 5-minute ischemia in mongolian gerbils. I. Blood-brain barrier, cerebral blood flow, and local cerebral glucose utilization changes. *Acta Neuropathol.* **60,** 207–216 (1983).
41. Gross, P. M., Teasdale, G. M., Angerson, W. J., and Harper, A. M. H2-receptors mediate increases in permeability of the blood–brain barrier during arterial histamine infusion. *Brain Res.* **210,** 396–400 (1981).
42. Ashani, Y., and Levi, D. A quaternary anticholinesterase probe for determining the integrity of the blood–brain barrier. (In preparation.)
43. Ashani, Y., and Catravas, G. N. Seizure-induced changes in the permeability of the blood–brain barrier following administration of anticholinesterase drugs to rats. *Biochem. Pharmacol.* **30,** 2593–2601 (1981).
44. Ashani, Y., Henry, F. H., and Catravas, G. N. Combined effects of anticholinesterase drugs and low-level microwave radiation. *Radiat. Res.* **84,** 496–503 (1980).
45. Petrali, J. P., Maxwell, D. M., Lenz, D. E., and Mills, K. R. *A Study of the Effects of Soman on the Rat Blood–Brain Barrier,* 4th Annu. Chem. Def. Biosci. Rev. U.S. Army Med. Res. Inst. Chem. Defense, Aberdeen, Maryland, 1984.
46. Ashani, Y., Catravas, G. N., and Alter, W. A. An irreversible anticholinesterase probe for studying increased permeability of the rat blood–brain barrier. *Biochem. Pharmacol.* **30,** 2585–2592 (1981).
47. Ashani, Y., Catravas, G. N., Hawkins, R. N., and Rogers, J. The application of a new anticholinesterase probe for studying radiation-induced changes in the blood–brain barrier of rats. (In preparation.)

CHAPTER 9

Acute Radiation Syndrome

ROBERT W. YOUNG*

Armed Forces Radiobiology Research Institute
Bethesda, Maryland 20814-5145

Exposure to a sufficient amount of ionizing radiation causes radiation sickness in man, which is manifest in characteristic clinical sequelae known as the acute radiation syndrome (ARS). The ARS is actually a combination of syndromes occurring in stages during a period of hours to weeks after exposure, as injury to various tissues and organs is expressed. The extent of injury is determined primarily by the total radiation dose received, how rapidly that dose is delivered, and how it is distributed in the body. In general, a dose to only part of the body causes less injury than the same dose to the whole body. Similarly, a dose delivered over a period of time is less harmful than the same dose delivered almost instantaneously. The most significant tissue damage manifest in the ARS from a high-level whole-body irradiation is to the bone marrow, gastrointestinal tract, cardiovascular system, central nervous system, gonads, and skin. The signs and symptoms that result from injury to these tissues constitute the ARS.

The various tissues of the body have different degrees of sensitivity to radiation. The bone marrow is the most sensitive, followed by the gastrointestinal tract, the cardiovascular system, and finally the central nervous system. This variation in tissue sensitivity to radiation causes the signs and symptoms of the ARS to occur in three successive phases: an initial prodromal phase, which develops within the first hours after irradiation; a subsequent latent period, which is relatively symptom free; and the main or manifest phase. The clinical pattern of the ARS has been compared to that of a viral disease (1) in which a delay occurs between the inoculation or exposure and the appearance of the prodromal stage of the illness. The prodromal phase lasts for a day or more and is followed by an incubation period or latent stage that can last for weeks. The latent phase is followed by the manifest illness. Recovery or death from the manifest radiation

*Present address: Science and Technology Biomedical Effects Directorate, Defense Nuclear Agency, Washington, D.C. 20305.

sickness occurs within 8 weeks of exposure, depending primarily on the radiation dose and the individual's susceptibility.

I. SOURCES OF DATA ON HUMAN RESPONSE TO PROMPT RADIATION

The "typical" response to prompt radiation has been carefully summarized by previous authors (2–6). Although these summaries have provided excellent generic clinical descriptions and summary statistics, they do not provide detailed quantitative data on the various signs and symptoms of the ARS. This chapter attempts to establish a complete quantitative profile for the incidence, severity, and duration of radiation sickness symptoms as a function of both radiation dose and time after exposure. As with the previous analyses, the data presented in this chapter were gathered from many sources, since no one source provides a comprehensive picture of the incidence, severity, and duration of radiation sickness in humans. Despite the large number of experimental studies with laboratory animals, those data were not used in order to avoid the uncertainty of extrapolating from animals that do not respond to radiation in exactly the same way as man. The sources of human data on radiation effects include case studies of nuclear accidents (7–22), records of patients given total-body radiation therapy for cancer and other diseases (3, 4, 6, 23–35), studies of the survivors of the Hiroshima and Nagasaki atomic bombings (36–41), and reaction of persons accidentally irradiated in a nuclear test in the South Pacific (23, 42–48). These data, together with previous analyses and expert opinions, serve as the human data base from which the frequency, severity, and duration of the symptoms of the ARS can be quantified.

II. ACUTE RADIATION SEQUELAE

The individual signs and symptoms for a person exposed to midlethal dose radiation are not specific for radiation injury, but collectively they are highly characteristic of the ARS. Irradiation, in the absence of the flash, detonation, and thermal pulse that accompany a nuclear detonation, occurs without any sensation. Even after a dose of several hundred centigrays (cGy), exposed individuals may remain perfectly normal and asymptomatic for 1–2 hr after irradiation.

A. Prodromal Phase of the ARS

The prodromal syndrome begins with a growing fatigue, which is described as being "washed out" or "worn out." In some cases, the listlessness and lack of initiative will progress to severe apathy, extreme weakness, or even prostration. Clinically irradiated patients simultaneously complain of headache, insomnia, dullness, dizziness, and occasionally vertigo. It is not uncommon for these signs to be accompanied by acute mental depression with bouts of hopelessness and despair (4). For doses of 1000 cGy or more, the latent period is almost nonexistent, and a transient prostration or incapacitation is the first sign of irradiation.

For persons who receive less than 1000 cGy, complaints of an ''upset stomach'' develop rather abruptly within the first 2 hr after exposure. This gastrointestinal distress is characterized by anorexia (loss of appetite), nausea, and malaise accompanied by listlessness, drowsiness, and fatigue. Vomiting soon follows, which increases in intensity until about 5–8 hr after irradiation, when it subsides almost as quickly as it began. On the second day after exposure, fatigue and episodes of nausea or vomiting may persist, but the patient's general condition is markedly improved. With the exception of a few who show some continued fatigue, anorexia, and mild nausea, most individuals are symptom-free on the third day after exposure. Fatigue is the basic symptom during the prodromal phase of the ARS with bouts of nausea and vomiting superimposed. Occasionally, fatigue is the only complaint. More rarely, vomiting may occur without fatigue.

Because of the predominance of fatigue and emesis and their apparent independence of one another, Gerstner (4) and Lushbaugh (6) divided the prodromal phase into a gastrointestinal and neuromuscular syndrome, which Court-Brown (27) has suggested arises from two separate etiologies. The gastrointestinal symptoms are anorexia, nausea, vomiting, diarrhea, intestinal cramps, salivation, fluid loss, dehydration, and weight loss. The neuromuscular symptoms include easy fatigability, apathy, or listlessness, sweating, fever, headache, and hypertension; after high doses, these are followed by hypotensive shock (6).

For doses above 750 cGy, some patients show decreased blood pressure and increased pulse rate due to circulatory hypovolemia. There are also some reports of acute myocardial insufficiency and death in patients with a history of myocardial disease. Some individuals also develop a painful mumps-like swelling of the parotid gland within a few hours of exposure. The pain usually subsides within 2 days, but the swelling sometimes persists for several more days. Dry mouth, accompanied by a metallic taste that persists for 1 week, contributes to the disinterest in food.

Approximately 10% of radiation therapy patients treated for leukemia with 750–1000 cGy develop diarrhea soon after irradiation. More develop diarrhea in 1–7 days after exposure. After 3–4 days, a sense of easy fatigability, apparently unrelated to fluid and electrolyte changes, becomes a major complaint for some patients. About 75% of patients develop oral infections, due to ulcerated mucosae in the presence of leukocytopenia and immunosuppression. These infections as well as fungal pharyngitis and herpes can appear within 3 days. Erythema can appear as early as 1 day after irradiation, but it usually occurs later and may persist for as long as 2 weeks. Beginning 7–10 days after exposure, loss of hair may occur, but it is incomplete and temporary. All of these signs and symptoms are not seen unless the exposure is in the supralethal range or unless the observation period is extended beyond the first 2 days postirradiation, which is usually considered to delineate the prodromal phase of the ARS. For potentially survivable doses, the primary symptoms are anorexia, nausea, vomiting, and easy fatigability. Prodromal diarrhea, fever, and hypotension are signs of exposure to supralethal doses.

The exact pathophysiology of the prodromal phase of the ARS is not known. Several causal factors have been suggested, including direct radiation effects on the central and autonomic nervous systems, disturbance of the endocrine balance, and production of various toxic substances (49). It is thought that chemical mediators are released from damaged cells of primarily lymphoid and bone marrow tissues (49, 50). This cellular debris reaches a maximum at 8–12 hr postirradiation, and is cleared by fixed and free macrophages by 24–48 hr after exposure. As such, phagocytic action may be largely responsible for the length of the prodromal phase of the ARS. Listlessness, nausea, and vomiting imply that structures of the nervous system are involved. However, since adult nervous systems are radioresistant in the midlethal range, likely causes seem to be indirect mediation through either increased intracranial pressure or a circulating substance acting on the chemoreceptor trigger zone of the medulla (51). Recent experimental work in animals has shown that a peptide released into the blood due to intestinal radiation injury can act on the area postrema to induce vomiting (52). These mechanisms are being actively investigated.

B. The Latent Period

The latent period, which follows the prodromal phase, is relatively asymptomatic. It can last until the nineteenth or twentieth day postirradiation. This quiescent period is the time between initial cell damage and its expression as interference with cell renewal in the affected organs (2, 53). Radiation interferes with the ability of mitotically active tissues to renew themselves. Bergonie and Tribondeau (54) first recognized this in 1906, when they stated the principle that the radiation sensitivity of tissue varies directly with the mitotic activity of its cells and inversely with the degree of differentiation of those cells. Thus different organs exhibit different sensitivities to the same radiation dose, and that sensitivity determines the type and severity of the manifest illness, the duration of the latent period, and the outcome of the illness. The number of surviving stem cells in these mitotically active tissues after irradiation determines the outcome of the manifest illness, since these cells are the only source for cell renewal in the affected tissues.

C. The Manifest Illness

The manifest illness of the ARS is classically divided into three major syndromes, which cause death due to damage of the marrow, small intestine, or brain after radiation exposure. These syndromes of radiation death are traditionally known as the hemopoietic, gastrointestinal, and central nervous system syndromes. The hemopoietic and gastrointestinal syndromes are still regarded as the major mechanisms of death for doses of less than 2000 cGy, but the current view replaces the classic central nervous syndrome with a neurovascular syndrome for doses of 2,000 cGy to approximately 10,000 cGy. Death due to direct injury of the brain is now thought to occur at doses of over 10,000 cGy.

D. Hemopoietic Syndrome

The hemopoietic stem cells are the most radiosensitive tissues in the body. As such, radiation doses of 100 cGy or more can significantly damage the blood-forming capability of the body and cause death for approximately 50% of individuals exposed to approximately 300 cGy. Radiation kills the mitotically active precursor stem cells, diminishing the subsequent supply of mature red cells, white cells, and platelets. As mature circulating cells die and the supply of new cells is inadequate to replace them, the physiological consequences of damage to the hemopoietic system become manifest. The physiological consequences of damage to bone marrow include increased susceptibility to infection, bleeding, anemia, and lowered immunity. For doses below 700 cGy, the hemopoietic syndrome begins at about 8–10 days postexposure with a serious drop in granulocyte and platelet counts (12, 23, 47, 55–60). Pancytopenia follows at about 3–4 weeks later, becoming complete at doses above 500 cGy (40, 61, 62). Purpura is evident, and bleeding may be uncontrolled, causing anemia. There may be fever as well as rises in pulse and respiratory rate due to endogenous bacterial and mycotic infections, with infections becoming uncontrolled due to impaired granulocyte and antibody production (6, 63, 64). If at least 10% of the hemopoietic stem cells remain uninjured, recovery is possible; otherwise, death occurs within 4–6 weeks (40, 62, 65).

E. Median Lethal Dose

As indicated above, approximately 50% of individuals exposed to 300 cGy of prompt whole-body ionizing radiation will die in 4–6 weeks after irradiation. The median lethal dose for death within 60 days from radiation exposure is designated as the $LD_{50/60}$. Individual differences in the status of the immune system and preexisting infection at the time of irradiation will produce considerable variability in this end point, which changes very rapidly as a function of dose. Given this variability and the limited number of human cases of whole-body prompt irradiations in this dose range, it is difficult to know the $LD_{50/60}$ with certainty. Numerous statistical analyses have provided estimates of the $LD_{50/60}$ value. Most of these analyses yield midline tissue values of approximately 300 cGy (450 cGy free in air).

F. Gastrointestinal Syndrome

At doses above 700 cGy, injury to the gastrointestinal tract contributes increasingly to the severity of the manifest-illness phase of the ARS. Irradiation inhibits the renewal of cells lining the digestive tract. These cells are short-lived and must be renewed at a high rate. Radiation of sufficient magnitude to interfere with the renewal of the cells leads to their depletion within a few days. Because the rate of cell turnover is highest in the small intestine, cell depletion occurs sooner in this part of the gastrointestinal tract.

The physiological consequences of gastrointestinal injury from radiation may vary, depending on the region and extent of damage. The responses of the stomach, colon, and rectum to radiation are similar to that of the small intestine, but they occur more slowly. Responses of the mouth and esophagus closely resemble that of the skin. Under normal conditions, the integrity of the intestinal mucosa prevents substantial escape of bacteria into the bloodstream. The few bacteria that do escape are soon inactivated by granulocytes or specific antibodies. The radiosensitive mucosal stem cells in the intestinal crypts have a rapid turnover rate (2, 65), producing mature, nondividing, differentiated cells that migrate to the functional mucosal lining. Those mature mucosal cells have a lifetime of several days. They are progressively shed and are not replaced when radiation kills the stem cells in the crypt. The result is breakdown of the mucosa and ulceration of the intestine. As the mucosa breaks down, bacteria can enter the bloodstream. These bacteria go unchallenged after irradiation because of the curtailed production of granulocytes (62, 66) due to bone marrow damage, resulting in fever and infections (65, 67).

At doses of 300–800 cGy, temporary injury to the tight junctions between epithelial cells of the mucosal lining reportedly permit the escape of bacterial endotoxins into the bloodstream (68). As dose increases, the epithelial lining is more extensively depleted. For doses of 700–1000 cGy, septicemia is more important than dehydration in causing death, which occurs in 2–3 weeks (2, 3, 40, 65, 68, 69). With doses of about 1000–1500 cGy, denudation of the mucosa, particularly in the small intestine, exacerbates the loss of fluid and electrolytes. Beginning at about 1250 cGy, early mortality occurs due to dehydration and electrolyte imbalance, because of leakage through the extensively ulcerated intestinal mucosa (58, 65, 70). These conditions develop over a few days and are characterized by cramping, abdominal pain, and diarrhea (which may become watery, more frequent, and more severe by the next week), followed by shock and death (41, 62, 64, 65, 69, 71).

G. Neurovascular Syndrome

The neurovascular syndrome is the least well understood of the radiation-induced deaths. The syndrome is unique in that death occurs very quickly (within 2 days) before failures in the gastrointestinal and hemopoietic systems become apparent. It is difficult to precisely define this syndrome due to (1) the lack of human exposures at the high radiation doses required to produce the syndrome and (2) confusion about the actual cause of death. This type of radiation death was originally designated central nervous system (CNS) syndrome because of the obvious CNS signs and symptoms, such as disorientation, loss of muscular coordination, respiratory distress, convulsive seizures, and coma associated with death. Histological evidence of microvascular damage (72–74) and evidence of cerebral edema [such as herniation of the cerebellar tonsils into the foramen magnum, the existence of pressure cones, and the narrowing of the sulci (75)] have led to the realization that the CNS failures that result in these early deaths are due to the buildup of intracranial pressure caused by edema. Lushbaugh (76) has termed this form of death a cardiovascular shock syndrome, wherein the massive loss of serum and electrolytes through leakage into the extravascular tissues causes extreme circulatory problems, edema, intracranial pressure, and cerebral anoxia, which can bring death within 2 days. The radiation threshold for this syndrome is equally unclear. Generally it is thought that 5,000 cGy is necessary for the neurovascular syndrome, and doses above 10,000 cGy are required for direct damage of the nervous system.

III. DETAILED DESCRIPTION OF THE ARS

The incidence, severity, and duration of the signs and symptoms of the ARS are shown in Tables 9-1 through 9-8. Each table provides a temporal summary of

TABLE 9-1

SYMPTOMS FOR DOSE RANGE 50–100 rads (cGy)

Symptom	Postexposure time: Hours (0, 4, 8, 12, 16, 20, 24)	Days (1, 2, 3, 4, 5, 6, 7)	Weeks (1, 2, 3, 4, 5, 6)
Nausea[a]	—5–30% mild—		
Vomiting (retching)	—5–20%— mild		
Anorexia	——15–50%—— -------------		
Diarrhea (cramps)			
Fatigue			
Weakness			
Hypotension			
Dizziness			
Disorientation			
Bleeding[b]			
Fever		—(c)—	——(c)——
Infection			
Ulceration			
Fluid loss/electrolyte imbalance			
Headache			
Fainting			
Prostration			
Death			

[a] References for this group of symptoms: 3, 6, 19, 22, 26, 30, 31, 42–45, 47, 60, 62–66, 68–69, 81–89. These symptoms not observed in American servicemen exposed to approximately 78 rads (cGy) of fallout radiation, according to Refs. 42 and 45.

[b] References for this group of symptoms: 2, 22, 26, 42–48, 58, 62, 63, 68, 69, 81, 85.

[c] Slight drop in lymphocyte, platelet, and granulocyte counts; no overt symptoms.

the ARS for the first 6 weeks after irradiation for a range of doses. The dose range from 50 to 3000 cGy has been divided into eight regions that correspond to major pathophysiological changes in response to irradiation. Table 9-9 summarizes those dose ranges and the pathophysiological events associated with them.

A. DOSES OF 50–100 cGy

An absorbed whole-body dose of less than 50 cGy will produce few if any acute effects in irradiated persons, and then only in those who are the most sensitive. As Table 9-2 indicates, the acute radiation effects at this level are mild, and they occur only during the first day after exposure. Blood cell counts may drop slightly, but survival of the individual is almost certain at this radiation dose level.

B. Doses of 100–200 cGy

The severity of the prodromal effects increases over the dose range of 100–200 cGy, with the incidence of vomiting increasing to 70% at 200 cGy (Table 9-3). Fatigue and weakness will be evident in approximately 30–60% of persons irradiated at this level. Significant destruction of bone marrow stem cells may lead to a 25–35% drop in blood cell production. As a result, mild bleeding, fever, and infection may occur during the fourth and fifth weeks postexposure. At 200 cGy, up to 5% may die by 5–6 weeks after irradiation.

C. Doses of 200–350 cGy

Over the dose range of 200–350 cGy, prodromal symptoms begin earlier and affect a greater number of exposed persons (Table 9-4). Moderate diarrhea at 4–8 hr may be experienced by 10% of the population. Most victims tire easily and experience mild to moderate weakness intermittently over the 6 weeks. Under normal conditions, the vomiting and diarrhea are not enough to cause serious fluid loss or electrolyte imbalance. However, in hot or humid conditions, the combined fluid loss and electrolyte imbalance could become serious.

Injury to the hemopoietic system is indicated by moderate bleeding, fever, infection, and ulceration at 3–5 weeks postexposure for more than 50% of those exposed. During the fourth and fifth weeks, diarrhea may complicate the condition. As the dose rises from 200 to 350 cGy, 5 to 50% of nontreated persons may die during the fifth week, with death coming earlier to those with preexisting infections, e.g., of the upper respiratory tract.

Five examples of the nonfatal bone marrow syndrome are contained in the 1958 Y-12 Oak Ridge accident. At that time, eight male workers (ages 25–56 years) were exposed to a mixed neutron–gamma field in a critical excursion that had been triggered by an unplanned transfer of enriched uranium to a 55-gal drum. Five of these men received doses of between 236 and 365 cGy. All experienced nausea and vomiting during the first 48 hr after exposure. In two individuals, nausea persisted for 3 and 5 days. One victim experienced headache and another complained of aches in his legs. All men began partial transient epilation on the seventeenth day after exposure. All developed the characteristic hematological pattern of radiation-induced bone marrow depression. The hemopoietic response was greatest in the man who received the highest dose (365

TABLE 9-2

Symptoms for Dose Range 100–200 rads (cGy)

Symptom	Postexposure time: Hours (0 4 8 12 16 20 24)	Days (1 2 3 4 5 6 7)	Weeks (1 2 3 4 5 6)
Nausea[a]	——30–70% mild to moderate——		
Vomiting (retching)	——20–50% mild to moderate——		
Anorexia	——50–90%——		
Diarrhea (cramps)[b]			
Fatigue[c]	——30–60% mild to moderate——	- - - - - - - -	- - - - Mild - - - -
Weakness	——30–60% mild to moderate——	- - - - - - - -	- - - - Mild - - - -
Hypotension			
Dizziness			
Disorientation			
Bleeding[d]			—(e)— —10%— mild
Fever		—(f)—	—(f)— ——— 10–50% mild to moderate
Infection		—(g)———	——— 10–50% mild to moderate
Ulceration			
Fluid loss/electrolyte imbalance			
Headache			
Fainting			
Prostration			
Death[h]			≤5%——

[a]References for this group of symptoms: 3, 6, 8, 9, 11, 15, 17, 19, 21, 22, 26, 27, 31, 33, 40–45, 49, 55, 60–62, 64–66, 68–69, 71, 82–84, 86, 87, 90–93.

[b]Ten percent of the Marshallese victims exposed to 175 rads (cGy) experienced diarrhea during the first postexposure day according to Ref. 55.

[c]References for this group of symptoms: 17, 38, 40, 46, 62, 65, 68, 71, 94.

[d]References for this group of symptoms: 2, 8, 21, 22, 31, 38, 40–46, 56–62, 65, 66, 69, 84, 89, 91, 95, 96.

[e]Slight to moderate drop in platelets from $3 \times 10^5/\text{mm}^3$ to $1.8–0.8 \times 10^5/\text{mm}^3$.

[f]Slight to moderate drop in granulocytes from $6 \times 10^3/\text{mm}^3$ to $4.5–2.0 \times 10^3/\text{mm}^3$.

[g]Slight to moderate drop in lymphocytes from $3 \times 10^3/\text{mm}^3$ to $2.0–1.0 \times 10^3/\text{mm}^3$.

[h]References for this event: 2, 5, 40, 55.

TABLE 9-3

SYMPTOMS FOR DOSE RANGE 200–350 rads (cGy)

Symptom	Postexposure time: Hours (0 4 8 12 16 20 24)	Days (1 2 3 4 5 6 7)	Weeks (1 2 3 4 5 6)
Nausea[a]	——70–90% moderate——		
Vomiting (retching)	——50–80% moderate——		
Anorexia	——90–100%——————		–60%——-------
Diarrhea (cramps)	— ← ~10% moderate		Moderate → –40——------- 60%
Fatigue[b]	——60–90% moderate——————	------Mild------------	------Moderate-------
Weakness	——60–90% moderate——————	------Mild------------	------Moderate-------
Hypotension			
Dizziness			
Disorientation			
Bleeding[c]		—(d)—	—10–50%——------- moderate
Fever		——(e)——	——————------- 10–80% moderate
Infection		————(f)——	
Ulceration	(g)		——30%——------- moderate
Fluid loss/electrolyte imbalance			
Headache			
Fainting			
Prostration			
Death[h]			≤5–50% →———

[a]References for this group of symptoms: 3, 4, 7, 8, 10, 11, 17, 19, 22, 26, 32, 38, 40, 41, 47–49, 55–57, 59, 60–62, 64–66, 68, 69, 71, 82–84, 86–89, 90, 92, 93, 97–100.

[b]References for this group of symptoms: 2, 4, 21, 30, 40, 46, 60, 62, 81, 89, 100.

[c]References for this group of symptoms: 2, 11, 22, 23, 26, 30, 32, 36, 38, 40, 41, 46, 47, 57–62, 65, 69, 84, 88, 89, 91, 93, 101.

[d]Moderate drop in platelets from $3 \times 10^5/mm^3$ to $0.8–0.1 \times 10^5/mm^3$.

[e]Moderate drop in granulocytes from $6 \times 10^3/mm^3$ to $2.0–0.5 \times 10^3/mm^3$.

[f]Moderate to severe drop in lymphocytes from $3 \times 10^3/mm^3$ to $1.0–0.4 \times 10^3/mm^3$.

[g]Epilation.

[h]References for this event: 2, 5, 40, 55, 62.

TABLE 9-4

SYMPTOMS FOR DOSE RANGE 350–550 rads (cGy)

Symptom	Postexposure time: Hours (0, 4, 8, 12, 16, 20, 24)	Days (1, 2, 3, 4, 5, 6, 7)	Weeks (1, 2, 3, 4, 5, 6)
Nausea[a]	—90–100%— severe moderate		60–100% moderate → —— --
Vomiting (retching)	—80–100%— severe moderate		(60–100% moderate) → —— --
Anorexia	——100%——		——100%—--
Diarrhea (cramps)	— ←~10% moderate to severe		——60–100%—— -- moderate to severe
Fatigue[b]	—— 90–100% moderate to severe ——	—— 90–100% moderate to severe ——	—— 90–100% moderate to severe ——
Weakness	—— 90–100% moderate to severe ——	—— 90–100% moderate to severe ——	—— 90–100% moderate to severe ——
Hypotension[c]			
Dizziness			Moderate → ——60%—— --
Disorientation			Moderate → ——60%—— --
Bleeding[d]			—(e)–50–100%—— -- moderate to severe
Fever		—(f)—	—80–100%—— -- moderate to severe
Infection		—(g)—	
Ulceration			—50% mild to—— -- moderate (h)
Fluid loss/electrolyte imbalance[i]	——50% mild—— to moderate		(j) Moderate: ——50%—— --
Headache	——50% mild—— to moderate		Moderate: ——50%—— --
Fainting			Moderate: ——50%—— --
Prostration			Moderate: ——50%—— --
Death[k]			——50–99%——

[a]References for this group of symptoms: 2, 3, 8, 10, 11, 13, 17, 19, 22, 26, 40, 41, 46, 47, 55, 60–62, 64, 65, 68, 69, 81–84, 86, 88–93, 96–98, 102.

[b]References for this group of symptoms: 2, 10, 17, 31, 40, 46, 61, 62, 65, 81, 90, 99.

[c]References for this group of symptoms: 96, 102.

[d]References for this group of symptoms: 2, 13, 17, 22, 26, 30, 36, 40, 46, 47, 56–62, 64–66, 69, 70, 81, 84, 89, 93, 102, 103.

[e]Severe drop in platelets from $3 \times 10^5/\text{mm}^3$ to 0.1×10^5–$0/\text{mm}^3$.

[f]Severe drop in granulocytes from $6 \times 10^3/\text{mm}^3$ to 0.5×10^3–$0/\text{mm}^3$.

[g]Severe drop in lymphocytes from $3 \times 10^3/\text{mm}^3$ to 0.4–$0.1 \times 10^3/\text{mm}^3$.

[h]Epilation.

[i]References for this group of symptoms: 2, 30, 40, 46, 62, 96.

[j]Mild intestinal damage.

[k]References for this event: 2, 5, 40, 55.

cGy); second greatest in the patient with the lowest dose (236 cGy); and third, fourth, and fifth greatest in the patients who received 270, 339, and 327 cGy, respectively. In the fourth week, the patients who received 365 and 339 cGy exhibited a mild hemorrhagic phenomenon associated with thrombocytopenia (platelets 10–20 $\times$ $10^3/ml^3$). The patients at 365 and 270 cGy developed mild upper respiratory infections, which did not clearly correlate with a minimal granulocyte level. These infections responded promptly to antibiotics. Spontaneous recovery of platelet and granulocyte counts occurred between the fifth and eighth weeks. All of the patients were released from the hospital on day 44 after exposure (77).

D. Doses of 350–550 cGy

When exposure is in the dose range of 350–550 cGy, almost everyone experiences severe prodromal symptoms (Table 9-5). Severe and prolonged vomiting can adversely affect the electrolyte balance. Most persons show moderate to severe fatigue and weakness for many weeks. If untreated, 50–99% of cases will die of extensive injury to the hemopoietic system. This injury will be manifest in overwhelming infections and bleeding during the third to sixth weeks. Nausea, vomiting, and anorexia may recur during the manifest illness, with approximately half the individuals experiencing diarrhea, electrolyte imbalance, and headache. Terminal conditions may be complicated by dizziness, disorientation, fainting, prostration, and symptoms of infection.

E. Doses of 550–750 cGy

As depicted in Table 9-6, almost all persons exposed to radiation doses of 550–750 cGy will experience severe nausea and vomiting on the first day. Those symptoms will moderate over the next day or two. Dizziness and disorientation are likely to accompany these prodromal reactions.

Untreated persons will lose their defense against infection because bone marrow stem cells and granulocytes will be almost completely eliminated. By the end of the first week postexposure, infection will be rampant from endogenous bacteria from the injured gastrointestinal tract into the immunocompromised host.

The combination of hemopoietic and gastrointestinal damage reduces the survival of all untreated persons to 2–3 weeks. During this entire time they suffer from severe fatigue and weakness. Toward the end of the first week, nausea, vomiting, and anorexia recur. Moderate to severe bleeding, headaches, hypotension, dehydration, electrolyte imbalance, and fainting complicate the condition of all such cases during their last days.

TABLE 9-5

SYMPTOMS FOR DOSE RANGE 550–750 rads (cGy)

Symptom	Postexposure time: Hours (0 4 8 12 16 20 24)	Days (1 2 3 4 5 6 7)	Weeks (1 2 3)	Weeks (4 5 6)
Nausea[a]	—severe 100% moderate to severe——		——-----	100% moderate to severe
Vomiting (retching)	—severe 100% moderate to severe——		-----	100% moderate to severe
Anorexia	——100%——	—100%—	——-----	
Diarrhea (cramps)	— 10% moderate to severe	——100%—— moderate to severe	——-----	
Fatigue[b]	——100% severe——	————	——-----	
Weakness	——100% severe——	————	——-----	
Hypotension[c]			——-----	100% severe
Dizziness	——100% severe——		——-----	100% severe
Disorientation	——100% severe——		——-----	100% severe
Bleeding[d]		(e)——	——-----	100% severe
Fever		(f)——	——-----	100% severe
Infection		——(g)——		
Ulceration			——-----	100% severe
Fluid loss/electrolyte imbalance[j]	——80% moderate——	(i)	(h) ——-----	80% moderate to severe
Headache	——80% moderate——		——-----	100% moderate to severe
Fainting			——-----	80% moderate to severe
Prostration				
Death[k]			——	100%

[a]References for this group of symptoms: 2, 3, 8–10, 12, 17, 19, 22, 26, 30, 35, 40, 41, 46, 47, 49, 55, 60–62, 64, 65, 68–70, 81–82, 84, 86, 88, 90–93, 97–99.

[b]References for this group of symptoms: 2, 10, 17, 31, 40, 46, 61, 62, 70, 81, 90, 99, 104.

[c]References for this group of symptoms: 2, 5, 17, 22, 26, 30, 46–48, 60–62, 65, 68, 69, 84, 88, 93, 97, 103.

[d]References for this group of symptoms: 2, 7, 9, 13, 17, 22, 26, 30, 40, 46, 47, 57–62, 65, 69, 81, 84, 91, 92, 103.

[e]Platelet count drops nearly to zero.

[f]Granulocyte count drops nearly to zero.

[g]Lymphocyte count drops nearly to zero.

[h]Epilation.

[i]Moderate intestinal damage.

[j]References for this group of symptoms: 2, 7, 17, 22, 26, 30, 40, 46, 47, 57–62, 65, 67, 69, 81, 84, 91, 92.

[k]References for this event: 2, 5, 40, 55, 70, 102.

TABLE 9-6

SYMPTOMS FOR DOSE RANGE 750–1000 rads (cGy)

Symptom	Postexposure time: Hours (0, 4, 8, 12, 16, 20, 24)	Days (1, 2, 3, 4, 5, 6, 7)	Weeks (1, 2, 3, 4, 5, 6)
Nausea[a]	100% severe		100% moderate to severe
Vomiting (retching)	100% severe		100% moderate to severe
Anorexia	100%	100%	
Diarrhea (cramps)	10% severe	100% severe	
Fatigue[b]	100% severe		
Weakness	100% severe		
Hypotension[c]	~80% mild[d]		100% severe
Dizziness	100% severe		100% severe
Disorientation	100% severe		100% severe
Bleeding[e]		(f)	100% severe
Fever	~30–45% moderate[d]	(g)	100% severe
Infection		(h)	
Ulceration		(i)	100% severe
Fluid loss/electrolyte imbalance[j]	100% moderate	(k)	100% severe
Headache	100% moderate to severe		100% severe
Fainting			70% moderate to severe
Prostration			70% moderate to severe
Death[l]			100%

[a]References for this group of symptoms: 2, 8, 10, 12, 17, 19, 26, 30, 34, 40, 41, 46, 47, 49, 60–62, 65, 68, 69, 82, 84, 86, 90–93, 98–99.

[b]References for this group of symptoms: 2, 10, 31, 46, 61, 62, 65, 81, 90, 97, 99, 103.

[c]References for this group of symptoms: 2, 5, 22, 26, 30, 34, 41, 46, 47, 60–62, 65, 68, 69, 84, 86, 91, 92, 96.

[d]Blood pressure drops 25%; temperature increases to 102°F according to Ref. 69.

[e]References for this group of symptoms: 2, 5, 8, 10, 17, 26, 30, 34, 40, 41, 46, 47, 61–63, 65, 69, 70, 84–86, 90–93, 96.

[f]Platelet count drops to zero.

[g]Granulocyte count drops to zero.

[h]Lymphocyte count drops to zero.

[i]Epilation.

[j]References for this group of symptoms: 2, 5, 26, 30, 40, 41, 46, 47, 60–62, 65, 68, 69, 84–86, 91, 96.

[k]Moderate to severe intestinal damage.

[l]References for this event: 2, 5, 40, 55, 102.

F. Doses of 750–1000 cGy

The survival time for untreated persons after exposure to doses of 750–1000 cGy diminishes from 2 to $2\frac{1}{2}$ weeks (Table 9-7). Symptoms resemble those experienced over the preceding dose range with the following notable differences: severe nausea and vomiting may continue into the third day, hypotension affects 80% and moderate fever between 30 and 45% during the first day, electrolyte imbalance is a persistent problem from the sixth hour on, all cases have moderate to severe headache during the first day, and nearly three quarters of cases are prostrate before the end of the first week.

G. Doses of 1000–2000 cGy

With doses of 1000–2000 cGy, severe nausea and vomiting affect everyone within 30 min of exposure, and the symptoms continue intermittently until death in the second week (Table 9-8). Severe headaches begin after about 6 hr and continue for 2–3 days. The severity of symptoms may diminish somewhat during days 3–5. Gastrointestinal injury predominates, which is manifest 4–6 days after exposure by the abrupt return of severe nausea, vomiting, anorexia, and diarrhea, along with high fever, abdominal distention, and peristalsis. During the second week, severe dehydration, hemoconcentration, and circulatory collapse, compounded by septicemia, will lead to coma and death.

The gastrointestinal phase of the ARS is illustrated by the reaction of a scientist who was exposed at Los Alamos when an experimental nuclear assembly unexpectedly became critical. In this case a 32-year-old man received a prompt exposure estimated at 1114 cGy, composed of 1000 cGy of neutrons and 114 cGy of gamma rays (78). He experienced nausea within a few minutes and vomited before reaching the hospital about an hour later. During the next few hours he vomited with increasing frequency and had one diarrheal stool. After 12 hr the prodromal signs and symptoms ceased completely, and his general condition began to improve. The patient remained in satisfactory condition until the sixth day, when his white blood cell count fell steeply. The low-grade fever that had existed since the accident suddenly rose to above 102°F. Nausea, vomiting, and diarrhea reappeared, and signs of severe paralytic ileus developed. The patient's major complaint was abdominal distention. Ten liters of green fluid with fecal odor and appearance were aspirated by gastric suction within a 24-hr period. In spite of continuous gastric suction, parenteral fluid supply, blood transfusions, and antibiotic treatment, the patient deteriorated swiftly to the beginning of circulatory collapse on the eighth day, and finally to death on the ninth day after exposure. At autopsy, the most striking changes were found in the small intestine. Hempelmann and colleagues (79) reported the following:

TABLE 9-7

SYMPTOMS FOR DOSE RANGE 1000–2000 rads (cGy)

Symptom	Postexposure time		
	Hours	Days	Weeks
	0 4 8 12 16 20 24	1 2 3 4 5 6 7	1 2 3 4 5 6
Nausea[a]	—100% severe—	------ —	← 100% severe
Vomiting (retching)	—100% severe—	---------- —	←
Anorexia	—100%—	—100%—	
Diarrhea (cramps)	— 20% severe	—100% severe—	
Fatigue[b]	—100% severe—	—	
Weakness	—100% severe—	—	
Hypotension[c]	—100% moderate to severe—	—	
Dizziness	—100% severe—	—	
Disorientation	—100% severe—	—	
Bleeding[d]		(e)—	100% severe
Fever	—45–80% moderate to severe—	{ —(f)—	100% severe
Infection		{ —(g)—	
Ulceration		—	100% severe
Fluid loss/electrolyte imbalance[h]	—100% moderate to severe—	--------- (i) —	100% severe
Headache	—100% severe—	—	100% severe
Fainting		—	80% severe
Prostration		—	80% severe
Death[j]		—(k)— —100%—	

[a]References for this group of symptoms: 2, 8, 17, 22, 26, 30, 40, 41, 46, 47, 60–62, 65, 68–70, 82, 84, 86, 91–93, 99, 103.

[b]References for this group of symptoms: 2, 10, 17, 30, 31, 46, 61, 62, 65, 99.

[c]References for this group of symptoms: 2, 5, 22, 26, 30, 40, 41, 46, 47, 60–62, 65, 68, 69, 84, 86, 91, 92, 96.

[d]References for this group of symptoms: 2, 8, 17, 22, 26, 30, 40, 41, 46, 47, 60–62, 65, 68, 69, 82, 84, 86, 91–93, 99, 103.

[e]Platelet count drops to zero.

[f]Granulocyte count drops to zero.

[g]Lymphocyte count drops to zero.

[h]References for this group of symptoms: 2, 5, 22, 26, 40, 41, 46–48, 59, 60, 62, 65, 68, 69, 76, 86, 91, 93.

[i]Severe intestinal damage.

[j]References for this event: 2, 5, 35, 40, 55.

[k]Renal failure according to Ref. 35.

TABLE 9-8

SYMPTOMS FOR DOSE RANGE 2000–3000 rads (cGy)

Symptom	Postexposure time: Hours (0 4 8 12 16 20 24)	Days (1 2 3 4 5 6 7)	Weeks (1 2 3 4 5 6)
Nausea[a]	——————100% severe——	——————	
Vomiting (retching)	——————100% severe——	——————	
Anorexia	——————100%——	——————	
Diarrhea (cramps)	—— ~30% severe	——	100% severe
Fatigue[b]	——————100% severe——	——————	
Weakness	——————100% severe——	——————	
Hypotension[c]	——————100% severe——	——————	
Dizziness	——————100% severe——	——————	
Disorientation	——————100% severe——	——————	
Bleeding[d]			
Fever	————80–90%———— moderate to severe	{ —(e)	
Infection		—(f)—	
Ulceration			
Fluid loss/electrolyte imbalance[g]	—————100% severe——	——————	
Headache	—————100% severe——	——————	
Fainting		——100%—— severe	
Prostration		——100%—— severe	
Death[h]		—100%—	

[a]References for this group of symptoms: 2, 3, 8, 17, 26, 30, 40, 41, 46–49, 60–62, 65, 68, 82, 84, 86, 91–93.

[b]References for this group of symptoms: 2, 8, 17, 18, 22, 26, 30, 40, 41, 46–48, 60–62, 65, 68, 82, 84, 86, 91–93.

[c]References for this group of symptoms: 2, 5, 18, 22, 30, 41, 46, 48, 60, 65, 68, 86, 91–93.

[d]References for this group of symptoms: 2, 5, 18, 22, 30, 40, 41, 46–48, 59, 60, 62, 65, 68, 86, 91–93.

[e]Granulocyte count drops to zero.

[f]Lymphocyte count drops to zero.

[g]References for this group of symptoms: 2, 5, 22, 26, 30, 41, 46, 48, 59, 60, 65, 68, 69, 84, 86, 91–93.

[h]References for this event: 2, 5, 40, 55, 102.

> It [the small intestine] was distended, flabby and filled with dark brown semi-liquid material. The vessels were intensely congested, and there were numerous petechial hemorrhages on the serosal surface. The mucosal surface was edematous and deep red, particularly in the region of the jejunum where, in addition, the surface was covered with membranous gray-green exudate that could be stripped off in sheets.

Histological examination found complete erosion of the jejunal and iliac epithelium with loss of the superficial submucosal layers. The denuded surfaces were covered by an exudate containing masses of bacteria that, in the ulcerated portions, had invaded the intestinal wall. The anatomical diagnosis was diffuse membranous and ulcerative enterocolitis associated with aplasia and depletion of bone marrow, lymph nodes, and the lymphatic system in the spleen and gastrointestinal tract.

H. Doses of 2000–3000 cGy

Gastrointestinal and cardiovascular injury predominate in the dose range of 2000–3000 cGy (Table 9-9). Prodromal effects, including severe headache and drowsiness, appear almost immediately after exposure and may persist as the gastrointestinal syndrome develops. Severe dehydration and electrolyte imbalance are manifest several hours after exposure. Initially, fluids and electrolytes are lost by vomiting, but in time the greater loss is from severe diarrhea. The increased permeability of capillaries throughout the body releases fluids into the interstitial spaces.

A prompt 4500-cGy whole-body exposure of an individual engaged in a plutonium recovery procedure occurred at Los Alamos in 1958. In this case the patient experienced an early transient incapacitation shortly after exposure. The signs and symptoms of both central nervous system damage and cardiovascular damage were prominent in the postirradiation picture, including fatigue, nausea, vomiting, disorientation, and fever. In contrast to what one may have predicted, vomiting ended after 12 hr, although intensive sedation was used to make the patient manageable and comfortable. There was a period of temperature rise which also ended at 12 hr after exposure. With the exception of one watery bowel movement at about 1–2 hr after the accident, no diarrhea was observed; however, the patient complained of severe abdominal distress. Early erythema of most of the face and trunk was present minutes after exposure. This progressed rapidly, and edematous swelling of the hands and forearms became prominent. The erythema had apparently disappeared by the time of death. The cardiovascular changes included hypertension, cardiac damage, and shock. Death occurred at 32 hr after exposure (65).

TABLE 9-9

DOSE RANGES AND ASSOCIATED PATHOPHYSIOLOGICAL EVENTS

Dose range rads (cGy)	Pathophysiological events		
	Prodromal effects	Manifest illness effects	Survival
50–100	Mild	Slight decrease in blood cell count	Virtually certain
100–200	Mild to moderate	Beginning symptoms of bone marrow damage	Probable (>90%)
200–350	Moderate	Moderate to severe bone marrow damage	Possible— Bottom third of range: $LD_{5/60}$ Middle third: $LD_{10/60}$ Top third: $LD_{50/60}$
350–550	Severe	Severe bone marrow damage	Death within 3½–6 weeks— Bottom half: $LD_{90/60}$ Top half: $LD_{99/60}$
550–750	Severe	Bone marrow pancytopenia and moderate intestinal damage	Death within 2–3 weeks
750–1000	Severe	Combined gastrointestinal and bone marrow damage; hypotension	Death within 1–2½ weeks
1000–2000	Severe gastrointestinal damage; early transient incapacitation; gastrointestinal death		Death within 5–12 days
2000–3000	Gastrointestinal and cardiovascular damage		Death within 2–5 days

IV. MODIFYING FACTORS

While the effects manifest in the ARS depend primarily on the radiation dose received by an individual, other factors influence the severity and course of the illness. First, the time of onset and the severity of response during the prodromal phase vary widely, depending on the person's radiosensitivity. Hypersensitive individuals exhibit a low threshold and a severe reaction, while hyposensitive individuals exhibit a high threshold and a mild response. Gerstner (4) indicates that, for a dose of 200 cGy, the following distribution of severity of prodromal symptom emerges: completely asymptomatic, 20%; mild reaction not exceeding nausea, 20%; moderate reaction with vomiting and marked weakness, 50%; and severe reaction leading to profuse vomiting and prostration, 10%. While there is a significant difference in individual sensitivity for the prodromal phase, the same is not true for damage to the hemopoietic system or the gastrointestinal tract. The dramatic differences in individual sensitivity are typical of only the initial reactions, which allows few prognostic conclusions about the severity of the later clinical sequelae. The difference in sensitivity is pointed out in two observations. First, the Y-12 accident victim exposed to 339 cGy was completely asymptomatic during the first day after exposure, but he experienced other symptoms including severe bone-marrow depression several weeks later. Second, of the Hiroshima population classified into a dose group of which everyone had severe forms of the ARS, only 32% vomited during the prodromal phase of the illness.

Responses to whole-body irradiation differ from the responses to partial-body irradiation. From radiation therapy experience, substantial evidence indicates that the epigastric region is the most sensitive to radiation, with decreasing sensitivity for exposure toward the head and toward the thighs. The same dose that induces prodromal reactions in 50% of patients when applied to the abdomen, produces only 33% incidence in the thorax, 25% incidence in the head and neck, and no incidence in the extremities (4).

The age and sex of the individual have also been shown to significantly affect the incidence of radiation sickness. This can be seen in an analysis of the response of over 6000 individual 20-day survivors at Hiroshima in whom vomiting clearly decreased with progressing age (41). Similarly, females at Hiroshima were found to be more likely to experience emesis than males receiving the same radiation dose (41). While these studies did not address the issue, the results suggest that females and the young may be more likely to vomit as a psychosomatic response to the bombing's aftermath than would the older persons and the males at Hiroshima. Certainly, psychological trauma associated with being the victim of a nuclear bombing, especially for unwarned and inexperienced individuals, can contribute to the incidence and severity of prodromal symptoms. While these factors account for some of the differences in prodromal sensitivity to radiation, they do not explain the range of differences in observed reactions and do not address at all the variability in radiation-induced lethality.

Preexisting conditions at the time of irradiation have been shown to influence radiation lethality. Furchtgott (80) and others have pointed out that postirradiation survival is linked to several preexposure factors. Preirradiation stress or fatigue, decreased immune competence, existing infection, and vitamin deficiencies will decrease the postirradiation survival. In contrast, alcohol content, lowered body temperature, and anoxia may increase the survival. These factors can alter the prodromal and manifest phases of radiation sickness, but they do not alter the types of symptoms or the clinical sequelae that follow suprathreshold radiation doses.

Although variability exists in individual responses to a given radiation dose and although certain factors can modify the ARS somewhat, the medical and operational significance of the ARS for military personnel is not in doubt. Even nonlethal radiation doses may pose significant medical and operational problems, since those doses can be accompanied by prodromal symptoms, severe fatigability, and weakness, which can render a person incapable of any sustained activity involving physical exertion. Because of these considerations, coping with the medical and operational implications of the acute radiation syndrome will be one of the most significant problems to be faced if nuclear weapons are ever used again.

REFERENCES

1. Saenger, E. L. *Radiation Accident Preparedness: Medical and Managerial Aspects.* Science-Thru Media, New York, 1980.
2. Bond, V. P., Fliedner, T. M., and Archambeau, J. O. *Mammalian Radiation Lethality: A Disturbance in Cellular Kinetics.* Academic Press, New York, 1965.
3. Gerstner, H. B. Acute radiation syndrome in man. *U.S. Armed Forces Med. J.* **9,** 313 (1958).
4. Gerstner, H. B. Reaction to short term radiation in man. *Annu. Rev. Med.* **11,** 289–302 (1960).
5. Langham, W. H., ed. *Radiobiological Factors in Manned Space Flight,* Publ. 1487. Natl. Acad. Sci., Natl. Res. Counc., Washington, D.C., 1967.
6. Lushbaugh, C. C. *Recent Progress in Assessment of Human Resistance to Total-Body Irradiation,* Conf. Pap. 671135. Natl. Acad. Sci., Natl. Res. Counc., Washington, D.C., 1968.
7. Blakely, J. *The Care of Radiation Casualties.* Thomas, Springfield, Illinois, 1968.
8. Brucer, M. B. (comp.) *The Acute Radiation Syndrome: A Medical Report on the Y-12 Accident, June 16, 1958,* Rep. ORINS-25. U.S. At. Energy Comm., Washington, D.C., 1959.
9. Gilberti, M. V. The 1967 radiation accident near Pittsburgh, Pennsylvania, and a follow-up report. In *The Medical Basis for Radiation Accident Preparedness* (K. F. Hubner and S. A. Fry, eds.). Elsevier/North Holland, New York, 1980.
10. Ingram, M., Howland, J. W., and Hansen, C. L., Jr. Sequential manifestation of acute radiation injury vs. "acute radiation syndrome" stereotype. *Ann. N.Y. Acad. Sci.* **114,** 356–367 (1964).
11. Jammet, H. P. Treatment of victims of the zero-energy reactor accident at Vinca. In *Diagnosis and Treatment of Acute Radiation Injury.* IAEA/World Health Organ., Int. Doc. Ser., New York, 1961.
12. Jammet, H. P., *et al.* Etude de six cas d'irradiation totale aigue accidentale. *Rev. Fr. Etud. Clin. Biol.* **4,** 210–225 (1959).
13. Jammet, H. P., Gongora, R., Le Go, R., and Doley, M. J. Clinical and biological comparison of two acute accidental irradiations: Mol (1965) and Brescia (1975). In *The Medical Basis for Radiation Accident Preparedness* (K. F. Hubner and S. A. Fry, eds.). Elsevier/North-Holland, New York, 1980.
14. Martinez, R. G., *et al.* Observations on the accidental exposure of a family to a source of cobalt-60. *Rev. Med. Inst. Mex., Seguro Soc.* **3,** 14–68 (1964) (transl. by F. V. Comas).
15. McCandless, J. B. Accidental acute whole-body gamma irradiation of seven clinically well persons. *J. Am. Med. Assoc.* **192,** 85–88 (1965).
16. Pendic, B. The zero-energy reactor accident at Vinca. In *Diagnosis and Treatment of Acute Radiation Injury.* IAEA/World Health Organ. Int. Doc. Ser., New York, 1961.
17. Saenger, E. L., ed., *Medical Aspects of Radiation Accidents.* U.S. At. Energy Comm., Washington, D.C., 1963.

18. Shipman, T. L. Acute radiation death resulting from an accidental nuclear critical excursion. *J. Occup. Med.* **3**(3), 145–192 (1961).
19. Thoma, G. E., Jr., and Wald, N. The diagnosis and management of accidental radiation injury. *J. Occup. Med.* **1,** 421–447 (1959).
20. Vodopick, H., and Andrews, G. A. *The Clinical Effects of an Accidental Radiation Exposure.* Oak Ridge Associated Universities, Oak Ridge, Tennessee, 1973.
21. Vodopick, H., and Andrews, G. A. The University of Tennessee Comparative Animal Research Laboratory accident in 1971. In *The Medical Basis for Radiation Accident Preparedness* (K. F. Hubner and S. A. Fry, eds.). Elsevier/North-Holland, New York, 1980.
22. Wald, N., and Thoma, G. E., Jr. *Radiation Accidents: Medical Aspects of Neutron and Gamma-Ray Exposures,* Rep. ORNL-2748, Part B. Oak Ridge Natl. Lab., Oak Ridge, Tennessee, 1961.
23. Adelstein, S. J., and Dealy, J. B., Jr. Hematologic responses to human whole-body irradiation. *Am. J. Roentgenol., Radium Ther. Nucl. Med.* **93,** 927–934 (1965).
24. Anno, G. H., Brode, H. L., and Washton-Brown, R. *Initial Human Responses to Nuclear Radiation,* Note 477. Pacific-Sierra Research Corporation, Los Angeles, California, 1982.
25. Barret, A. Total body irradiation (TBI) before bone marrow transplantation in leukemia: A cooperative study from the European group for bone marrow transplantation. *Br. J. Radiol.* **55,** 562–567 (1982).
26. Bond, V. P., Fliedner, T. M., and Cronkite, E. P. Evaluation and management of heavily irradiated individuals. *J. Nucl. Med.* **1,** 221–238 (1960).
27. Court-Brown, W. M. Symptomatic disturbance after single therapeutic dose of X-rays. *Br. Med. J.* **1,** 802–805 (1953).
28. Court-Brown, W. M., and Doll, R. *Leukemia and Aplastic Anemia in Patients Irradiated for Ankylosing Spondylitis,* Spec. Rep. Ser. 1-50. Br. Med. Res. Counc., H. M. Stationery Office, London, 1957.
29. Court-Brown, W. M., and Doll, R. Mortality from cancer and other causes after radiotherapy for ankylosing spondylitis. *Br. Med. J.* **2,** 1327–1332 (1965).
30. Lushbaugh, C. C., *et al.* Clinical studies of radiation effects in man. *Radiat. Res., Suppl.* **1,** 398–412 (1967).
31. Miller, L. S., Fletcher, G. H., and Gerstner, H. B. Radiobiologic observations on cancer patients treated with whole-body X-radiation. *Radiat. Res.* **8,** 150–165 (1958).
32. Rider, W. D., and Hasselback, R. The symptomatic and hematological disturbance following total body radiation at 300-rad gamma-ray irradiation. In *Guidelines to Radiological Health,* pp. 139–144. U.S. Public Health Serv., Washington, D.C., 1968.
33. Saenger, E. L., *et al. Metabolic Changes in Humans Following Total Body Irradiation,* Rep. 1633. Defense Atomic Support Agency, Washington, D.C., 1964.
34. Salazar, O. M., *et al.* Systemic (half-body) radiation therapy: Response and toxicity. *J. Radiat. Oncol. Biol. Phys.* **4,** 937–950 (1978).
35. Thomas, E. D., *et al.,* Allogeneic marrow grafting for hematologic malignancy using HL-A matched donor–recipient sibling pairs. *Blood* **38**(3), 267–287 (1971).
36. Ishida, M., and Matsubayashi, I. *An Analysis of Early Mortality Rates Following the Atomic Bomb in Hiroshima,* Tech. Rep. 20-61. Atomic Bomb Casualty Commission, Hiroshima, 1948.
37. Ishimaru, T., *et al.* Leukemia in atomic bomb survivors, Hiroshima and Nagasaki, 1 October 1950–30 September 1966. *Radiat. Res.* **45,** 216–233 (1971).
38. Kumatori, T. Hematological effects on heavily irradiated Japanese fisherman. In *Biological Aspects of Radiation Protection* (T. Sugahara and O. Hug, eds.). Igaku Shoin, Tokyo, 1971.
39. Levin, S., *et al., Early Biological Effects From Initial Nuclear Radiation in Hiroshima and Nagasaki.* Armed Forces Radiobiol. Res. Inst., Bethesda, Maryland (in preparation).

40. Ohkita, T., II A review of thirty years study of Hiroshima and Nagasaki atomic bomb survivors. II. Biological effects. A. Acute effects. *J. Radiat. Res. (Tokyo)* **16,** Suppl., 49–66 (1975).
41. Oughterson, A. W., and Warren, S. *Medical Effects of the Atomic Bomb in Japan.* McGraw-Hill, New York, 1956.
42. Conard, R. A. *A Twenty-Year Review of Medical Findings in a Marshallese Population Accidentally Exposed to Radioactive Fallout,* Rep. BNL 50424 (TLD-4500). Brookhaven Natl. Lab., Long Island, New York, 1975.
43. Conard, R. A., *et al. Medical Survey of Rongelap People Eight Years After Exposure to Fallout,* Rep. BNL 780 (T-296). Brookhaven Natl. Lab., Long Island, New York, 1963.
44. Conard, R. A., *et al. Review of Medical Findings in a Marshallese Population Twenty-six Years after Accidental Exposure to Radioactive Fallout,* Rep. BNL 51261 (TLD-4500). Brookhaven Natl. Lab., Long Island, New York, 1980.
45. Cronkite, E. P., *et al.* Response of human beings accidentally exposed to significant fallout radiation. *J. Am. Med. Assoc.* **159,** 430–434 (1955).
46. Upton, A. C. Effects of radiation on man. *Annu. Rev. Nucl. Sci.* **18,** 495–528 (1968).
47. Wald, N., Thoma, G. E., and Broun, G. Hematologic manifestations of radiation exposure in man. *Prog. Hematol.* **3,** 1–5 (1962).
48. Warren, S. You, your patient, and radioactive fallout. *N. Engl. J. Med.* **266,** 1123–1125 (1962).
49. Gerstner, H. B. Practical implication of the initial reaction to penetrating ionizing radiation. U.S. Air Force School of Aerospace Medicine, Brooks Air Force Base, Texas, 1970 (unpublished).
50. Edsall, D. L., and Pemberton, R. The nature of the general toxic reaction following exposure to X-rays. *Am. J. Med. Sci.* **133,** 426–431 (1970).
51. Young, R. W. Mechanism and treatment of radiation-induced nausea and vomiting. In *Mechanisms and Treatment of Emesis* (D. Graham-Smith, *et al.*, eds.). Springer-Verlag, Berlin and New York, 1985.
52. Harding, R. K., McDonald, T. J., Hagenholtz, H., and Kucharczyk, J. Characteristics of a neo emesis-producing intestinal factor. *Physiologist* **27**(4), 279 (1984).
53. Patt, H. M., and Quastler, H. Radiation effects on cell renewal and related systems. *Physiol. Rev.* **43,** 357–396 (1963).
54. Bergonie, J., and Tribondeau, L. Interpretation de quelque résultat de la radiothérapie et essai de fixation d'une technique rationnùlle. *C. R. Hebd. Seances Acad. Sci.* **143,** 938–985 (1906).
55. Alpen, E. L. *Radiological Hazard Evaluation: A Critical Review of Present Concepts and a New Approach Thereto,* Rep. USNRDL-TR-186. U.S. Naval Radiol. Def. Lab., San Francisco, California, 1957.
56. Court-Brown, W. M., and Abbatt, J. D. The effect of a single dose of X-rays on the peripheral blood count of man. *Br. J. Haematol.* **1,** 75–85 (1955).
57. Dienstbier, Z., Arient, M., and Pospisil, J. Hamatologische Veranderung bei der Strahlenkrankheit-IV; Hamokoagulations-veranderungen. *Atompraxis* **9,** 189–194 (1963).
58. Liebow, A. A., Warren, S., and De Coursey, E. Pathology of atomic bomb casualties. *Am. J. Pathol.* **25,** 853 (1949).
59. McFarland, W., and Pearson, H. A. Hematological events as dosimeters in human total-body irradiation. *Radiology* **80,** 850–855 (1963).
60. McLean, A. S. Early adverse effects of radiation. *Br. Med. Bull.* **29,** 69–73 (1973).
61. Andrews, G. A. Total-body irradiation in the human being. *Int. Congr. Ser.—Excerpta Med.* **105,** 1583–1589 (1965).
62. Andrews, G. A., and Cloutier, R. J. Accidental acute radiation injury. *Arch. Environ. Health* **10,** 498–507 (1965).

63. Cronkite, E. P., Bond, V. P., and Dunhan, C. L. *Some Effects of Ionizing Radiation on Human Beings,* Rep. TID 5358. At. Energy Comm., Washington, D.C., 1956.
64. Zellmer, R. W., and Pickering, J. E. *Biological Effects of Nuclear Radiation in Primates,* Tech. Rep. 60-66. U.S. Air Force School of Aviation Medicine, Brooks Air Force Base, Texas, 1960.
65. Prasad, K. N. *Human Radiation Biology.* Harper & Row, Hagerstown, Maryland, 1974.
66. Warren, S. The early changes caused by radiation. *J. Mt. Sinai Hosp. (N.Y.)* **19,** 443–455 (1952).
67. Bond, V. P. Radiation mortality in different mammalian species. In *Comparative Cellular and Species Radiosensitivity* (V. P. Bond and T. Sugahara, eds.). Igaku Shoin, Tokyo, 1969.
68. Howland, J. W. Injury and recovery from ionizing radiation exposure. *Annu. Rev. Med.* **7,** 225–244 (1956).
69. Cronkite, E. P., and Bond, V. P. *Radiation Injury in Man.* Thomas, Springfield, Illinois, 1960.
70. Cronkite, E. P., and Bond, V. P. Acute radiation syndrome in man. *U.S. Armed Forces Med. J.* **9,** 313–324 (1958).
71. *Radiological Factors Affecting Decision-Making in a Nuclear Attack,* Rep. 42. National Council on Radiation Protection and Measurement, Bethesda, Maryland, 1974.
72. Haymaker, W., Lagner, G., Nanta, W. J. H., Pickering, J. E., Sloper, J. C., and Vogel, F. S. The effects of barium-140, lanthanum-140 (gamma) radiation on the central nervous system and pituitary gland of macaque monkeys. *J. Neuropathol. Exp. Neurol.* **17,** 12–57 (1958).
73. Vogel, F. S. Changes in the fine structure of cellebellar neurons following ionizing radiation. *J. Neuropathol. Exp. Neurol.* **18,** 580–589 (1959).
74. Wilson, S. G. Radiation induced central nervous system death: A study of the pathologic findings in monkeys inadeated with massive doses of cobalt-60 (gamma) radiation. *J. Neuropathol. Exp. Neurol.* **19,** 195–215 (1960).
75. Clemente, C. D., and Holst, E. A. Pathological changes in neurons, neuroglia and blood brain barrier induced by X-irradiation of heads of monkeys. *AMA Arch. Neurol. Psychiatry* **71,** 66–79 (1954).
76. Lushbaugh, C. C. Theoretical and practical aspects of models explaining "gastrointestinal death" and other lethal radiation syndromes. In *Comparative Cellular and Species Radiosensitivity* (V. P. Bond and T. Sugahara, eds.), pp. 288–297. Igaku Shoin, Tokyo, 1969.
77. Andrews, G. A., Hubner, K. F., Fry, S.A., Lushbaugh, C. C., and Littlefield, L. G. Report on 21-year medical follow-up on survivors of the Oak Ridge Y-12 accident. In *The Medical Basis for Radiation Accident Preparedness* (K. F. Huber and S. A. Fry, eds.), pp. 59–79. Elsevier/North-Holland, New York, 1979.
78. Lawrence, J. N. P. *Dosimetry Evaluation Handbook,* Internal Memorandum on Los Alamos Criticality Accidents, 1945–1946, Personnel Exposure, HI-78. Los Alamos Sci. Lab., Los Alamos, New Mexico, 1978.
79. Hempelmann, L. H., Lisco, H., and Hoffman, J. G. The acute radiation syndrome: A study of nine cases and a review of the problem. *Ann. Intern. Med.* **36,** 279–510 (1952).
80. Furchtgott, E. Behavioral effects of ionizing radiation. In *Pharmacological and Biophysical Agents and Behavior* (E. Furchtgott, ed.), pp. 1–64. Academic Press, New York, 1971.
81. Adams, R., and Cullen, S. *The Final Epidemic: Physicians and Scientists on Nuclear War.* Educational Foundation for Nuclear Science, Chicago, Illinois, 1981.
82. Cairnie, A. B., and Robitaille, H. A. *Arguments for the Greater Importance of the Prodromal Syndrome Than Incapacitation (Involving Early Transient Incapacitation) in the Consideration of Radiation Effects in Irradiated Military Personnel, Together With a Proposal to Simulate the Prodromal Effects Using Lithium Carbonate,* Rep. 836. (Canadian) Defense Research Establishment, Ottawa, 1980.

83. Ellinger, F., *et al.* A clinical study of radiation sickness. *Am. J. Roentgenol., Radium Ther. Nucl. Med.* **68,** 275–280 (1952).
84. Laumets, E. *Time History of Biological Response to Ionizing Radiation,* Rep. USNRDL-TR-905. U.S. Naval Radiobiol. Def. Lab., San Francisco, California, 1965.
85. Martin, E. J., and Rowland, R. H. *Castle Series, 1954,* Rep. DNA 6035F. Defense Nuclear Agency, Washington, D.C., 1982.
86. *Nuclear Weapons Employment Doctrine and Procedures,* FM 101-31-1. U.S. Army, Washington, D.C., 1977.
87. Saenger, E. L., *et al. Radiation Effects in Man: Manifestations and Therapeutic Efforts,* Rep. 2751. Defense Nuclear Agency, Washington, D.C., 1971.
88. Thoma, G. E., Jr., and Wald, N. The acute radiation syndrome in man. In *Epidemiology of Radiation Injury,* Postgraduate Course Syllabus. St. Louis University, School of Medicine, St. Louis, Missouri, 1961.
89. Zellmer, R. W. Human ability to perform after acute sublethal radiation. *Mil. Med.* **126,** 681–687, (1961).
90. Howland, J. W., Ingram, M., Mermagen, H., and Hansen, C. L., Jr. The lockport incident: Accidental partial body exposure of humans to large doses of X-irradiation. In *Diagnosis and Treatment of Acute Radiation Injury,* pp. 11–26. IAEA/World Health Organ., Int. Doc. Serv., New York, 1961.
91. *NATO Handbook on the Medical Aspects of NBC Defensive Operations,* Rep. AMED P-6. U.S. Departments of the Army, Navy, and Air Force, Washington, D.C., 1973.
92. R&D Associates. Collateral damage implications of low radiation dose criteria for battlefield nuclear operations. Unpublished manuscript (1980).
93. Storb, R. Total-body irradiation and marrow transplantation. *Transplant. Proc.* **9,** 1113–1119 (1977).
94. Upton, A. C. *Radiation Injury: Effects, Principles, and Perspectives.* Univ. of Chicago Press, Chicago, Illinois, 1969.
95. Conard, R. A. Acute myelogenous leukemias following fallout radiation exposure. *J. Am. Med. Assoc.* **232,** 1356–1357 (1975).
96. Rubin, P., and Casarett, G. W. *Clinical Radiation Pathology.* Saunders, Philadelphia, Pennsylvania, 1968.
97. Cronkite, E. P., *et al.* Diagnosis and therapy of acute radiation injury. In *Atomic Medicine,* 3rd ed., Chapter 2. Williams & Williams, Baltimore, Maryland, 1959.
98. Gerstner, H. B. Acute clinical effects of penetrating nuclear radiation. *J. Am. Med. Assoc.* **168,** 381–388 (1958).
99. Grant, G. A., *et al. A Predictive Study of the Incidence of Vomiting in Irradiated Military Personnel,* Rep. 817. (Canadian) Defense Research Establishment, Ottawa, 1979.
100. Yochmowitz, M. G., and Brown, G. C. Performance in a 12-hour 300-rad profile. *Aviat. Space Environ. Med.* **48,** 241–247 (1977).
101. Finch, S. C. Recognition of radiation-induced late bone marrow changes. *Ann. N.Y. Acad. Sci.* **145,** 748–754 (1967).
102. Messerschmidt, O. *Medical Procedures in a Nuclear Disaster.* Verlag Karl Thieming, Munich, 1979.
103. Lushbaugh, C. C. Reflections on some recent progress in human radiobiology. *Adv. Radiat. Biol.* **3,** 277–315 (1969).
104. Porvaznik, M. Tight junction disruption and recovery after sublethal gamma irradiation. *Radiat. Res.* **78,** 233–259 (1979).

CHAPTER 10

The Combined Injury Syndrome

GARY J. BOWERS*

Armed Forces Radiobiology Research Institute
Bethesda, Maryland 20814-5145

The modern proliferation of nuclear materials for civilian and military use has increased the risk of human injury from exposure to these inherently toxic substances. Transport vehicles carrying radioactive sources or nuclear weapons may become involved in unexpected mishaps, such as a train derailment, airplane crash, or motor vehicle accident, with subsequent radioactive contamination of the surrounding environment and nearby persons. Conceivably, a nuclear device(s) may be detonated by a terrorist group or by third-world nations involved in a regional conflict. Unfortunately, there is always the possibility of a nuclear world war, in which tactical and/or strategic weapons may be exchanged by the superpowers, with catastrophic results.

In these situations, the number of casualties could range from a few to millions. Our health care systems must be prepared to cope with this horrid prospect, although we know that a nuclear world war would essentially eliminate all ability to provide modern medical care. On the other hand, terrorist and third-world nuclear explosions would find many worldwide health care facilities (civilian or military) relatively intact, and would present a large number of victims for care. Modern Med-Evac capabilities would enable the evacuation of these injured and irradiated persons out of the zone of destruction and their transport to widely distributed hospitals throughout the world.

In many situations involving radiation accidents and/or explosion of a nuclear device, a large proportion of the injured will have a variety of injuries, including blunt or penetrating trauma, crush injuries, fractures, burns, or any combination of these in addition to radiation injury. Some estimates place them at 65–70% of the expected casualties (1). For such clinical conditions, the term "combined injury" [which evolved in the early German literature (2)] is used to designate the combination of conventional trauma and radiation injury. The clinicopatho-

*Present address: Department of Surgery, Boston University Medical Center, Boston, Massachusetts 02118.

logical features resulting from this are here referred to as the combined-injury syndrome (CIS).

A large volume of morbidity and mortality statistics, derived mostly from animal models, documents the various harmful effects of radiation. Some of these data suggest that the combination of conventional trauma and radiation injury has adverse synergistic effects on the victim. This evidence predicts that the resulting morbidity and mortality associated with the CIS will be greater than the sum of both injuries (3–8). So combined injury is a potentially severe and lethal form of multiple trauma. To understand this, one must be familiar with the many effects of radiation on human systems, and how these effects influence the management of trauma victims.

I. PATHOPHYSIOLOGY

Much of the lethality associated with CIS is caused by the physiologic derangements due to acute radiation damage (described in the preceding chapters). Radiation injury is manifest primarily in those organs whose structural and/or functional integrity depends on a large pool of actively dividing stem cells. Therefore, organs such as the bone marrow, gastrointestinal tract, skin, and gonads are said to be more sensitive to the effects of radiation than are those composed of more mature and stable cell populations such as found within most of the parenchymal organs. Even so, the principal adverse effects of radiation on these radiation-sensitive organs, for the most part, are not immediately seen; they develop over the days to weeks following the accident. In CIS, the initial resuscitation and necessary surgical treatment of the victim's wounds will have been dealt with before many of these problems arise. But many of the early management decisions must be weighed in light of the knowledge that the individual has been irradiated and that subsequent wound healing will be delayed.

In addition to interfering with cellular replication, radiation exerts many of its injurious effects by inducing widespread inflammatory responses. Inflammation has local and systemic components that are mediated by the release of vasoactive and inflammatory substances. In excess, these agents are swept into the systemic circulation, where they can influence distant targets. Although they evolve as defense mechanisms, many of these reactions can be detrimental, especially when stimulated in excess or allowed to proceed without appropriate regulation. Some of the systemic manifestations are quite dramatic, such as anaphylaxis. Also, "bystander" injury to normal surrounding tissue frequently occurs in inflammatory reactions. These inflammatory events are generally seen early, and they may account for many of the early prodromal signs and symptoms of acute radiation exposure and other forms of trauma. Nevertheless, many of these early events may have prolonged sequelae, which can result in significant and diffuse compromise of organs. One such sequela is development of the adult respiratory distress syndrome (ARDS) following traumatic injury, blood transfusions, anesthesia, or sepsis. Much of the pathophysiology of radiation-induced toxicity may result from the actions of an inflammatory mediator such as the various ecosinoids, kinins, oxygen radicals, and complement factors.

The CIS is further complicated by a sequence of events initiated by conventional trauma. The primary effect of trauma on the host is the physical disruption of tissues and organs, which compromises the structural and functional integrities of the affected systems. In addition, trauma, like radiation, unleashes a cascade of secondary inflammatory and neurohormonal events that have systemic ramifications. Many of these events may produce signs and symptoms that mimic or obscure those that are attributable to acute radiation intoxication. Some of these too may be detrimental to the patient. Stress ulceration, bleeding diatheses, metabolic disturbances, hypoxemia, etc., frequently complicate the recovery course of trauma victims. Furthermore, the immunocompetency of an injured person is often jeopardized by multiple trauma.

Following any major trauma, a catabolic state develops. This is in part a response to the activation of repair mechanisms, but it is also a consequence of the loss of body constituents via the wound. In the prolonged absence of a normal energy input, these metabolic consequences of trauma can rapidly exhaust the energy stores needed for continuation of the repair processes. Also, the functional status of other nontraumatized organs may begin to deteriorate. The result is a downward spiral, yielding to multiple-organ failure and eventually death.

Superimposed on the systemic manifestations of trauma are the consequences of radiation. The combined effect of these two insults on the host's morbidity and mortality appears to be synergistic. Many studies in a variety of different animal models have documented the enhanced killing effect of combined injury. The nonlethal irradiation of animals subjected to experimental wounds varying from low to moderate lethality will invariably produce death in almost all subjects. This holds for thermal burns (3, 6–9), soft-tissue trauma (5, 6, 10, 11), and varying degrees of hemorrhagic shock (12).

The temporal relationship of the two insults, radiation and trauma, does influence the subsequent mortality (3, 5, 8, 10, 11, 13, 14). Synergism occurs when the two occur together, and it increases when the injury follows the radiation insult. This effect peaks when wounding occurs during that period of time when the radiation effects on the bone marrow are most prominent; then the effect declines as the marrow begins to recover its function (5, 9, 10). Surprisingly, in some animal series, mortality appears to be less when a small injury occurs before the radiation insult (3, 5, 8, 14, 15). This "protective" effect of prior trauma generally occurs when the injury is delivered 3–4 days before the radiation. The protective effect decreases as the time span between the two insults either increases or approaches concurrent injury. This effect appears to correlate with the fact that, at the time of irradiation, the marrow and reticuloendothelial system are at a higher state of activation. These protective effects of trauma generally hold for minor injury and diminish as the severity of the traumatic insult increases. Large burns (greater than 30% body-surface area), grade III or IV hemorrhagic shock, or major crush injuries involving the pelvis or an entire extremity, because of their magnitude and severity, are not expected to result in any survival advantage for anyone receiving those injuries and concurrent radiation. Because of the added radiation injury, such injuries, which otherwise might be survivable, will probably be uniformly fatal despite vigorous medical therapy. Of course, no protection can be expected in individuals who sustain very large doses of radiation, regardless of the extent of traumatic injury or its temporal relation to the irradiation.

Resuscitation and cardiovascular stabilization are important for the survival of a multiple-trauma victim. In patients with CIS, various degrees of cardiovascular instability will exist. Conventional trauma of any significance is nearly always associated with a hypovolemic state. Following burn, for instance, hypovolemia results from the copious loss of proteinaceous fluids through the burn wounds. These patients require large volumes of fluids to compensate for their losses. A variety of treatment schemes have been recommended for the proper fluid resuscitation of burn victims, but the bottom line for all is adequate fluid replacement. Traumatic injuries, on the other hand, generally produce hemorrhage. Treatment of hemorrhage requires the quick and satisfactory control of bleeding along with the rapid replenishment of lost blood with fluids and/or blood products, depending on the severity of the hemorrhage. Only rarely is hypotension in a multiple-trauma patient associated with anything other than hypovolemic shock. For this reason, fluids and not pressor agents are indicated for the proper management of trauma. The aggressiveness of the fluid resuscitation is predicated on clinical parameters, including blood pressure, pulse, mental status, urine output, skin color, and presence of obvious blood losses.

Acute radiation in doses exceeding 1000 cGy may also produce a prodromal hypotensive state. This hypotension results from significantly lowered systemic vascular resistance and excessive transudation of fluids from damaged microcapillary beds. The hypotension may be superimposed on the acute fluid losses resulting from the traumatic injury. Various vasoactive inflammatory substances, including histamine, have been implicated in the mediation of radiogenic shock (16, 17). But this form of shock tends to resist pressor therapy (16, 18, 19), presumably because of primary disturbances at the level of the capillary sphincters by as-yet-undefined mechanisms. Large volumes of fluids are required to support these patients. Even so, after lethal irradiation (doses greater than 1000 cGy), massive infusions of crystalloids still fail to adequately resuscitate them.

The patients remain in refractory shock and eventually succumb to cardiopulmonary arrest, usually within 36–48 hr of injury.

Only with proper hemodynamic resuscitation and stabilization, with restoration of satisfactory tissue perfusion, can multiple-trauma victims be expected to enter any kind of recovery phase. It is the inability to control fluid losses from the vascular spaces that kills during the initial few hours after injury, be it from massive trauma or lethal-to-supralethal doses of radiation.

Following stabilization, the combined-injury patient should enter a recovery phase. In general, the recovering patient is one whose traumatic injury was amenable to surgical correction and/or repair and whose radiation exposure was below outright lethal thresholds. The recovery period in such a patient may be stormy and involve many complications, particularly in the more severe trauma or higher dose of radiation. Speaking of radiation alone, the higher the dose, the more extensive the injury and the more organ systems that will be involved. Hence, following very high levels of exposure, patients may exhibit a variety of central nervous system (CNS), cardiopulmonary, gastrointestinal (GI), and hematopoietic symptomatology, whereas only hematopoietic dysfunction will be seen at the lower doses.

Ultimately, recovery in combined-injury patients will depend on the victim's capacity to repair his/her injuries and defend his/her internal milieu from infectious organisms. Radiation and conventional trauma can cause physiological disturbances that impair an individual's capacity to recover from injury. The nature and extent of the individual's wounds as well as the degree of radiation-induced hematopoietic and GI injuries are particularly important variables that influence survival in CIS. Deficits in wound healing, hematopoiesis, and GI function (all prominent features of CIS) have pivotal roles in prolonging and profoundly complicating the recovery of these patients. Death in these patients who survive beyond the initial 24–48 hr is generally related to problems associated with these three systems.

II. HEMATOPOIETIC–IMMUNE DYSFUNCTION

The degree of hematopoietic dysfunction following combined injury appears to be the critical determinant of survival. Normal metabolic processes, hemostasis, and immunocompetency are each vitally dependent on cellular elements derived directly or indirectly from the marrow.

The CIS imparts a protean insult to hematopoiesis. Trauma depletes the circulating blood volume and places enormous demands on the splenic, marrow, and venous reserves. Radiation, on the other hand, disturbs stem cell replication and therefore impedes the marrow's ability to replenish its stores or meet new demands. Both insults initiate widespread inflammatory responses that require and consume a large number of marrow-derived elements. The combination of these insults, if large enough, will severely compromise hematopoietic function and its subservient systems. It is not surprising, therefore, that the manner in which both experimental animals and patients die is sepsis (described further in the next chapter).

Besides compromising the oxygen transport and clotting mechanisms, radiation imposes profound immunosuppression on its victims. In brief, immunocompetency relies on the presence and functional capabilities of a variety of leukocyte subpopulations, which express varying sensitivities to radiation. Once the structural barriers are breached by injury (i.e., the epidermal or mucous membranes), these cells are mobilized to deal with foreign substances and invading organisms. Classically, inflammation evokes the early recruitment of circulating polymorphonuclear leukocytes to the site of injury or infection. This is followed by an infiltration of mononuclear cells. Activated tissue macrophages will phagocytize the debris and then process many of the foreign antigens. Besides clearing the wound and circulation of these agents, the macrophage presents to T cells the processed portions of the antigen, which results in the proliferation of immunocompetent T cells. These in turn activate and modulate B-cell maturation, which results in the expression of immunoglobulins by mature B or plasma cells. Many of the immunologically important steps in antigen recognition and lymphocyte proliferation occur within regional lymph nodes and various other lymphoid tissues throughout the body, all of which are sensitive to the effects of radiation.

Ultimately, all of the immunologically important cells are derived from either the bone marrow precursors or progenitors within the lymphoid tissues. Radiation injury to the marrow and lymphoid tissues, such as the thymus and spleen, can thus be expected to significantly impair the overall functionality of the immune system. Because of this, irradiated hosts that are challenged with an infection have a limited capacity to effectively combat the pathogens; this is an immensely important limitation when confronting, for example, fecal peritonitis after traumatic disruption of the large bowel. The extent of this dysfunction is generally proportional to the dose of radiation absorbed by these immunologically important tissues.

Immune and hematopoietic dysfunctions are also features of many of the conventional forms of trauma, including burn, crush, and hemorrhage. If the insult is severe enough, impairments of the marrow and the reticuloendothelial system (RES) may develop. The factors responsible for these immunosuppressive effects of traumatic injury are not fully understood.

Most forms of conventional trauma are associated with some degree of hemorrhage or fluid loss, resulting in a hypovolemic state. Hypovolemic shock of various etiologies brings about the phenomenon of RES blockade (20–26). Traumatized and/or ischemic tissues appear to release factors into the circulation that have inhibitory effects on the immunocompetency of the individual. One such factor, although poorly defined, has been isolated from the portal blood of animals subjected to hemorrhagic and intestinal ischemic shock (21, 22). In a burn model, stroma hemoglobin has been shown to decrease the RES ability to clear intravenously injected carbon particles (27). It appears that this substance is produced at the injury site in amounts sufficient to generate the blood levels required to elicit systemic effects (27). The conclusions from these and similar studies (20, 23, 28) are that conventional trauma has the capacity to generate factors that produce functional deficits in the mononuclear phagocytic cells that comprise the RES. These cells are vitally important for monitoring the circulation, and they serve to eliminate toxins, foreign antigens, immune complexes, organisms, and debris from the blood. A defect in this elimination function, or a blockade, has been shown in several animal models to increase the mortality associated with any given traumatic event (24–26).

Pancytopenia frequently complicates the clinical course of multiple-trauma patients. At the time of injury, hemorrhage expends peripheral blood elements. The size of this loss is proportional to the amount of blood lost by the circulation. Clotting systems are activated, which in themselves consume red cells in addition to platelets. Splenic and marrow reserves are mobilized to restore this deficit. Also, neurohormonal mechanisms are activated to vasoconstrict the affected regions in order to stem the loss of blood. In conjunction with this, the circulating blood volume is contracted to maintain the flow to vital areas. Following mild injury, such mechanisms can quickly stabilize and recover the peripheral circulation. But such compensatory mechanisms are not adequate for most forms of major trauma. Severe trauma will greatly diminish—if not completely exhaust—the peripheral blood pool and all readily available splenic, marrow, and venous reserves. Nevertheless, if the patient is properly resuscitated with exogenous fluids and blood components, a recovery phase will ensue, and during this time the body will attempt to restore the blood volume.

During the recovery phase after any form of hemorrhage, the marrow continues to receive signals to enhance the maturation of the preformed progenitors and to add these to the peripheral circulation. A differential examination of the patient's peripheral blood smear may reveal a variety of immature forms, including bands and nucleated red blood cells. This activity will persist until the peripheral pools are replenished. Following major trauma, the demands on the marrow to provide cellular elements for oxygen transport, inflammation, clotting, and repair are great, and the demands may continue for a prolonged period of time. The pool of preformed progenitors may itself become exhausted as a result of these unrelenting demands, despite substantial initial reserve. Until these depleted stores are replenished to a critical level by stem cell mitosis, the maturation processes will be slowed. This will later decrease the release of mature elements into the peripheral circulation. Anything that hinders the repletion of this progenitor pool will perpetuate a state of peripheral pancytopenia. Further demands on the peripheral pools by renewed injury or invasion by pathogenic organisms may worsen the situation. It may be that such demands are ultimately not met.

Substrate and/or energy deficits will impede the recovery of the progenitor pool. Such energy deficiencies frequently arise in multiple-trauma patients, especially those who experience protracted recovery periods as a result of their primary injury or the complications of it. For a variety of reasons, nutrient intake may be restricted in many of these patients due to the nature of their injury. Thus, a prolonged state of negative nitrogen balance will impair hematopoiesis. The same applies to combined-injury victims. But in addition, the ability of the stem cells to repopulate the progenitor stores will be directly compromised by the direct and indirect effects of radiation on these cells. Thus the peripheral and central pancytopenia of CIS should be more severe and prolonged, and should develop earlier than with either radiation or conventional injury alone. This will immensely influence the patient's ability to recover.

The symptoms associated with hematopoietic suppression of any etiology are as expected. The combined effect of radiation and trauma on hematopoiesis may exaggerate these symptoms. Besides immunosuppression with concomitant problems of infection, thrombocytopenia and anemia can be expected. Thrombocytopenia will predispose the patient to varying degrees of bleeding complications. Petechias and purpura within the soft tissues, hemarthrosis, and GI bleeding may develop. Low platelet counts may also jeopardize surgical repairs. Hematomatous extravasation between reapproximated wound edges can disrupt suture lines. Raw surface areas resulting from the primary injury and/or surgical therapy may continue to ooze blood in the presence of a coagulopathy. In general, these problems should not be too severe, as long as the peripheral platelet count remains above 50,000 (which may require intermittent platelet transfusion).

Thus, trauma-related blood losses and induced coagulopathies together with a radiation-induced decrease in erythropoiesis will contribute to unrelenting anemia. This may be even more pronounced and protracted than that seen after either injury alone (29). This anemia may produce various cardiopulmonary abnormalities, including flow murmurs, arrhythmias, and low oxygen-carrying capabilities. Decreased oxygen delivery to tissues will compromise cellular respirations and will contribute to a variety of metabolic disturbances, including persistent acidemia.

In summary, CIS profoundly compromises normal hematopoietic function. Fortunately, acute losses from the peripheral pools can be transiently replaced with exogenous blood products. During the later recovery period, the patient may require additional transfusions of platelets and erythrocytes. However, true recovery will ultimately depend on the ability of the stem cells to repopulate the marrow. If a sufficient reserve of stem cells survives and is provided with an adequate substrate base and energy supply, then marrow reconstitution will occur. If the marrow does reconstitute and infectious complications are avoided, the patient is expected to regain a normal hematologic profile, although it may be months before the peripheral and marrow indices fully recover. On the other hand, if hematopoietic function is so extensively altered by the combined stresses of radiation and trauma that the number of functional stem cells falls below the critical number necessary for repopulation of the marrow, then effective hematopoiesis will not recover.

III. GASTROINTESTINAL DYSFUNCTION

Any major trauma affects the bowel. If the abdominal cavity is directly involved, the severity of the resulting bowel disturbances will be even more pronounced. The functional integrity of the GI tract is altered by both acute radiation exposure and trauma. Diffuse motility disorders are common to both. Direct injury to the mucosal and submucosal layers tends to be regional after traumatic injury. Radiation injury, on the other hand, produces diffuse damage that involves all layers and encompasses the entire length of the bowel, particularly the small intestine.

Vomiting often accompanies many forms of injury. At times it may be difficult to distinguish this vomiting from that often arising as part of an individual's emotional response to stress or injury. Nonetheless, vomiting is characteristically an early feature of acute radiation intoxication, and its onset and severity are related to the dose of radiation (30, 31). (This is explored in later chapters.) However, it is important to recognize that trauma in addition to radiation can produce vomiting. Because of this, triaging cases of suspected radiation injury will be very difficult when the cases are associated with trauma and when reliable dosimetry is absent.

Following whole-body irradiation, many patients also experience severe abdominal cramping followed by watery, often explosive, diarrhea. These events are generally associated with higher doses of radiation. As with vomiting, the onset of diarrhea is related to the radiation dose: the higher the absorbed dose, the earlier the manifestations, and consequently, the more guarded the prognosis. With time, hypermotility may yield to an ileus. Abdominal distention secondary to aerophagia and increased accumulation of fluids within the gut will develop. Patients may again begin to vomit. Pharmacologic agents have no role in the management of such vomiting; nasogastric intubation is the preferred treatment.

Nasogastric decompression will generally be the treatment of choice in CIS. Many combined-injury patients will require early nasogastric intubation as a preoperative maneuver. Victims of head trauma or blunt and penetrating thoracicoabdominal trauma frequently develop a posttraumatic ileus. Postoperative patients, particularly those who have undergone abdominal or retroperitoneal procedures, all develop an ileus. In any of the above, the likelihood of aspirating gastric contents is a real risk. A nasogastric tube will decrease this risk as well as decompress a distended upper intestinal tract. Further, this approach eliminates drug-induced changes in the mental status, which is important when head injury is suspected.

Nasogastric intubation of the second portion of the duodenum may aspirate enough bile and pancreatic secretions to afford some protection to the more distal bowel. Without an intact epithelium, the underlying basement membrane is exposed to the lumenal contents, including bile and pancreatic juices. Both trauma and radiation generate erosive mucosal changes, with radiation generally inducing injury that is more widespread and more severe. Bile and pancreatic enzymes are thought to potentiate radiation damage to the bowel, with a subsequent increase in mortality. In some irradiated animal models, diversion of the common bile and/or pancreatic ducts has improved survival (32, 33). But such maneuvers in humans are not feasible or technically practical in the absence of direct traumatic injury to these ducts. Binding agents such as cholestyramine may be beneficial in theory, but in reality they may not be practical or as thoroughly effective, because of motility disturbances that generally accompany these injuries.

Radiation enteropathy has many gross and microscopic features that are germane to the CIS (34). Grossly, the bowel becomes edematous, hyperemic, distended, and hemorrhagic. The glycocalyx is lost. Mucosal changes occur, ranging from punctate ulcerations to complete denudation of the mucosa. Many of these same mucosal changes are seen with the acute hemorrhagic enteropathy that accompanies other forms of hypovolemic shock (35). In any case, a hemorrhagic and ulcerated mucosal surface will extravasate blood and proteins into the gut lumen. Such bleeding (1) will aggravate the anemia associated with hematopoietic effects of radiation and (2) may potentially precipitate some degree of hepatic encephalopathy, especially in those individuals with underlying hepatic insufficiency or radiation-induced hepatitis. Included in the problems associated with hepatic encephalopathy are protean CNS dysfunctions. Loss of proteins into the lumen will also stress the individual's ability to maintain an adequate nitrogen balance. These losses will further accelerate the skeletal muscle proteolysis, lower the plasma oncotic pressures, and attenuate the levels of circulating immunoglobulins, primarily IgG.

In addition to mucosal injury, vascular changes occur after irradiation and/or trauma, which result in microperfusion abnormalities. These result either as a direct insult to the microvascular beds or, more likely, as a secondary effect following the release of vasactive substances by damaged splanchnic tissue and/or inflammatory cells. Vessels become irregularly dilated and tortuous, and develop microhemorrhages (36). Transcapillary leakage of fluids results in interstitial edema throughout all tissue layers of the bowel. This edema may contribute to propulsional, microprofusional, metabolic, and other functional disturbances within the irradiated gut.

Along with the structural abnormalities, functional disturbances arise. Radiation and trauma both perturb normal transport functions of the epithelium (32, 37–41). Disturbances occur within membrane-bound ionic exchange pumps, which may account for some of the earliest functional disturbances seen in an irradiated bowel (38,39). This is particularly true of sodium transport.

Many of the pathophysiologic changes in the irradiated and/or traumatized bowel result in accumulations of fluids and electrolytes within the gut lumen. An irradiated bowel loses its ability to absorb sodium (34,38,39), and it may in fact begin to excrete sodium (38,39). This, in combination with motility disturbances, can produce significant pooling of fluids and critical electrolytes within the gut. Similarly, fluids and electrolytes usually accumulate within the gut after many forms of conventional trauma. Such shifts can have pronounced effects on the overall homeostatic balance of the host. Within the gut, these fluids and electrolytes are lost to the circulation and may never be recovered, especially if the victim experiences vomiting and/or diarrhea and is subjected to nasogastric suctioning. If the fluids and electrolytes are not adequately replaced, profound and life-threatening electrolyte disturbances can arise. Continual GI loss of fluids will contribute to a secondary shock state and subsequent renal failure. Such fluid shifts will only compound the problem of maintaining hemodynamic stability in patients after trauma. These fluid and electrolyte losses must be taken into account when prescribing maintenance fluid and electrolyte therapy for CIS patients.

Overall, GI injury associated with combined injury has profound and protean effects on an individual who is already severely compromised by wounds and hematopoietic dysfunction. This injury will limit the acquisition of new energy and substrates, and also render the victim more susceptible to bacteremia of enteric origins.

IV. CHANGES IN ENDOGENOUS FLORA

During times of severe body stress (i.e., following trauma, surgery, burn, or radiation), the growth dynamics of the enteric flora may change. The gut normally harbors a vast array of microbes, including beneficial species and potentially harmful species. Proliferation of the latter is generally kept in check by various factors within the gut microenvironment. These pathogens are normally not associated with any significant illness or pathology. But chronic disease, acute injury, prolonged stress, diet, and various drugs may disturb this equilibrium, and favor the proliferation of opportunistic pathogens. For example, the majority of the bacteremia associated with trauma, stress, burn, or immunosuppression derives from the enteric flora. *Pseudomonas, Escherichia coli,* and gram-negative anaerobes are particularly prevalent in the gut and generally do not cause a healthy person any problem; but they are frequently seen in the cultured blood of patients with severe injury.

An individual with an irradiated bowel is particularly prone to infectious complications arising from the enteric flora (42–44). Radiation adversely affects all the enteric bacterial species (discussed in the next chapter). Gut lymphoid tissues are markedly depleted. Hypomotility increases the transit time, which in turn favors the local accumulation of toxins and bacteria. These agents then have easier access to the host, in which they are disseminated by the portal and lymphatic circulations. As a result of enteric derived bacteremia (45), microabscesses arise within the parenchyma of the liver, spleen, and kidneys, and are invariably culture-positive for enteric organisms (46).

V. METABOLIC REQUIREMENTS

If the radiation dose is sufficiently elevated, malnutrition may arise in irradiated individuals due to pathological changes within the bowel. These nutritional disturbances will persist until normal absorptive functions resume, even if the patient is able to tolerate an oral diet. Normal gut function can be expected to resume only after the gut epithelium is restored to an intact, normal mucosa. To what extent nutritional impairment develops will depend on the degree of injury to the gut. Such injury may be either marginal or extensive.

Following any major injury, patients will generally be in a profound catabolic state. The input of exogenous substrates may be profoundly restricted, as in patients with extensive maxillofacial trauma or a paralytic ileus from a variety of etiologies. Without adequate input, energy reserves will be rapidly depleted by the increased metabolic demands accompanying injury. Following trauma, immediate energy demands are met by the mobilization of skeletal glycogen stores and muscle proteins. The latter provides substrates for gluconeogenesis to proceed within the liver. As part of the endocrine response to trauma, levels of insulin decrease while levels of epinephrine and norepinephrine increase. These hormonal events favor catabolism. If these processes are prolonged, profound wasting of mucles may occur, including intercostal muscles. Atrophy of these muscles by the catabolic energy demands will eventually compromise respiratory function. To reverse this problem, aggressive nutritional therapy is warranted once initial stabilization has been achieved. In both elective surgical and trauma patients, the success of operative procedures, wound healing, recovery, and septic complications are related to the nutritional status of the individual. One can assume that the same is true for CIS.

Because of the extensive nature of their injuries, the nutritional needs of combined-injury patients may be even more pronounced, perhaps greater than 3000 cal/day. If enteral nutrition is feasible, either by mouth or a feeding catheter, a high-calorie, high-protein diet should begin early in the patient's hospital course. Unfortunately, this may not be possible in a significant number of combined-injury patients, especially those with significant radiation enteropathy or traumatic abdominal injury. In these individuals, the early institution of total parenteral nutrition (hyperalimentation) should be considered. One cannot afford to wait for the return of GI function before nutritionally supplementing these patients, especially since the irradiated bowel may require a longer recovery time than the bowel after the usual posttraumatic or postoperative ileus. Most of the major problems associated with this kind of nutritional support (i.e., fluid and electrolyte abnormalities, catheter insertion, line sepsis, etc.) can be satisfactorily controlled if scrupulous attention is paid to the details of care and management in this technique.

As long as the patient remains hypermetabolic and continues to leak plasma proteins into the interstitial tissue, gut lumen, or external environment, maintaining a satisfactory nitrogen balance will be difficult. Serum albumin levels will fall and remain low. Nutritional therapy must be designed to provide necessary substrates for meeting metabolic needs. In doing so, one hopes to spare the body's endogenous energy sources, promote recovery, and provide resources for combating infection.

VI. HEALING OF TRAUMATIC WOUNDS

Wounds associated with the CIS may range from simple cuts, contusions, and abrasions to extensive burns, crush, and degloving injuries. All wounds involve a break in continuity of the host's epidermal barrier, which exposes internal compartments to the effects of the external world. Because of the pervasiveness of microbes in our environment, wounds quickly become colonized by an array of microbiological organisms. Such an organism generally arises from the normal skin flora, or it may be a contaminant on the penetrating projectile, or it may arise rom secondary contamination of the wound by soil. A wound is an abnormal port of entry through which pathogens, toxins, and foreign matter gain easy access to the body. Also, a wound allows the egress of body constituents (i.e., fluids, electrolytes, and proteins) to the external environment. This places another metabolic burden on the host to replace these lost substances while also attempting to provide new substrates for repairing the injury.

Experience has shown that radiation impairs the normal healing of soft-tissue injuries (5, 47–51). A delay in or even an arrest of the normal repair processes occurs. Wounds that seem to be minor and not life threatening often become major problems. Japanese survivors of the atomic attacks on Hiroshima and Nagasaki during the 1940s who suffered penetrating wounds and/or thermal burns experienced this phenomenon (52). Granulation beds in these patients were seen to stop proliferating and even regress. Recurrent bleeding from persistently open wounds often occurred. Infections were rampant. In those who survived and finally did heal, significant keloid formation was often seen. Alterations in normal wound healing have similarly been observed in numerous animal and clinical models after experimental and therapeutic irradiations.

Alterations in the wound-repairing mechanisms after acute exposure to radiation may be correlated with both the degree of bone marrow injury and unchecked bacterial growth in the wound. Normal wound healing is a dynamic process that involves a rapid turnover of many marrow-derived cells. Platelet and red cells form clots, which stem the loss of fluids. Polymorphonuclear leukocytes and mononuclear cells infiltrate and eliminate debris and foreign matter, including most microbes. Fibroblasts proliferate, deposit collagen fibers, and stimulate wound contracture and reepithelialization. The availability of all these cells depends on an active bone marrow.

Normally, injury results in the release of colony-stimulating factors (CSF) that enhance the maturation of preformed progenitors and their subsequent release into the circulation. These cells are actively recruited to the site of injury. Radiation injury to the bone marrow with subsequent hematopoietic suppression interferes with the supply of cells for wound repair. Once the readily available supply is exhausted, further repair is arrested. Such wounds remain open, and they continue to lose fluids and proteins and also provide access to bacteria.

All wounds are quickly contaminated with bacteria. The milieu of a wound is an excellent broth for the growth and proliferation of bacteria. This process is accompanied by the release of toxins and enzymes by the bacteria, which facilitate the dissemination of the bacteria into the surrounding soft tissues. An otherwise healthy individual is well equipped with a variety of mechanisms to cope with these organisms and their metabolites. But massive injury may overwhelm those capabilities. Prior disease may impede the proper processing of bacteria. Immunosuppression limits the host's ability to effectively mount any resistance to these pathogens.

Leukopenia is a hallmark of CIS. Lymph nodes and lymphoid tissues are vastly depleted of the important cell mediators of immunity. Functional deficiencies within the RES appear. In essence, CIS results in widespread immunosuppression and interferes with nearly every aspect of the body's defense capabilities, including local inflammatory reactions (47). So invading wound bacteria meet little resistance to their growth and proliferation. Their toxins and metabolic wastes accumulate locally, which may further impair wound healing. The process is analogous to a full-thickness burn which, because of thermal coagulation of blood vessels, is isolated from circulating immune mechanisms. Thus leukocytes, immunoglobulins, and systemic antibiotics fail to reach the eschar. Bacterial proliferation proceeds unabated, generating a wound abscess. But unlike the thermal model, bacteria and toxins in a combined-injury wound are relatively unimpeded as they enter the lymphatics and venous drainage. In dogs, for instance, both gamma and beta hemolytic streptococci were isolated in the blood of 75% of the wounded and irradiated animals (8). Others have observed similar findings (53). Also, CIS victims tend to harbor higher bacterial counts than do their nonirradiated counterparts (54). So wound sepsis is a likely consequence in these immune suppressed individuals. Defects in granulation formation will delay the closure and subsequent reepithelialization of the wound. Until the epithelial barrier is restored, the wound will continue to harbor high numbers of bacteria and will continue to serve as a nidus for seeding infectious organisms into the circulation.

The importance of bacterial colonization in influencing the morbidity and mortality associated with combined-injury wounds can be demonstrated in germ-free rodents (55). Gnotobiotic rodents, when subjected to traumatic injury and subsequent radiation, have a lower mortality rate than their normal counterparts who harbor the usual commensal organisms. Furthermore, these animals do heal their wounds, although some delay occurs, likely due to hematopoietic suppression. These studies indicate that if pathogens are eliminated or at least controlled, then combined-injury victims should survive and heal their injuries.

VII. MANAGEMENT OF WOUNDS

The initial approach to the treatment of victims of CIS does not differ from that for any multiple-trauma patient. The basics of trauma management must prevail. Following resuscitation and stabilization of the patient, one can make a full assessment of the situation. This entails a complete inquiry into the nature of the accident and a thorough physical examination of the patient, with appropriate laboratory and radiographic studies.

Assessment of the extent and nature of the radiation exposure is important for anticipating future clinical events, particularly in the case of accidents. One must discern if the victim was indeed exposed to clinically significant levels of radiation and if body contamination with radioactive materials did occur. Few, if any, patients will have dosimeters in their possession. Even so, a dosimeter will generally not provide any descriptive information on the nature of the radiation beam. The exact amount and type of radiation energies involved in the accident may never be known, but they are critically important variables in determining the extent of radiation injury. With the assistance of radiation safety personnel, a crude estimate can be derived from the nature of the source and the spatial relationship between the victim and the source at the time of exposure. Witnesses, especially paramedics, may also provide valuable information. The patient's clinical course over the subsequent 48–72 hr, along with serial analysis of blood specimens and cytogenetic studies when available, may provide the only reliable index of the extent of the victim's radiation injury.

Many of the early signs and symptoms that develop in combined-injury patients may be attributed to either the trauma or the radiation. It is important that the physician not be too quick to attribute early findings merely to the effects of radiation alone. One must maintain a high index of suspicion for underlying traumatic injury, which may not be immediately apparent. Conversely, if a patient with minor to moderate injury but few other symptoms is known to have possibly significant radiation injury, he/she should be admitted to the hospital for a period of observation unless the specifics of the patient's radiation exposure are known. Even then, this patient must be followed closely with repeated clinic visits and serial hematologic examinations.

A patient contaminated with radioactive debris must be decontaminated as thoroughly and quickly as the patient's clinical status allows. Nonetheless, decontamination must not stand in the way of the proper administration of medical support in patients who require immediate therapy. A significant portion of the radioactive debris can usually be removed by fully undressing and showering the patient. (This subject is more fully discussed in Chapter 13, "Internal Contamination.")

Proper fluid therapy is vital for the management of those patients with CIS. Following the initial resuscitation and stabilization, further fluid management is dictated by the patient's condition and clinical course. In general, these patients require fluids. Depending on the magnitude of the syndrome, the initial maintenance requirements can be expected to be high because of the nature of combined injury. The consequences of conventional trauma and the surgical therapy required for managing it may produce high ongoing fluid losses. This includes fluid sequestration by raw peritoneal surfaces, large wound defects resulting in increased evaporative losses, multiple drains, nasogastric losses, and other GI losses. Radiation effects also contribute to persistent fluid losses. Irradiated microcapillary beds become leaky and allow the outpouring of proteinaceous fluids into the interstitial spaces. Gastrointestinal losses will be pronounced following doses large enough to produce the gastrointestinal syndrome. Irradiated dogs, for example, within 24–48 hr postirradiation experience profuse radiation-induced diarrhea. Without supplemental fluids, they quickly become dehydrated and show significant hemodynamic alterations (17). So combined injuries have the potential to produce large and probably synergistic changes in the victim's fluid status (56). Several studies have demonstrated the beneficial effect of fluid therapy on survival in both the acute radiation syndrome (17, 44, 57–60) and CIS (12, 56).

Infectious complications will only aggravate fluid requirements, and may produce a secondary shock-like state in these patients. Septic shock will probably have many causes in these individuals, but presumably will be associated with severe myocardial suppression, decreased vascular tone, and further third spacing of fluids. A high rate of infusion may be required to replace the patient's intravascular volume in order to restore and maintain an acceptable tissue profusion pressure. Even so, to fully reverse this problem, one must gain control of the aggravating infection.

Few combined-injury patients will be able to consume sufficient amounts of fluids to meet their needs. Parenteral supplementation with appropriate fluids and electrolytes will be needed. Daily determinations of the patient's weight and 24-hr fluid inputs and outputs should guide therapeutic decisions. Even so, for many severely ill patients, one may have to resort to invasive hemodynamic monitoring to accurately assess and replenish the intravascular volume.

Inadequate replacement must be avoided. The persistent hypovolemia can be associated with a low flow state, which results in inadequate perfusion of critical organs. The kidneys are particularly sensitive to the effects of poor perfusion and will quickly exhibit signs of failure. Acute renal failure is a particularly ominous event because of its inherent morbidity and particularly high mortality (60–70%) in posttrauma and postsurgery patients. If sufficient fluid replacement is provided, a stable hemodynamic profile should ensue. This will assure the necessary profusion of all tissues with the appropriate delivery of oxygen and energy substrates.

Proper wound management is paramount in the treatment of combined-injury patients. With any multiple-trauma patient, wound management begins in the emergency room (or sooner, if possible) with a full evaluation of the extent of injury, halting of excessive blood loss, removal of foreign matter, and debridement of all nonviable tissues. It may be necessary to repeat this process over the next several days as the full extent of the injury, including the host's reaction, becomes apparent. This process, in conjunction with irrigations, topical and systemic antimicrobials, and dressing changes, is designed to limit bacterial proliferation within the wound and to minimize subsequent tissue invasion. None of these modalities is as effective as debridement. The final outcome, however, relies on a competent immune system.

Whenever possible, wounds should be closed primarily. They heal faster and with fewer complications. Within 24–48 hr an epithelial barrier is reestablished, which impedes further bacterial contamination and decreases losses by evaporation. Nonetheless, one should consider for primary closure only those wounds that are minimally contaminated with bacteria, debris, or devitalized tissues, after a thorough surgical evaluation, Unfortunately, most wounds cannot be closed primarily, and alternatives must be used. Wounds in a multiple-trauma patient are generally extensively contaminated. In other cases, after the initial surgical treatment of the wound, the resulting defect may be too large to reapproximate. Attempts to close such a wound will place undue tension on the suture line. Ischemic necrosis of the wound edge will ensue, with subsequent breakdown of the repair, enlargement of the wound, and reinfection.

Time is a critical variable in wound management. As discussed earlier, bacteria proliferate rapidly within a wound, increasing in number proportionally with time. Once the bacterial counts exceed a given threshold, wound infection occurs. This threshold depends on a variety of factors, including the treatment schemes applied to the wound, the host's tissue response, presence of pathogens, and accessibility of antimicrobials to the organisms involved.

Wounds closed a short time after injury generally heal without infectious complications. Once cleansed and properly debrided, these wounds resemble surgical incisions, in that the residual bacterial counts are below the critical threshold for infection. The host's defense mechanisms are capable of eliminating these few organisms. On the other hand, wounds that remain untreated for long periods of time (greater than 6–8 hr) will invariably harbor bacterial counts in excess of the critical threshold. Debridement and irrigations are usually not thorough enough in this setting to reduce the bacterial levels to the point where primary closure can succeed without complications. Also, these wounds have associated edematous changes that distort the tissue planes and, because of subdermal swelling, retard reapproximation of the skin edges.

Primary closure of bacterially contaminated wounds produces a clinically significant wound infection. The contaminating organisms within a closed wound are relatively isolated from the host's defense mechanisms. The vicinity of a wound is generally avascular. The fatty tissues normally have a poor blood supply, which in a wound is compounded by traumatic disruption of the microvasculature. Serosanginous fluids accumulate. Oxygen tensions are reduced. These factors favor the growth and proliferation of bacteria. The utility of topical agents is reduced by the closure. Thus, within the confines of a closed space, various factors interact to lower the threshold for clinical infection to arise.

The threshold for infection is also influenced by the competence of the immune response. An effective inflammatory response may contain the infectious process, leading to a localized abscess. But even a "localized" process may adversely influence distant organ systems. On the other hand, depending on the pathogen or the presence of a defective immune response, the infectious thresholds may be further reduced. Because combined injury is associated with widespread disturbances in immune responsiveness, a closed contaminated wound will pose an even greater risk for a clinically significant wound infection in the CIS.

There are two alternatives to primary closure of a contaminated and/or large wound. The more conservative approach involves debriding and cleansing the wound and then allowing the healing process to proceed by secondary intent. As long as the wound remains open and is tended to daily, bacterial proliferation can usually be controlled enough to minimize the number of bacteria. In the presence of a competent host defense system, granulation will proceed and closure will eventually occur. However, wounds that remain open (1) require a long time to fully heal, (2) often produce unacceptable cosmetic results, and (3) until reepithelialized, continue to exude body fluids, electrolytes, and proteins. Also, the open wounds remain a nidus for persistent bacteremia and/or toxemias. An individual with a normal, competent immune system, who is nutritionally balanced, can generally recover without serious complications. Immunosuppressed, malnourished, and massively injured persons do not tolerate such long-term metabolic demands and bacteremic challenges. This includes patients with CIS.

A dirty contaminated wound can be converted into a clean wound by surgical extirpation of the wound. This differs from debridement. Surgical extirpation totally excises the wound and adjacent traumatized tissues using aseptic techniques. Such a procedure, in effect, eliminates almost all microbes from the site of injury. This surgically created, clean incision is then primarily closed without tension. This approach may be accompanied by fairly significant blood loss, but the wound generally heals with satisfactory results.

Large avulsive injuries are not amenable to primary closure because of the unacceptable tension required to bring these wounds together. For similar reasons they cannot be excised and simply closed. Nonetheless, such injuries and many thermal burn injuries can be treated successfully by excising *in toto* the primary injury, in staged procedures if necessary, followed by grafting of the defect with autologous or heterologous biologic coverings. Using this approach, burn and avulsive injuries can be dealt with early in the patient's hospital course, resulting in speedier recoveries and fewer complications. The alternative is long-term wound treatment until granulation tissue fills the defect and reepithelialization occurs, all by secondary intent. During this time, the patient remains in a metabolically stressed state and is subject to frequent complications, mostly infections or fluid loss. Primary excision and grafting have been used successfully in treating such injuries at the Massachusetts General Hospital and the Boston University Medical Center.

Wounds present a special problem to the surgeon involved in the care of patients with CIS. Left open, wounds become sources of systemic infection and produce unacceptably high incidences of complications and death. Few of these wounds are amenable to primary closure. Primary excision with subsequent grafting may be the treatment of choice for these persons. This approach restores a functional biological barrier in these immunologically suppressed and metabolically stressed individuals. Indeally, one prefers to use heterologous or synthetic materials. These technologies are rapidly advancing for burn injuries, and they may be adaptable to combined injuries. In lieu of this, autologous covering with a split-thickness skin graft from a nonaffected portion of the victim's body may be usable. Of major concern in the use of autologous coverings is that the donor site may place an unacceptable burden on the host, if it becomes a source of bacteriologic complications. The widespread availability of synthetic materials should alleviate this concern.

Support for this therapeutic approach to combined-injury wounds is available in the European literature (5, 54, 61–63). In experimentally wounded and irradiated animals, Messerschmidt (5) and colleagues demonstrated that survival could be enhanced by closing the wound following injury. Although it is not entirely clear, presumably the wound either was excised before closure or was closed shortly after the injury. Nonetheless, a definite advantage in survival was demonstrated over those animals in which the wounds were left open. Of further interest is the German observation that wound coverage also reduced the bacterial counts in these irradiated wounds (54).

In general, one wants to avoid, if at all possible, inflicting further injury (i.e., a surgical wound) on combined-injury patients, especially those whose traumatic injuries are blunt in nature. Eventual problems with wound healing, immunosuppression, anemia, and thrombocytopenia will only complicate postsurgical recovery. However, many injuries require surgical intervention early in the course of the patient's illness. Such injuries may be readily apparent at the time of presentation (such as an evisceration or major arterial disruption), or they may become apparent only later in the hospital course. Intraabdominal injury following blunt trauma may not be as readily obvious. Such injuries may require prolonged observation and thorough radiographic investigation before they can be detected.

Injuries associated with blunt abdominal wounds vary as to their treatment. Some may require immediate operative therapy, for example, a fractured liver or an avulsed tail of the pancreas. Others may be merely observed, allowing the patient to heal of his/her own accord. Frequently, parenchymal injuries to the spleen and kidney can be managed by close observation and restricted patient activity. Unfortunately, some of these individuals may later decompensate as a result of renewed hemorrhage or infectious complications involving the contused tissues. Such infections presumably arise from hematogenous seeding of devitalized and/or severely contused tissues. In these cases, the risks remain high for days to weeks following the initial injury. Young, healthy individuals have been known to present to emergency rooms for intraabdominal bleeding after blunt trauma to the spleen that occurred several days to weeks before their current intraabdominal catastrophe.

Persons with defects in their healing abilities are at increased risk for developing secondary complications, such as renewed hemorrhage and infections, arising from blunt organ injury. Besides such healing disorders, patients with CIS are at significant risk because of radiation-induced immunodeficiencies and coagulopathies. Following whole-body irradiation in humans, hematopoietic suppression reaches its nadir within 1–2 weeks, depending on the initially absorbed dose. During this time, the exploration of these CIS patients for secondary infectious and/or bleeding problems associated with contused organs will carry an inordinately high mortality rate. So after early detection, the removal of badly contused or fractured organs might be justifiable in these patients. This aggressive approach may ultimately result in less morbidity and mortality. Few parameters exist in the surgical literature to indicate how extensive an injury must be before the risks of developing secondary complications from the contused organ *in situ* outweigh the risks associated with removal of the injured (but perhaps salvageable) organ in normal patients, much less in irradiated persons. In pediatric populations, splenectomy appears to increase the risks for future septic complications; however, the picture in adults is not so clear.

When evaluating combined-injury patients, one must suspect an underlying but not immediately obvious traumatic injury. These injuries need to be pursued aggressively when present, especially if significant hematopoietic and GI disturbances seem imminent. It is imperative to identify and treat traumatic injuries before the onset of significant marrow disturbances. In humans, this grace period extends no longer than 36–72 hr after irradiation. Beyond this, invasive procedures must be kept to a minimum until hematopoietic recovery begins.

As with any patient, but particularly the combined-injury patient, one must devote particular attention to technical details. One does not want to reexplore these patients later because of an undrained collection, nonviable tissue, an anastomotic leak resulting from a misplaced stitch, undue tension on a suture line, or reanastomosis in the face of intraabdominal sepsis. Most indications suggest that small bowel anastomosis (64, 65) and major arterial repairs, at least those using autologous materials (65), will heal satisfactorily in spite of recent irradiation. On the other hand, salvaging a highly vascularized organ such as the spleen has to be weighed against the consequences of possible recurrent hemorrhage at a later, less desirable time. Penetrating injuries to the left colon should be exteriorized as a colostomy when feasible; if not feasible, the left colon should be resected, with a subsequent end colostomy and mucous fistulae or Hartmann's procedure. Right colon injuries are generally resected and primarily reanastomosed without difficulties, even in the presence of gross contamination. This precludes the fluid and electrolyte problems associated with high-output ileostomies and cecostomies. Nonetheless, in those individuals who exhibit florid manifestations of significant GI injury secondary to ongoing radiation enteropathy, one should probably consider an ileostomy. Good surgical margins may not exist in these cases as a result of a diffusely edematous and/or hemorrhagic irradiated ileum. This situation is analogous to dealing with a terminal ileum that has been compromised by inflammatory bowel disease. Primary anastomosis of such a bowel will invariably result in a leak with subsequent fistulization, abscess, and/or generalized peritonitis. Later, after the marrow depression subsides and recovery ensues, usually 2–3 months postirradiation, GI continuity can be restored in these and in left colon injuries.

In addition to delaying the healing of soft-tissue injuries, radiation interferes with the regeneration of normal bone after traumatic fracture (66, 67). In rabbits, nonunion and delayed mineralization of experimentally fractured long bones were evident long after control fractures had healed (66). Both osteoblasts and osteoclasts, which are important cellular actors in bone remodeling and repair, derive from the bone marrow. These mature cells, found within the bone matrix, appear to be relatively radioresistant. However, their marrow progenitors, much like other marrow constituents, may be very radiosensitive. Thus, without the repletion of damaged and/or senescent cells within the bone matrix by marrow progenitors, a delayed healing of fractures can be expected. Such delays prolong the incapacitation of the individual, especially one with long-bone fractures. The extended bedrest and immobilization required for proper reunion and refusion of the fractured bony ends only worsen the complication rate of these patients. These complications include pulmonary and embolic events.

Incomplete immobilization will encourage nonunion. Patients with delayed healing capabilities will require longer immobilization and longer periods with plaster casts or splints. Plaster may not provide sufficiently adequate immobilization over the long period of time required for healing to occur. Very early internal fixation using Ender's rods (or their equivalent) before the full effects of radiation-induced marrow suppression arise might be beneficial in patients with closed fractures, especially compound fractures. Such an approach allows earlier mobilization than otherwise expected.

Open fractures present a more complex problem. Radiation impairs both soft-tissue and bone repair processes. It also predisposes these wounds to an increased risk of infectious complications. Such soft-tissue injuries require increased surveillance in these immunocompromised patients. Diligent debridement of all devitalized tissues is a must in managing these injuries for a period of 24–48 hr following the initial injury. Plaster casting obviously interferes with management of the soft-tissue component of the injury and, as mentioned, may not provide sufficient immobilization of the orthopedic injury. An alternative is the use of a Hoffman external fixator. This device, which requires minimal wounding for proper insertion of the pins, minimizes the amount of foreign material within the patient, in contrast to intramedullary rods. Plaster is not required, and rigid and constant immobilization at any desired angle can be achieved. The pin sites and the soft-tissue injury can be observed daily and frequently cleansed and dressed. Most important, this device allows early skin grafting of the soft-tissue injury. In numerous multiple-trauma cases, the Hoffman device has produced good clinical results, and it probably should be used for all open fractures in those patients with CIS. Such an approach is most likely superior to that recommended by Messerschmidt (5). In reviewing the East European literature, he concluded that intramedullary pins in conjunction with antibiotics, because they appeared to enhance bone healing and survival in experimental animals, was the procedure of choice for open compound fractures. Unfortunately, intramedullary stabilization is technically more difficult to perform, it still requires plaster immobilization to control rotation motion at the injury, and it may seed the intramedullary canal with infection because it is a large foreign body. The Hoffman device is a far superior technique.

The major determinant of survival after initial resuscitation in CIS is the ability to prevent or control established infections in affected patients. Wound management as proposed here is one possible method to achieve this. Other methods should also be used to reduce the levels of bacteria in the patient's environs. Included are isolation protocols, dietary manipulations, and scrupulous attention to personal hygiene.

Antibiotics currently play a prominent role in the treatment of multiple-trauma victims. Patients with traumatic wounds are routinely given tetanus prophylaxis and begun on a course of systemic chemotherapy. The duration of the chemotherapy often depends on the nature of the injury and the patient's subsequent clinical course. In patients who later become immunosuppressed, the duration may be long. But antibiotic therapy is not the panacea of medicine. It must be combined with the drainage of all infected fluid and hemorrhagic collections, debridement of devitalized tissues, and copious irrigation of wounds followed by good nursing care of the wound. For antibiotics to be fully effective, however, some functional level of the immune system must be present. Herein lies the crux of the problem of the CIS.

The little evidence currently available in the literature suggests that systemic and perhaps oral antibiotics do have some beneficial role in CIS (7, 8, 17, 42, 45, 68–71). Following the atomic blasts in Japan sulfonamides were available and were used extensively only in Nagasaki (68). Compared with Hiroshima, the survival rate for victims in Nagasaki was higher. Although other factors may contribute to these differences, some positive benefits from the use of the sulfonamides are suggested. The efficacy of antibiotics in irradiated animals seems to be potentiated when combined with other forms of treatment, most importantly fluid therapy (12, 17, 57–60). Intuitively, one suspects both antibiotics and fluids to be of value, based on clinical observations in patients with other immunosuppressive states and multiple trauma. Oral agents will probably have limited use in most cases because of motility disturbances inherent in a large proportion of combined-injury situations.

Injudicious use of antibiotics is unacceptable. Numerous drug and allergic reactions are associated with various agents. Many antibiotics, notably the aminoglycosides, have significant toxicities. Of particular concern in hematopoietically suppressed individuals is the fact that many antibiotics (most notably chloroamphenicol) have marrow-suppressing effects. The prolonged use of antibiotics frequently results in the emergence of resistant strains of bacteria. Many parenteral antibiotics (including cleocin, the ampicillins, and most of the cephalosporins) are associated with the overcolonization of the gut with *Clostridium difficile,* the causative agent for pseudomembranous colitis. Long-term antibiotic use may also predispose to the proliferation and dissemination of yeast, particularly in immunosuppressed patients. Fungemias are very lethal complications in this setting. Even in the absence of fungemia, mucocutaneous candidiasis frequently develops. This most uncomfortable complication may be controlled with oral nystatin, perineal hygiene, and topical antifungal agents.

REFERENCES

1. Geiger, K. *Grundlagen der Militarmedizin,* p. 220. Deutscher Militarverlag, Berlin, 1964.
2. Wintz, H. Die vor und nachbehandlung bei der rotgenbestrahlung. *Ther. Gegenw.* **64,** 209 (1923).
3. Alpen, E. L., and Sheline, G. E. The combined effects of thermal burns and whole body X-irradiation on survival time and mortality. *Ann. Surg.* **140,** 113–118 (1954).
4. Parr, W. H., Daggs, V. M., O'Neil, T. A., and Bush, S. M. *A Study of Combined Thermal Radiation Burn and X-Irradiation in Mice,* Report No. 94. Army Med. Res. Lab., Fort Knox, Kentucky, 1952.
5. Messerschmidt, O. Combined injury caused by nuclear explosions. In *Present Day Surgery* (R. Zenker, F. Deucher, and W. Schink, eds.), vol. 4, pp. 1–102. Urban & Schwarzenberg, Munich, 1976.
6. Schildt, B. E. Mortality rate in quantified combined injuries. *Strahlentherapie* **144,** 440–449 (1972).
7. Baxter, H., Drummond, J. A., Stephens-Newsham, L. G., and Randall, R. G. Reduction of mortality in swine from combined total body radiation and thermal burns by streptomycin. *Ann. Surg.* **137,** 450–455 (1954).
8. Brooks, J. W., Evans, E. I., Han, W. T., and Reid, R. D. The influence of external body radiation on mortality from thermal burns. *Ann. Surg.* **136,** 533–545 (1952).
9. Messerschmidt, O., Birkenmayer, E., Bomer, H., and Koslowski, L. *Radiation Sickness Combined with Burns,* IAEA-SM 119/34, pp. 173–179. IAEA, Vienna, 1970.
10. Messerschmidt, O., and Kolowski, L. Studies on surgical-radiological combined injuries. *Hefte Unfallheild.* **87,** 269–274 (1966).
11. Stromberg, L. W. R., Woodward, K. T., Maehn, D. T., and Donati, R. M. Combined surgical and radiation injury. The effect of timing of wounds and whole body gamma irradiation on 30 day mortality and rate of wound contracture in the rodent. *Ann. Surg.* **167**(1), 18–22 (1968).
12. Perman, V., Sorenson, D. K., Usenik, E. A., Bond, V. P., and Cronkite, E. P. Hematopoietic

regeneration in control and recovered heavily irradiated dogs following severe hemorrhage. *Blood* **19,** 738–742 (1962).
13. Messerschmidt, O., and Stahler-Michelis, O. Effects of ionizing radiation combined with other injuries. *Acta Anaesthesiol. Belg.* **3,** 308–314 (1968).
14. Ledney, G. D., Exum, E. D., and Sheehy, P. A. Survival enhanced by skin wound trauma in mice exposed to ^{60}Co radiation. *Experientia* **37,** 193–194 (1981).
15. Messerschmidt, O. Untersuchungen uber kombinationsschaden. 6. Mitterlung: Uber die lebenserwartung von Mausen, die met ganzkorperbestralungen in combination mit offenen oder geschlossenen oder kompressionsschaden belastet wurden. *Strahlentherapie* **131,** 298–310 (1966).
16. Haley, T. J., Riley, R. F., Williams, I., and Andem, M. R. Presence and identity of vasoactive substances in blood of rats subjected to acute whole body roentgen ray irradiation. *Am. J. Physiol.* **168,** 628–636 (1952).
17. Mount, D., and Bruce, W. R. Local plasma volume and vascular permeability of rabbit skin after irradiation. *Radiat. Res.* **23,** 430–445 (1964).
18. Kundel, H. L. The effect of high energy proton irradiation on the cardiovascular system of rhesus monkey. *Radiat. Res.* **28,** 529–537 (1966).
19. Phillips, R. D., and Kemildorf, D. The effect of whole-body X-irradiation on blood pressure in the rat. *Radiat. Res.* **18,** 86–95 (1963).
20. Loegering, D. J., and Carr, F. K. Humoral factor activity and carbon clearance rate during the early stages of hemorrhagic shock. *J. Reticuloendothel. Soc.* **21**(4), 263–269 (1977).
21. Loegering, D. J. Circulating reticuloendothelial depressing substance following thermal injury and intestinal ischemia. *Proc. Soc. Exp. Biol. Med.* **166,** 515–521 (1981).
22. Blattberg, B., Levy, M. N., and Strong, I. Effect of reticuloendothelial depressing substance on survival from shock. *RES, J. Reticuloendothel. Soc.* **3,** 65–70 (1966).
23. Ollodart, R., and Masberg, A. R. The effect of hypovolemic shock on bacterial defenses. *Am. J. Surg.* **110,** 302–307 (1965).
24. McKenna, J. M., and Weifach, B. W. Reticuloendothelial system in relation to drum shock. *Am. J. Physiol.* **187,** 263–268 (1956).
25. Zweifach, B. W. Q., Benacerraf, B., and Thomas, L. The relationship between the vascular manifestation of shock produced by endotoxin, trauma, and hemorrhage. II. The possible role of the reticuloendothelial system in resistance to each type of shock. *J. Exp. Med.* **106,** 403–414 (1975).
26. Altura, B. M., and Hershey, S. G. Sequential changes in reticuloendothelial system function after acute hemorrhage. *Proc. Soc. Exp. Biol. Med.* **139,** 935–939 (1972).
27. Loegering, D. J. Intravascular hemolysis and RES phagocytes and host defense functions. *Circ. Shock* **10,** 383–395 (1983).
28. Sanford, J. P., and Evans, J. R. Effect of hemorrhagic shock on certain antibacterial defense mechanisms in the rabbit. *J. Appl. Physiol.* **10**(1), 88–92 (1957).
29. Messerschmidt, O., Birkenmayer, E. E., and Koslowski, L. Untersuchungen uber kombinationsschaden. Veranderungen dei weissen und roten blutbildes von ratten bei ganzkorperbestrahlung in kombintion mit offenen hautwunden. *Strahlentherapie* **135,** 586–596 (1968).
30. Middelton, G. R., and Young, R. W. *Postirradiation Vomiting,* Sci. Rep. SR74-23. Armed Forces Radiobiol. Res. Inst., Bethesda, Maryland, 1974.
31. Dubois, A., Jacobus, J., and Grissom, M. Altered gastric emptying during radiation induced vomiting. *J. Nucl. Med.* **24,** 56A (1983).
32. Sullivan, M. F., Hulse, E. F., and Mole, R. H. The depleting action of bile in the small intestine of the irradiated rat. *Br. J. Exp. Pathol.* **46,** 235–244 (1965).
33. Jackson, K. L., and Enteman, C. The role of bile secretion in the gastrointestinal radiation syndrome. *Radiat. Res.* **10,** 67–79 (1959).
34. Conrad, R. A., Cronkite, E. P., Brecher, G., and Strome, C. P. A. Experimental therapy of the

gastrointestinal syndrome produced by lethal doses of ionizing radiation. *J. Appl. Physiol.* **9,** 227–233 (1956).
35. Robbins, S. L., and Angell, M. *Basic Pathology,* 2nd ed., pp. 217–218. Saunders, Philadelphia, Pennsylvania, 1976.
36. Law, M. Radiation induced vascular injury and its relation to late effects in normal tissues. *Adv. Radiat. Biol.* **9,** 37–73 (1981).
37. Detrich, L. E., Upham, H. C., Higby, D., Debley, V., and Haley, T. J. Influence of X-ray irradiation on glucose transport in the rat intestine. *Radiat. Res.* **2,** 483–489 (1955).
38. Jackson, K. L., Rhodes, R., and Enteman, C. Electrolyte excretion in the rat after severe intestinal damage by X-irradiation. *Radiat. Res.* **8,** 361–373 (1958).
39. Goodner, C. J., Moore, T. E., Bowers, J. Z., and Armstrong, W. D. Effects of acute whole body irradiation on the absorption and distribution of ^{22}Na and $H^3{}_2O$ from gastrointestinal tract of the fasted rat. *Am. J. Physiol.* **183,** 475–484 (1955).
40. Moss, W. T. The effect of irradiating the exteriorized small bowel on sugar absorption. *Am. J. Roentgenol., Radium Ther. Nucl. Med.* **78,** 850–854 (1957).
41. Sullivan, M. F. Bile salt absorption in the irradiated rat. *Am. J. Physiol.* **209**(1), 158–164 (1965).
42. Miller, C. P., Hammond, C. W., Tompkins, M. Reduction of mortality from X-radiation by treatment with antibiotics. *Science* **111,** 719 (1950).
43. Miller, C. P., Hammond, C. W., and Tompkins, M. The role of infection in radiation injury. *J. Lab. Clin. Med.* **38,** 331–335 (1951).
44. Taketa, S. T. Water-electrolyte and antibiotic therapy against acute (3 to 5 day) intestinal radiation death in the rat. *Radiat. Res.* **16,** 312–326 (1962).
45. Webster, J. B. The effect of oral neomycin therapy following whole body X-irradiation of rats. *Radiat. Res.* **32,** 117–124 (1967).
46. Walker, R. I. Hematologic contribution to increases in resistance or sensitivity to endotoxin. In *Experimental Hematology Today* (S. J. Baum and G. D. Kedney, eds.), pp. 55–62. Springer-Verlag, Berlin and New York, 1979.
47. Rovnov, A. S. Besonderhecten des verlaufs von thermischen verbrennungen bei strahlenkrankheit. *Khirurgiia (Moscow)* **32**(4), 88–94 (1956).
48. Bystnova, V. V., and Sakolov, S. S. Morphologische sharakteristik des wundheilungsprozesses bei strahlenkrankheit. *Med. Radiol.* **3**(5), 71–77 (1958).
49. Raventos, A. Wound healing and mortality after total body exposure to ionizing radiation. *Proc. Soc. Exp. Biol. Med.* **87,** 165–167 (1954).
50. Radakovich, M., Dutton, A. M., and Schelling, J. A. The effect of total body irradiation on wound closure. *Ann. Surg.* **139,** 186–194 (1954).
51. Lawrence, W., Jr., Nickerson, J. J., and Warshaw, L. M. Roentgen rays and wound healing: Experimental study. *Surgery* **33,** 376–384 (1953).
52. Oughterson, A. W., and Warnen, S. *Medical Effects of the Atomic Bomb in Japan,* Natl. Nucl. Ser., Div. VII, vol. 8. McGraw-Hill, New York, 1956.
53. Korlof, B. Infection in burns. *Acta Chir. Scand., Suppl.* **209,** 1–144 (1956).
54. Werdan, K., Schlick, P., Zimmer, P., Messerschmidt, O., and Maurer, G. Untersuchungen uber kombinationsschaden. 23. Mitteilung: Uber die pathogenetische bedeutung der bakteriellen wundbesiedlung bei mausen, die durch ganzkorperbestrahlung in komination mit offenen hautwurden gelastet wurden. *Strahlentherapie* **154,** 342–348 (1978).
55. Donati, R. M., McLaughlin, M. M., and Stromberg, L. W. R. Combined surgical and radiation injury. VII. The effect of the gnotobiotic state on wound closure. *Experientia* **29,** 1388–1390 (1972).
56. Messerschmidt, O., Kettemann, K. B., and Koslowski, L. Untersunchungen uber kombinationsschaden. Vergleichenden elektrolytuntersuchungen bei ratten, die durch ganzkorperkerrtrahlungen und offene Hautwurden belastet wurden. *Strahlentherapie* **141,** 705–711 (1971).

57. Sorenson, D. K., Bond, V. P., Cronkite, E. P,. and Perman, V. An effective therapeutic regime for the hematopoietic phase of the acute radiation syndrome in dogs. *Radiat. Res.* **13,** 669–685 (1960).
58. Coutler, M. P., Furth, F. W., and Howland, J. W. Therapy of the X-irradiation syndrome with terramycin. *Am. J. Pathol.* **28,** 875–881 (1952).
59. Allen, J. G., Moulder, P. V., and Emerson, D. M. Pathogenesis and treatment of the post-irradiation syndrome. *J. Am. Med. Assoc.* **145,** 704–711 (1951).
60. Hammond, C. W. The treatment of post-irradiation infection. *Radiat. Res.* **1,** 448–458 (1954).
61. Burmistrov, V. M., and Slinko, V. G. Derlokale verlauf von verbrennungen deitten grades bei strahlenkrankheiten. *Vestn. Khir. im. I. I. Grekova* **80**(6), 74–78 (1958).
62. Petrov, V. I. Freies hauttransplantat bei strahlenkrankheit. Vestn. Khir. im. I. I. Grekova **77**(9), 85–90 (1956).
63. Razgovorov, B. L. Primare wundnaht bei strahlenkrankheit. *Eksp. Khir.* **2,** 47–50 (1957).
64. Gustafson, G. E., and Cebul, F. A. *The Effect of Surgery on Dogs Following Whole Body X-irradiation,* At. Energy Rep. No. NYO-4017. Western Reserve University School of Medicine, Cleveland, Ohio, 1953.
65. Tepper, J. E., Sindelar, W., Travis, E. L., Terrill, R., and Padikal, T. Tolerance of canine anastomosis to intraoperative radiation therapy. *Int. J. Radiat. Oncol. Biol. Phys.* **9,** 987–992 (1983).
66. Zemlianoi, A. G. Healing of fractures and distribution of radioactive phosphorus in bone marrow following preliminary whole body irradiation of experimental animals. *Vestn. Khir. im. I. I. Grekova* **77**(6), 59–64 (1956).
67. Antipina, A. N., and Zeulianoi, A. G. Die allgemeine morphologische charakteristik der heilung offenen frakturen bei strahlenkrankheit. *Med. Radiol.* **2**(1), 70–79 (1957).
68. Cronkite, E. P. *Critical Evaluation of the Value of Antibiotics in Prophylaxis and Therapy of Combined Radiation and Traumatic Injuries,* Prog. Rep., Contract No. DE-AC02-76CH00016. U.S. Department of Energy, Washington, D.C., 1980.
69. Coutler, M. P., and Miller, R. W. *Treatment with Successive Antibiotics of Dogs Exposed to Total Body X-irradiation,* At. Energy Comm. Proj. UR-276. University of Rochester, Rochester, New York, 1953.
70. Wise, D., and George, R. E. *An Effective Readily Available Treatment for Acute Radiation Injury in Beagles,* Sci. Rep. SR71-14. Armed Forces Radiobiol. Res. Inst., Bethesda, Maryland, 1971.
71. Nutter, J. E., Graw, R. G., Jr., and Baum, S. J. *Therapy of Postirradiation Marrow Hypoplasia with Blood Components and Antibiotics,* Sci. Rep. SR73-5. Armed Forces Radiobiol. Res. Inst. Bethesda, Maryland, 1973.

CHAPTER 11

Mechanisms and Management of Infectious Complications of Combined Injury

RICHARD I. WALKER AND JAMES J. CONKLIN

Armed Forces Radiobiology Research Institute
Bethesda, Maryland 20814-5145

Microbial infections have long been recognized as a serious, often fatal, complication of major wounds and trauma. In fact, when patients with severe conventional injuries survive longer than 5 days, infection is second only to head injury as the cause of mortality. Trauma-associated infections are due to opportunistic pathogens from endogenous or exogenous sources, which colonize the susceptible epithelial surfaces of hosts with depleted defenses. These pathogens, often presented as mixed infections, include *Pseudomonas aeruginosa, Staphylococcus aureus, Klebsiella pneumoniae,* and *Candida albicans.*

A relationship has also been established between infections and deaths after whole-body radiation doses around the midlethal range (1). Invading microorganisms are often of enteric origin, and a curve drawn from mouse data depicts the increasing incidence of infection as parallel to but preceding the mortality curve. Furthermore, all mice that had developed bacteremia after irradiation died, whereas those that had not developed bacteremia survived. Studies such as this demonstrate that infection is an important cause of death in animals that die from the hematopoietic syndrome. Evidence also exists for postirradiation infections in man. Histological specimens of spleen, liver, lymph nodes, intestinal wall, and other tissues were taken from Japanese patients dying from the effects of the atomic blasts at Hiroshima and Nagasaki. Many of those specimens revealed microscopic bacterial colonies of both gram-positive and gram-negative bacteria growing freely in the tissues.

When radiation injury is superimposed on a victim with another type of trauma, the result is a phenomenon known as combined injury. A characteristic feature of these injuries is that insults that are sublethal or minimally lethal when given alone act synergistically to give a much greater mortality than could be

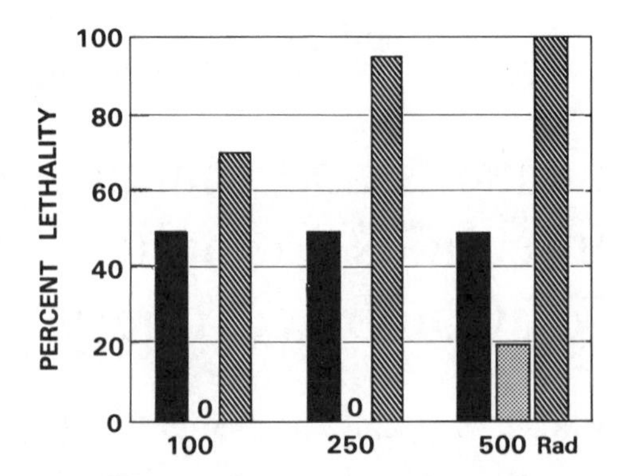

FIG. 11-1 Percentage mortality of rats given a 33% burn (solid bars), radiation (dotted bars), or a combination of the two injuries (hatched bars).

achieved by additive effects. For example (Fig. 11-1), almost 100% mortality was observed in rats given both 250 rad gamma radiation (no mortality) and a 33% burn (50% mortality) (2). The importance of the two types of injury acting simultaneously is illustrated in another set of experiments with mice (3). Five hundred ten roentgens of radiation caused 26% mortality; in combination with an open wound, it caused 90% mortality (Fig. 11-2). When the wound was closed, this synergistic effect was reduced. Then the number of deaths was similar to the number that died from radiation alone, suggesting that some signal(s) or mediator(s) was interrupted.

A role for endogenous microorganisms in deaths following combined injury has been established by determining the survival in germfree rats and in conventional rats undergoing combined radiation and wound injuries (4). No mortality

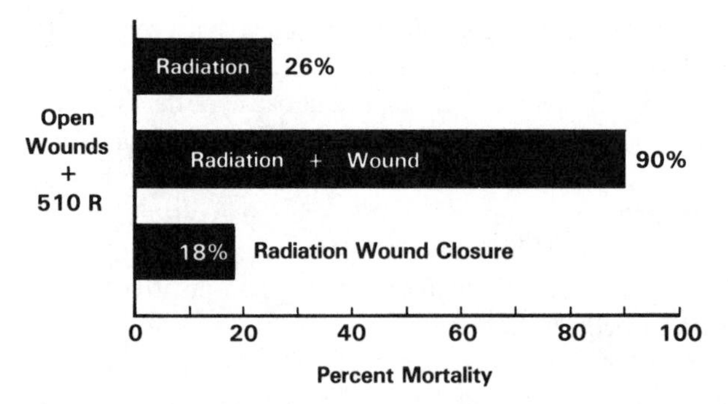

FIG. 11-2 Effect of wound closure on percentage mortality of mice given a minimally lethal dose of radiation.

TABLE 11-1

EFFECT OF ENDOGENOUS FLORA ON MORTALITY IN ANIMALS UNDERGOING COMBINED INJURY

	Positive cultures/total			
	Germfree		Conventional	
	Single injury	Combined injury	Single injury	Combined injury
800 rads TBI	+	+	+	+
3-cm wound	−	+	−	+
Percentage mortality	0	17	55	100
Survival (days)	30	22 ± 3	21 ± 3	10 ± 4

was seen in germfree rats given 800 rads total-body irradiation (TBI), compared to 55% mortality in conventional rats (Table 11-1). If a 3-cm wound was also inflicted, mortality was 17% and 100% in germfree and conventional animals, respectively.

I. ABNORMAL COLONIZATION OF EPITHELIAL SURFACES

We are gaining a better understanding of the events leading to postinjury infection through further studies with rodents. Mice receiving 850 rads X rays or 1000 rads ^{60}Co are neutropenic by 3 days after exposure, but infections that precede death do not occur until almost 1 week later (Fig. 11-3). This observa-

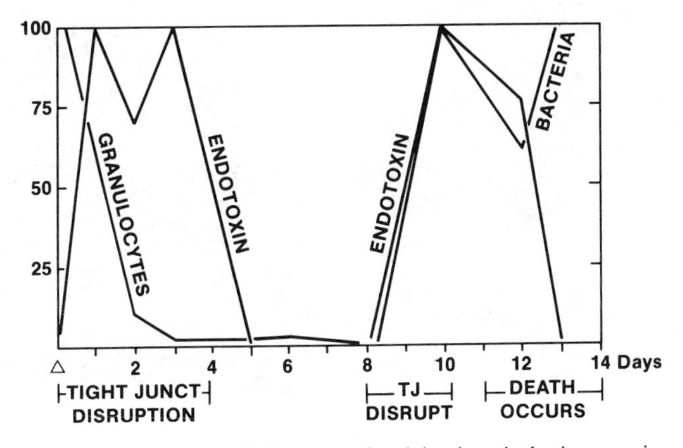

FIG. 11-3 Association of postirradiation events involving intestinal microorganisms in mice.

tion raises the question of what events affecting colonization by microorganisms occur before systemic infection. Significant changes must have occurred in gut physiology after irradiation, because biphasic changes in gut permeability were detected. Between days 1–4 and 8–12 after irradiation, endotoxin, presumably from cell walls of bacteria in the intestine, was detected in mouse livers. This leakage of substances from the intestine was accompanied by disruption of the tight junctional permeability barriers between adjacent epithelial cells.

The postirradiation findings described above led to the examination of changes in populations of microorganisms in the intestine. The ilea of normal rodents are colonized by unusual bacteria known as segmented filamentous microflora (SFM), which cannot be cultivated but whose numbers can be discerned easily with scanning electron microscopy. These organisms are intimately associated with the epithelium of the intestine. Their role in the intestine is unknown, but since they are associated with well-being, we used SFM as indicator organisms of injury to the intestine that could alter bacterial populations. Twenty-four hours after sublethal irradiation (500 rads), SFM disappeared from the rat ilea (Fig. 11-4) (5). By 4 days, however, the SFM populations were back to normal. In rats given lethal radiation (1000 rads), SFM did not return at 4 days and was still absent from the ilea at 11 days after irradiation. Potential opportunistic pathogens, measured by cultivation of dilutions of intestinal homogenates, declined in number shortly after exposure to 500–1000 rads of radiation. At the lower radiation dose, their numbers began to increase after return of the SFM, but they were still subnormal at 11 days after irradiation. In contrast, rats given 1000 rads showed increasing numbers of facultative flora after the first few days postir-

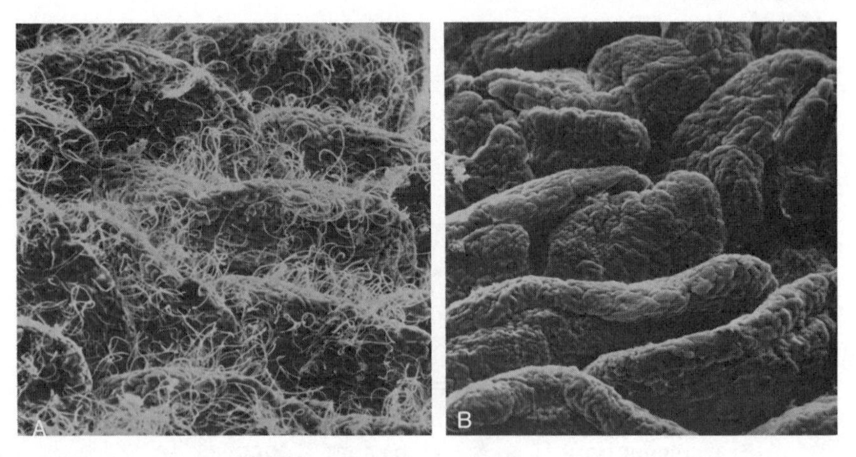

FIG. 11-4 Loss of segmented filamentous microflora from normal rat ileum (A) 24 hr after exposure to 500 rads ^{60}Co radiation (B).

radiation, and by day 11, these organisms had colonized the ilea in numbers far above normal. This event correlated with systemic infection and preceded death.

It is apparent that an event associated with higher doses of radiation is responsible for the abnormal colonization preceding infection. At present, little is known about the pathophysiology of this injury. Intestinal epithelial cells are almost as sensitive to radiation as are marrow cells, and even sublethal doses cause injury that could affect bacterial colonization. Recent evidence suggests that in trauma victims, some epithelial surfaces can lose fibronectin and become prone to attachment by gram-negative bacteria such as *P. aeruginosa* (6). This may also occur after irradiation and enhance the possibility that pathogens will successfully colonize the host.

Abnormal colonization after trauma could also be due to factors such as altered mucous barriers or loss of immunologic capabilities. Studies of posttrauma permeability barriers of the intestine indicate that mucin production could be altered. Increased permeability of intestinal epithelium to endotoxin has previously been associated with a variety of injuries (7). More recent studies (Fig. 11-3) demonstrate that this change in permeability is due to disruption of the tight junction (zonula occludens), a gasket-like structure that prevents such passage in normal animals. Disruption of this barrier occurs after a variety of traumas such as radiation or full-thickness dermal wounds and also during endotoxin shock. Furthermore, when the status of the tight junctions was examined at intervals after irradiation, their disruption coincided with the periods when endotoxin was detected in the liver. An important finding regarding tight junction disruption is that breaks occur only between adjacent goblet cells or goblet cells and absorptive epithelial cells, never between only absorptive epithelial cells. This phenomenon may indicate direct or mediated injury to cells that are responsible for mucin production. Since most intestinal microorganisms reside in the mucin layer, any qualitative or quantitative changes in the production of this substance could significantly alter the patterns of microbial colonization in the intestine. The effectiveness of the mucin barrier depends in part on the secretory immunoglobulin A (SIgA) it contains. How trauma or combined injury affects SIgA levels in the intestine remains to be determined.

Infection associated with combined injury could be due in part to the exacerbation of effects on goblet cells found after single injuries. Full-thickness dermal wounds administered to rats 2 days after irradiation with 500 rads (sublethal dose) ^{60}Co caused a much greater incidence of tight junction disruptions in ileal samples than would be expected from either injury alone.

Although altered mucosal surfaces are undoubtedly a source of infection in compromised patients, the direct contamination of wounds with soil and debris containing microorganisms must also be considered. Foreign material in wounds can increase the susceptibility to infection by as much as a millionfold. Therefore, an essential factor of wound management is debridement. The removal of particles upon which microcolonies can form and the elimination of necrotic tissue will significantly reduce a major source of infection.

II. INTERACTION OF HOST AND PARASITE DURING INFECTION

When microorganisms penetrate normally sterile tissues, inflammatory responses are called on to control the microorganisms. Humoral components of inflammation (e.g., complement) interact with microbial antigens and become activated to induce cellular responses such as vasoconstriction and exudation of polymorphonuclear leukocytes (PMNL). PMNL phagocytize and kill many microorganisms. Later, macrophages enter the inflammatory site where they contribute to the removal of microorganisms and debris, and they secrete factors that promote tissue repair.

Microorganisms are also removed from the circulation by the reticuloendothelial system (RES). In the presence of proper opsonins, RES macrophages (such as those found in the liver and spleen) sequester and kill microorganisms and also secrete mediators that help augment host defenses. Failure in systemic host defenses after trauma may be mediated in part by a deficiency of circulating opsonic protein (plasma fibronectin). Infusion with this substance in persons depleted by trauma has been associated with enhanced RES activity (8). Immunoglobulin G is depleted after burns, and this depletion can be associated with susceptibility to infection (9).

In traumatized subjects, an adequate inflammatory response is often not obtained due to the generation of immunosuppressive factors or the loss of functional cells. For example, the mortality rate from infection varies directly with the degree of granulocytopenia. Dogs given gentamicin plus granulocytes survived longer and cleared *P. aeruginosa* better than dogs given the antibiotic alone (10). Although granulocytes are undoubtedly important in controlling gram-negative sepsis, it is questionable whether transfusion with these cells is essential for effective therapy. In patients with infections due to antibiotic-sensitive organisms, granulocyte replacement is probably not necessary.

The uncontrolled multiplication of microorganisms in tissues of compromised hosts leads to the accumulation of microbial toxins. Little is known at present about the lethal mechanisms of most bacteria. Gram-negative bacteria are the microorganisms most frequently responsible for the consequences of posttrauma infections. A characteristic feature of these bacteria is a lipopolysaccharide cell wall component known as endotoxin. If administered in sufficient quantities, this substance induces a variety of toxic host responses that mimic many of those associated with overwhelming bacterial infections. Microbial pathogenesis, however, is probably not due to any single factor. Instead, multiple factors of virulence that can amplify or potentiate each other must be considered in order to

TABLE 11-2

ASSOCIATION OF VARIOUS *Aeromonas* CHARACTERISTICS WITH VIRULENCE FOR MICE

	Positive cultures/total	
Factors	High virulence	Low virulence
Cytotoxin	15/15	0/9
Elastase	11/15	0/9
Lipase	13/15	1/9
Protease	15/15	7/9
Piliation	9/15	2/8
Hemolysin	8/9	1/7

understand the mechanism of disease. For example, in a survey of 24 *Aeromonas* isolates that were grouped according to high virulence and low virulence (Table 11-2), a variety of factors were more often associated with the high-virulence group.

The various virulence factors (e.g., pili, toxins, capsules, and outer cell-surface components) of opportunistic pathogens are all potential targets for vaccines. Although the development of specific immunoprotection is complicated by the variety of potential opportunistic pathogens, vaccines against the more common pathogens or an antigen common to many pathogens is feasible. For example, a major virulence factor of *P. aeruginosa* is exotoxin A. Immunization against this protein protects 80% of burned mice challenged with live *Pseudomonas,* and this protection is 100% if gentamicin is given (11).

Vaccines can be directed against the bacterial cell itself. Surface antigens can be opsonized by immune animals. Furthermore, antibodies against pili can interfere with the bacterial attachment to host cells and enhance phagocytosis. In recent years, we have seen an unprecedented explosion in immunological techniques. The ability to produce unlimited quantities of pure microbial antigens through recombinant DNA techniques or to make synthetic or modified antigens with new immunochemical processes should accelerate the production of effective vaccines.

Antiserum against the J5 mutant of *Escherichia coli* offers the possibility of specific passive immunoprotection against a variety of gram-negative bacteria. This mutant has no O-antigenic side chains attached to its endotoxin. Therefore, the core portion of the molecule, which is much less specific, is exposed. Antisera against this substance can neutralize endotoxin and opsonize bacteria. Recently bacteremic patients given J5 antiserum had only 14% mortality versus 26% in control patients. Furthermore, in severely ill patients with prolonged shock, many more patients given antiserum recovered than the controls (82 versus 29% recovery). This antiserum also protects experimental animals against *Pseudomonas* infections (12).

The development of human monoclonal antibodies opens new possibilities for passive immunization against microbial infections. Passive antibody therapy was widely used in the preantibiotic era for the treatment of infectious diseases. Although it may be very effective in specific instances, this therapy has suffered from the technical difficulty that appropriate stocks of specific antibodies with high-affinity constants are expensive and very hard to prepare in large amounts by conventional methods. Furthermore, such antibodies are usually obtained from animals (particularly horses), and humans often mount major immune reactions to the injection of foreign immunoglobulin. This results in the syndrome of serum sickness, which obviates further use of this specific therapy in that patient. The new technique of production of antibodies *in vitro* by somatic cell fusion (hybridoma technology) has the potential to generate the desired amounts of specific human antibodies. Hybridoma technology could make available high concentrations of pure antibody preparations, which can be administered to counter the opportunistic pathogens detected in host tissues.

Less specific immunoglobin therapy may also be used successfully to control posttrauma infection in the future. Injury from burns, and possibly other insults, is characterized by significant decrements in immunoglobulin G (IgG) (9), and replacement could be protective. In support of this, Davis (13) has reported that a commerical preparation of immune serum globulin administered intravenously to mice affords significant protection against *Pseudomonas* infections. This material seems to act as an opsonin rather than an antitoxin. In the future, nonspecific immunoglobulins could be spiked with synergistic combinations of specific monoclonal antibodies.

Nonspecific immunoenhancement, which can be induced by a variety of microbial preparations and pharmacological agents, may someday be a major procedure in the treatment of trauma patients. Nonspecific immunoenhancers can stimulate macrophages and other cells that modulate the resistance to microbial pathogens and also enhance tissue repair. A cell wall component of *Mycobacterium,* known as muramyl dipeptide (MDP), and its derivatives have been successfully used to enhance nonspecific resistance to *C. albicans* and a variety of bacterial pathogens. Endotoxin is also a natural immunomodulator, which in sublethal amounts or detoxified form increases the granulopoiesis and nonspecific resistance to infection, the production of interferon and antibodies, and the repair of tissue. Soluble glucan has also shown promise as a safe, effective immunomodulator. Today a new armamentarium of immunoregulators such as thymosin and interleukin 1 promise future means to restore resistance to opportunistic infections.

Another approach to the control of infection after trauma is the regulation of host inflammatory processes. For example, baboons have been effectively protected against *E. coli*-induced shock by combined treatment with methylpred-

TABLE 11-3

ANTIBIOTIC THERAPY

SYSTEMIC

I. Types of agents
  A. Aminoglycosides, such as gentamicin, netilimicin, tobramycin, and amikacin, are the most effective
  B. Ureido-penicillins, such as ticarcillin and piperacillin, are less effective than the aminoglycosides, but are synergistic with them against gram-negative enterics
  C. Monobactams are effective against gram-negative enterics, to a lesser degree than aminoglycosides, but have no renal toxicity as aminoglycosides
  D. β-lactam-resistant penicillins, such as methicillin or dicloxicillin, are effective for therapy of *S. aureus*; vancomycin can be administered for the therapy of methacillin-resistant *S. aureus*
  E. Imipenem (combined with cilastalin) is the only single agent that is effective against aerobic gram-positive and gram-negative organisms as well as anaerobic bacteria; however, some strains of *Pseudomonas* may be resistant

II. Combination therapy
  Several combinations have been advocated for therapy of mixed aerobic-anaerobic infection or gram-negative infections in the compromised host
  A. Gram-negative infection
    1. Aminoglycoside plus ureido-penicillins
    2. Aminoglycoside plus a cephalosporin (second or third generation)
    3. Aminoglycoside plus a monolactam
  B. Gram-positive infection
    Combinations of β-lactam-resistant penicillin and an aminoglycoside
  C. Mixed aerobic–anaerobic infections
    An aminoglycoside plus either clindamycin, cefoxitin, or metronidazole

SELECTIVE DECONTAMINATION[a]

  A. Trimethoprim-sulfamethoxazole (co-trimoxazole) (six regular tablets/day)
  B. Polymixin (800 mg/day) + co-trimoxazole (six regular tablets/day)
  C. Polymixin + nalidixic acid[b] (800 mg + 6 g/day)
  D. Antifungal agents may also be used such as amphotericin B (2 g/day) or nystatin (6×10^6 IU/day)

[a]These combinations are administered orally, and are proved most effective, provided colonization resistance flora are present.

[b]Could be replaced by pipemedinic acid (800 mg/day) and perhaps norfloxacin (800 mg/day).

nisolone sodium succinate and gentamicin (14). Adrenal corticosteroids, in contrast to other antiendotoxin agents, have also been shown to prevent mortality in mice challenged with *E. coli, Proteus mirabilis,* or *K. pneumoniae*. Recent findings have shown a protective effect of indomethacin in rats with *E. coli* peritonitis. Opiate antagonists, particularly in conjunction with methylprednisolone, may also be useful in the treatment of sepsis (15).

Most new approaches to the control of infection in compromised patients are still tentative. Once infections are established, antibiotic therapy may be the only means of treatment now available (Table 11-3). Unfortunately, systemic antibiotics alone are often not sufficient to control infection in a compromised host. Antibiotics, however, may enhance the effectiveness of other treatments. The future use of antibiotics may also be more successful if applied in synergistic combinations. For example, it has been demonstrated that successful therapy requires at least two effective antibiotics (9). Current studies also indicate that new antibiotics such as third-generation cephalosporins may control pathogens that have previously resisted treatment. Another approach uses substances such as clavulanic acid (16), which inhibits β-lactamase activity and, in combination with penicillin derivatives, produces a very effective antimicrobial preparation.

Further consideration must be given to identifying the organisms against which antibiotic therapy should be directed. For example, *Bacteroides* and facultative gram-negative microorganisms often occur together. One organism could produce substances that protect the other from phagocytosis or antimicrobial therapy (i.e., capsular material and β-lactamase, respectively). Also one organism could create an anaerobic environment that is essential to the other, or it could produce nutritional growth factors. Many slowly growing and fastidious pathogens that are important in infections in severely compromised patients may be still unrecognized. Studies of these organisms will be necessary to select appropriate antimicrobial therapy.

Antibiotics can also be used to reduce the colonization of intestinal mucosa by opportunistic pathogens. Total intestinal decontamination is difficult to achieve, and it creates a vulnerability to colonization by antibiotic-resistant pathogens. However, selective decontamination with oral antibiotics has already been tested clinically, and it offers significant promise for the management of mass casualties receiving midlethal radiation in nuclear warfare. This approach uses the oral administration of specific antibiotics that eliminate opportunistic pathogens but leave intact the relatively benign intestinal flora (17). These benign flora increase colonization resistance by occupying binding sites and creating an environment that is inhospitable to pathogens. Selective decontamination eliminates the need for elaborate methods of isolation. In patients with aplastic anemia, leukemia, or burns, selective decontamination with antibiotics (such as oral nalidixic acid, co-trimoxazole, or colistin and amphotericin B) significantly reduced the number of infectious episodes.

III. SUMMARY

Combined injuries compromise a host's normal defensive processes and induce changes in the colonization resistance of the host's epithelial surfaces. These two events permit opportunistic infections to occur, which can become lethal complications of injury.

TABLE 11.4

POTENTIAL APPROACHES TO CONTROL INFECTIONS IN COMBINED-INJURY VICTIMS

Before injury	
Vaccination	
Radioprotective drugs	
After injury–before infection	
Surgical repair and administration of prophylactic antibiotics	
Drug modulation of injury	
Selective decontamination	
Replace:	Circulating fibronectin
	Granulocytes
	Bone marrow
	Immunoglobulin G
Nonspecific immunostimulation	
Enhance SIgA-mucous barriers	
During infection	
Monoclonal antibodies or antisera	
Antiinflammatory drugs	
Antibiotics	

Relatively little is known at present concerning the physiologic and immunologic mechanisms that are responsible for combined injury and its infectious complications. As the cellular and humoral factors regulating these phenomena become better understood, new prophylactic and therapeutic regimens should become available (Table 11-4). Since a period exists between injury and infection, there should be time to initiate a variety of approaches to treatment once the patient is stabilized. Pharmacologic modulation of the synergistic effects of combined injury may become possible, or at least we should learn to better implement some of the means described above for controlling infection in injured subjects. It will be necessary to evaluate these regimens in animal models with combined injury to accurately assess their effectiveness. Also, it must be remembered that to be truly usable in mass-casualty situations, the approaches that are perfected will have to be simple in their requirements for personnel and facilities.

REFERENCES

1. Miller, C. P., Hammond, C. W., and Tompkins, M. The incidence of bacteria in mice subjected to whole body x-radiation. *Science* **3,** 540–551 (1950).
2. Alpen, E. L., and Sheline, G. E. The combined effects of thermal burns and whole body x-irradiation on survival time and mortality. *Ann. Surg.* **140,** 113–118 (1954).
3. Messerschmidt, O. Kombinationsschadin als Folge nuclearer Explosion. In *Sonderdrucans Chi-*

rurgie der Gegenwart: Vol. 4, Unfallchirurgie (D. F. Zenker and W. Schink, eds.), pp. 1–102. Urban & Schwarzenberg, Munich, 1976.

4. Donati, R. M., McLaughlin, M. M., and Stromberg, L. W. R. Combined surgical and radiation injury. VIII. The effect of the gnotobiotic state on wound closure. *Experientia* **23,** 1388–1390 (1973).
5. Porvaznik, M., Walker, R. I., and Gillmore, J. D. Reduction of the indigenous filamentous microorganisms in the rat ileum following radiation. *Scanning Electron Microsc.* **3,** 15–22 (1979).
6. Wood, E. E., Straus, D. C., Johanson, W. G., Jr., and Bass, J. A. Role of fibronectin in the prevention of adherence of *Pseudomonas aeruginosa* to buccal cells. *J. Infect. Dis.* **143,** 784–790 (1981).
7. Cuevas, P., and Fine, J. Production of fatal endotoxic shock by vasoactive substances. *Gastroenterology* **64,** 285–291 (1973).
8. Saba, T. M., and Jaffe, E. Plasma fibronectin (opsonic glycoprotein): Its synthesis by vascular endothelial cells and role in cardiopulmonary integrity after trauma as related to reticuloendothelial function. *Am. J. Med.* **68,** 577–594 (1980).
9. Munster, A. M., Hoagland, H. C., and Pruitt, B. A., Jr. The effect of thermal injury on serum immunoglobulins. *Ann. Surg.* **172,** 9565–9569 (1970).
10. Love, L. J., Schimpf, S. C., Schiffer, C. A., and Wiernick, P. H. Improved prognosis for granulocytopenic patients with gram-negative bacteremia. *Am. J. Med.* **68,** 643 (1980).
11. Wretlind, B., and Pavlovskis, O. R. The role of proteases and exotoxin A in the pathogenicity of *Pseudomonas aeruginosa* infections. *Scand. J. Infect. Dis.* **29,** Suppl., 13–19 (1981).
12. Ziegler, E. F., McCutchan, A. J., Douglas, H., and Braude, A. I. Prevention of lethal pseudomonas bacteremia with epimerase-deficient *E. coli* antiserum. *Trans. Assoc. Am. Physicians* **88** 101–109 (1975).
13. Davis, S. D. Efficacy of modified human immune serum globulin in the treatment of experimental murine infections with seven immunotypes of *Pseudomonas aeruginosa. J. Infect. Dis.* **134,** 717–721 (1975).
14. Hinshaw, L. B., Archer, L. T., Beller-Todd, B. K., Coalson, J. J., Flourney, D. J., Passey, R., Benjamin, B., and White, G. L. Survival of primates in LD_{100} septic shock following steroid/antibiotic therapy. *J. Surg. Res.* **28,** 151–170 (1980).
15. Weissglas, I. S., Hinchey, E. J., and Chiu, R. C. J. Naloxone and methylprednisolone in the treatment of experimental septic shock. *J. Surg. Res.* **33,** 131–135 (1982).
16. Ball, P., Watson, T., and Mehtar, S. Amoxycillin and clavulanic acid in intra-abdominal and pelvic sepsis. *J. Antimicrob. Chemother.* **7,** 441–444 (1981).
17. Mulder, N. H., Nieweg, H. W., Sleijfer, D. T., de-Vries-Hospters, H. G., van der Waaij, D., Fidler, V., and van Saene, H. K. F. Infection prevention in granulocytopenic patients by selective decontamination of the digestive tract. In *New Criteria for Antimicrobial Therapy: Maintenance of Digestive Tract Colonization Resistance* (D. van der Waaij and J. Verhoef, eds.), pp. 113–117. Exerpta Medica, Amsterdam, 1979.

CHAPTER 12

Diagnosis, Triage, and Treatment of Casualties

JAMES J. CONKLIN AND RICHARD I. WALKER

Armed Forces Radiobiology Research Institute
Bethesda, Maryland 20814-5145

Injuries associated with radiation pose new and significant challenges for medical management. Radiation affects many organ systems, often compounding the problems produced by conventional injuries. These challenges are great enough when the injuries are confined to a few patients, but the use of nuclear weapons raises the probability of large numbers of casualties. Mass-casualty medicine using simplified and standardized regimens will be required to accomplish what is now done by labor-intensive and resource-demanding means. Thus new concepts in treatment must be considered.

If nuclear weapons are ever used again, the basic principles of mass casualty (triage, evacuation, and use of standard procedures) will have to be followed. Difficulties in their application will be due to relative inexperience in dealing with these types of patients. Life-threatening doses of acute total-body radiation have been so infrequent that management policies must be derived in part from different but analogous clinical situations and from studies with experimental animals.

Conventional injuries should be treated first, since no immediate life-threatening hazard exists for radiation casualties who can ultimately survive. The patient with multiple injuries should be resuscitated and stabilized. During this process, the standard preparation for surgery will accomplish much decontamination of radioactivity. After surgery, a more definitive evaluation of radiation exposure can be initiated.

I. DECONTAMINATION OF THE PATIENT

Radiation injury per se does not imply that the patient is a health hazard to the medical staff. Studies indicate that the levels of intrinsic radiation present within the patient from activation (after exposure to neutron and high-energy photon sources) are not life threatening.

Patients entering a medical treatment facility should be routinely decontaminated if the monitoring of radiation is not available. Removal of the patient's clothing will usually reduce most of the contamination, and washing of exposed body surfaces will further reduce it. Both procedures can be performed in the field or on the way to the treatment facility. Once the patient has entered the treatment facility, his care should be based on the obvious injuries. Care for life-threatening injuries should not be delayed until completion of the decontamination procedures.

When radiation safety personnel are available, decontamination procedures will be established to assist in rendering care and to minimize the hazard from radioactive contaminants. A more extensive decontamination procedure is to scrub the areas of persistent contamination with a mild detergent or a diluted strong detergent. Caution should be taken to not disrupt the integrity of the skin while scrubbing, because disruption can lead to incorporation of the radioisotope into deeper layers of the skin. Contaminated wounds should be treated first, since they will rapidly incorporate the contaminant. Washing, gentle scrubbing, or even debridement may be necessary to reduce the level of contaminants.

Wearing surgical attire will reduce the possible contamination of health personnel. If additional precautions are warranted, rotation of attending personnel will further reduce the possibility of significant contamination or exposure. The prevention of incorporation is of paramount importance. The inhalation or ingestion of radioactive particles is a much more difficult problem, and resources to deal with it will not be available in a field situation.

II. EVALUATION OF RADIATION INJURY

Radiation complicates the requirements of patient management. Hematologic injuries cause anemia, infection, bleeding, and delayed wound healing. Performance decrements due largely to neuromediator release can also impact the patient. At higher doses of radiation, severe fluid and electrolyte losses occur through the intestinal wall.

The problem of the physician is to triage patients according to the severity of radiation injuries so that appropriate treatment can begin. This problem is compounded because the response of any given individual may vary greatly, and a nonhomogeneous exposure of radiation (especially if the bone marrow and gut are spared) may result in a markedly decreased effect. Also, United States forces do not at present carry personal dosimeters that measure neutron and photon exposures. Finally, dose rate effects can be very pronounced, especially in a fallout environment. In this situation, tactical dosimeters (two per platoon) may be useful to a commander who is deciding whether to commit exposed troops to battle, but they are less useful to the health-care provider. Other problems will also exist. Casualties will be numerous and resources certainly will be inadequate. Complicating this will be the occurrence of blast and thermal injuries (in addition to radiation injuries). Improved dosimetry is needed for triage because the goal of military medical personnel should be the allocation of precious resources to salvage the maximum number of causalties. But improved dosimetry is currently unavailable.

Consequently, the following guidelines apply to medical personnel operating in austere field conditions, based on recent recommendations (1). Lymphocyte levels may be used as a biologic dosimeter to confirm the presence of pure radiation injury, but not in combined injuries. If the physician has the resources of a clinical laboratory, additional information can be obtained to support the original working diagnosis suggested by the presence of prodromal symptoms. An initial blood sample for concentrations of circulating lymphocytes should be obtained as soon as possible from any patient classified as "radiation injury possible" or "radiation injury probable." After the initial assessment or at least no later than 24 hr after the event in question, additional blood samples should be taken for comparison. The samples may be interpreted as follows:

(1) *Lymphocyte levels in excess of 1500/mm*^{3}. The patient most likely has not received a significant dose that would require treatment.

(2) *Lymphocyte levels between 1000 and 1500/mm*^{3}. The patient may require treatment for moderate depression in granulocytes and platelets within 3 weeks postexposure.

(3) *Lymphocyte levels between 500 and 1000/mm*^{3}: The patient will require treatment for severe radiation injury. The patient should be hospitalized to minimize the complications from hemorrhage and infection that will arise within 2–3 weeks postexposure.

(4) *Lymphocyte levels of less than 500/mm*^{3}: The patient has received a radiation dose that may prove fatal. The patient needs to be hospitalized for the inevitable pancytopenic complications.

(5) *Lymphocytes not detectable:* The patient has received a supralethal radiation dose, and survival is very unlikely. Most of these patients have received severe injuries to their gastrointestinal and cardiovascular systems, and will not survive for more than 2 weeks.

A useful rule of thumb is, If lymphocytes have decreased by 50% and are less than 1000/mm^3, then the patient has received a significant radiation exposure. *In the event of combined injuries, the use of lymphocytes may be unreliable.* Patients who have received severe burns or trauma to more than one system often develop lymphopenia.

TABLE 12-1

PRELIMINARY TRIAGE OF CASUALTIES WITH POSSIBLE RADIATION INJURIES[a]

	Radiation injury		
Symptoms	Unlikely	Probable	Severe
Nausea	–	++	+++
Vomiting	–	+	+++
Diarrhea	–	±	± to +++
Hyperthermia	–	±	+ to +++
Hypotension	–	–	+ to ++
Erythema	–	–	– to ++
CNS Dysfunction	–	–	– to ++

[a]From Alter and Conklin (1).

It is difficult to establish an early definitive diagnosis. Therefore it is best to function within a simplified, tentative classification system based on the three possible categories of patients noted in Table 12-1 and discussed in the following.

(1) *Radiation injury unlikely.* If no symptoms appear after radiation injury, patients are judged to be at minimal risk for radiation complications. These patients should be triaged according to the severity of conventional injuries. If the patients are free of conventional injuries or disease states that require treatment, they should be released and returned to duty.

(2) *Radiation injury probable.* Anorexia, nausea, and vomiting are the primary prodromal symptoms associated with radiation injury. Priority for further evaluation will be assigned after all life-threatening injuries have been stabilized. Casualties in this category will not require any medical treatment within the first few days for their radiation injuries. In the absence of burns and trauma, evidence to support the diagnosis of significant radiation injury may be obtained from lymphocyte assays taken over the next 2 days. If the evidence indicates that a significant radiation injury was received, these casualties need to be monitored for pancytopenic complications.

(3) *Radiation injury severe.* These casualties are judged to have received a radiation dose that is potentially fatal. Nausea and vomiting will be almost universal for persons in this group. The prodromal phase may also include prompt explosive bloody diarrhea, significant hypotension, and signs of neurologic injury. These patients should be sorted according to the availability of resources. Patients should receive symptomatic care. Lymphocyte analysis is necessary to support this classification.

III. SYMPTOMS FREQUENTLY OCCURRING IN WHOLE-BODY IRRADIATED CASUALTIES WITHIN THE FIRST FEW HOURS POSTEXPOSURE

A. Nausea and Vomiting

Nausea and vomiting occur with increasing frequency as the radiation exceeds 100–200 cGy. The symptoms may last as long as 6–12 hr postexposure, but they usually subside within the first day. If vomiting occurs within the first 2 hr after exposure, the radiation dose may be classified as severe. Vomiting within the first hour, especially if accompanied by explosive diarrhea, is associated with doses that frequently prove fatal. Due to the transient nature of these symptoms, it is possible that the patient will have already passed through this initial phase of gastrointestinal distress before being seen by a physician. It will be necessary for the physician to inquire about these symptoms at the initial examination.

B. Hyperthermia

Casualties who have received a potentially lethal radiation injury show a significant rise in body temperature within the first few hours postexposure (2). This appears to be a consistent finding, although the number of such cases is few. The occurrence of fever and chills within the first day postexposure is associated with a severe and life-threatening radiation dose. Hyperthermia may occur in patients who receive lower but still serious radiation doses (200 cGy or more). Present evidence indicates that hyperthermia is frequently overlooked. Individuals wearing a chemical ensemble will normally be hyperthermic, so this will not be a useful sign.

C. Erythema

A person who received a whole-body radiation dose of more than 1000–2000 cGy will experience erythema over the entire body within the first day postexposure. An individual who received a comparable dose to a local body region will also experience erythema, but only at the affected area. With doses lower but still in the potentially fatal range (200 cGy or more), erythema is less frequently seen.

D. Hypotension

A noticeable and sometimes clinically significant decline in systemic blood pressure has been recorded in victims who received a supralethal dose of whole-body radiation (3). A severe hypotensive episode was recorded in one person who had received several thousand rads. In individuals who received several hundred rads, a drop in systemic blood pressure of more than 10% has been noted. Severe hypotension after irradiation is associated with a poor prognosis.

E. Neurologic Dysfunction

Experience indicates that almost all persons who demonstrate obvious signs of damage to the central nervous system within the first hour postexposure have received a supralethal dose (3). Symptoms include mental confusion, convulsions, and coma, probably accompanied by intractable hypotension. Despite vascular support, these patients succumb within 48 hr.

Casualties who have received a potentially fatal dose of radiation will most likely experience a pattern of prodromal symptoms that is associated with the radiation exposure itself. Unfortunately, these symptoms are nonspecific and may be seen with other forms of illness or injury, which may complicate the process of diagnosis. Therefore, the triage officer must determine the symptoms that have occurred within the first day postexposure, evaluate the possibility that they are indeed related to radiation exposure, and then assign the patient to one of the three categories "radiation injury unlikely," "radiation injury probable," "radiation injury severe." In the last two categories, the study of changes in circulating lymphocytes may either support or rule out the original working diagnosis. All combined-injury patients should be treated initially as if no significant radiation injury is present. Triage and care of any life-threatening injuries should be rendered without regard for the probability of radiation injury. The physician should make a preliminary diagnosis of radiation injury only for those patients for whom radiation is the sole source of the problem. This is based on the appearance of nausea, vomiting, diarrhea, hyperthermia, hypotension, and neurologic dysfunction.

IV. INITIAL TREATMENT FOR PATIENTS WITH WHOLE-BODY RADIATION INJURY

The primary determinants of survival among most patients receiving intermediate radiation doses (which are serious but not uniformly fatal, if treated) are the management of microbial infections and the stopping of any bleeding. If a high intermediate dose has been received, fluid and electrolyte losses may cause early death. If properly resuscitated, however, these patients can survive until the consequences of hematologic failure become apparent.

For a casualty who received a sublethal whole-body radiation dose, gastrointestinal (GI) distress will predominate in the first 2 days. Antiemetics (metoclopramide and dazopride) may be effective in reducing the GI symptoms, but the presently available drugs have significant side effects. Unless severe radiation injury has occurred, the symptoms will usually subside within the first day. For a patient who continues to experience gastrointestinal distress, parenteral fluids should be considered. If explosive diarrhea occurs within the first hour postexposure, fluids and electrolytes should be administered if available. For triage purposes, the presence of explosive diarrhea (especially bloody) is likely to be related to a fatal radiation dose.

Cardiovascular support for patients with clinically significant hypotension and neurologic dysfunction should be undertaken only when resources and staff allow. These patients are not likely to survive injury to the vascular and gastrointestinal systems combined with marrow aplasia.

V. DIAGNOSIS AND TREATMENT OF THE PATIENT WITH COMBINED INJURIES

In the event of a radiation accident or nuclear detonation, many patients will probably suffer burns and traumatic injuries in addition to radiation effects. The initial triage of combined-injury patients is based on these conventional injuries. Further reclassification may be warranted on the basis of prodromal symptoms associated with radiation injury. The prognosis for all combined injuries is worse than for radiation injury alone. Animal studies indicate that, when other injuries are accompanied by sublethal doses of radiation, infections are much more difficult to control, and wounds and fractures heal more slowly. Thus, potentially survivable burns and trauma will be fatal in a large percentage of persons who have also received significant injury from sublethal doses of radiation. Often with conventional injuries, staged reparative surgery is scheduled for 1–2 days after the initial surgery, and reconstructive surgery later still. Because of the delays in wound healing and the subsequent granulocytopenia and thrombocytopenia with injuries from nuclear weapons, most of the life-saving and reconstructive surgery must be performed within 36 hr after the exposure. Then, if possible, no surgery should be performed for the next 6–8 weeks postexposure.

VI. MANAGEMENT OF INFECTION

In spite of antibiotics, infections with opportunistic pathogens are still a major problem. The majority of these organisms today are gram-negative, including

TABLE 12-2

SYSTEMIC ANTIBIOTIC THERAPY

Types of agents
Aminoglycosides, such as gentamicin, netilimicin, tobramycin, and amikacin, are the most effective.
Ureido-penicillins, such as ticarcillin and piperacillin, are less effective than the aminoglycosides, but are synergistic with them against gram-negative enterics
Monobactams are less effective against gram-negative enterics than are aminoglycosides, but they have no renal toxicity whereas the aminoglycosides do
Beta lactam-resistant penicillins, such as methicillin or dicloxicillin, are effective for therapy of *Staphylococcus aureus;* vancomycin can be administered for therapy of methicillin-resistant *S. aureus*
Imipenem (combined with cilastalin) is the only single agent that is effective against aerobic gram-positive and gram-negative organisms as well as anaerobic bacteria; however, some strains of *Pseudomonas* may be resistant
Combination therapy
Several combinations have been advocated for the therapy of mixed aerobic–anaerobic infection or for the therapy of gram-negative infections in compromised hosts as follows:
Gram-negative infection
Aminoglycoside plus ureido-penicillins
Aminoglycoside plus a cephalosporin (second or third generation)
Aminoglycoside plus a monolactam
Gram-positive infection
Combinations of β-lactam-resistant penicillin and an aminoglycoside
Mixed aerobic–anaerobic infections
An aminoglycoside plus clindamycin, cefoxitin, or metronidazole

Escherichia coli and *Pseudomonas aeruginosa*. These infections are a consequence of profound immunosuppression and abnormal colonization of body surfaces and medical devices. Susceptible body surfaces include the oropharyngeal–respiratory tree and the intestine. Wound sites and artificial invasive devices such as catheters are also important sources of infection. Infections may be more prevalent and severe if patients are maintained for long periods in environments containing antibiotic-resistant pathogens.

Wound debridement, dressings, and (when necessary) antibiotics are key elements in controlling infection. Antibiotics, preferably in combination therapy (Table 12-2), should be used promptly to treat any new fever. When signs or symptoms of infection do appear in the granulocytopenic patient, tratment should be started without waiting for culture and sensitivity studies. Initial coverage should include gram-negative organisms and *Staphylococcus aureus*. The prevalent organisms and antimicrobial susceptibility patterns in the particular medical facility should also be considered. At present, the drugs most often used for initial treatment are the synthetic penicillins such as ticarcillin, combined with an aminoglycoside like tobramycin. It is recommended to either continue treatment until granulocytes return to more than 500, or treat for just 2 weeks and stop even if the white cell count is still low, as long as all signs of infection have cleared.

VII. FUTURE CONCERNS FOR MANAGEMENT OF RADIATION INJURIES

Treatable radiation-associated injuries include only those obtained with the hematologic syndrome and possibly the gastrointestinal syndrome. Combined injuries will shift the range of treatable injuries to only the lower radiation doses. Even for these ranges, very little definitive information is available now. Many approaches of treatment that are suitable for conventional injuries may be of little utility for irradiated patients.

The first actions in dealing with radiation casualties are to treat any conventional injuries, that is, to maintain ventilation and perfusion and to stop hemorrhage. Most decontamination will be accomplished through routine management of the patient. After patient stabilization, it will be necessary to triage for radiation injuries, followed by steps to prevent infection, fluid and electrolyte imbalance, and bleeding. Unfortunately, the military physician is severely limited in his ability to effect these treatments successfully, particularly on a large scale with limited resources.

New means of radioprotection and repair of radiation damage are on the horizon. Furthermore, immunomodulators are now under study, which may not only facilitate the regeneration of marrow but also help reduce the profound immunosuppression that is responsible for infections in severe injury. These agents may be used in combination with radioprotectors and antibiotics to further enhance survival. Leukopenia is a significant problem in irradiated casualties, but hazards exist with the transfusion of leukocytes into patients. Stimulation of the repair of surviving stem cells in selected populations probably offers the best opportunity to correct this deficiency. Although the transfusion of platelets is certainly desirable for radiation victims, it is presently not practical for mass-casualty situations. This is also true for the transplantation of bone marrow, although enormous progress is being made in autologous bone marrow transplantation. Again, stem-cell regeneration is probably the best near-term hope of solving this problem. Problems of the effective management of wounds and the replacement of fluids and electrolytes remain to be overcome in the neutropenic patient. Pharmacologic means to regulate performance decrements such as emesis and early transient incapacitation still are not available for use by military personnel.

The foregoing shows that much remains to be done in order to achieve effective treatment of radiation victims or combined-injury victims. However, progress is being made, and the concerns outlined above will be resolved.

REFERENCES

1. Alter, W. A., and Conklin, J. J. Radiation casualties. In *Disaster Medicine* (B. Walcott, F. M. Burkle, and P. Sanner, eds.), pp. 197–209. Medical Examination Publishing Co., Inc., New York, 1984.
2. Hempelmann, L. H., Lisco, H., and Hoffman, J. G. The acute radiation syndrome. *Ann. Intern. Med.* **36,** 279 (1952).
3. Fanger, A., and Lushbaugh, C. C. Radiation death from cardiovascular shock following a criticality accident. *Arch. Pathol.* **83,** 446 (1967).

CHAPTER 13

Internal Contamination with Medically Significant Radionuclides

ASAF DURAKOVIC

Armed Forces Radiobiology Research Institute
Bethesda, Maryland 20814-5145

In a nuclear explosion, over 400 radioactive isotopes are released in the biosphere (1). Of these, about 40 radionuclides are of potential hazard to man. Of particular interest to the field of medicine are the isotopes whose organospecificity and long half-life present a danger of irreversible tissue damage or the induction of malignant alterations.

Some incorporated radionuclides are partly diminished in the processes of radioactive decay and biological elimination, but other isotopes have a long half-life and are incorporated in firm tissue. These latter isotopes must be therapeutically removed from the contaminated individual.

The radiation effects of an internally deposited radionuclide depend on its chemical nature, solubility, half-life, type of radioactive decay, the tissue of incorporation, and the physiological factors determining its metabolic fate. High organospecificity of certain fission products will result in radiation damage to selective tissues; in contrast, other radionuclides that are uniformly distributed in the body fluids will result in the relatively uniform exposure of various organs to the radiation. Some radionuclides produce detectable tissue damage soon after their incorporation in tissues of high radiosensitivity, whereas other radioisotopes may result in induced somatic and genetic changes as late effects.

From the biomedical point of view, an approach to the problem of internal contamination should focus on radionuclide organospecificity, physical characteristics, and chemical characteristics. The fate of radioisotopes will depend on various factors, including their chemical and physical natures, solubility, particle size, homeostasis, type of decay, biological decorporation, and elimination from the contaminated individual.

Radioisotopes that have no specific target organs will be rapidly eliminated by the processes of natural clearance. But that is not true for the organospecific radioisotopes. Some highly organospecific radioisotopes (e.g., iodine) are concentrated in their natural target organs (thyroid), and other radionuclides that are not normally present in nature (e.g., plutonium) also show high organospecificity, with osteotropic characteristics (due to their metabolic pathways) similar to the normal constituents of the calcified tissue. Incorporation of such radiosotopes in their target organs can result in considerable tissue damage. This is because some of those isotopes have extremely long half-lives and involve decay by the particulate (alpha and beta) radiation, resulting in a high probability of malignancy in the target organs that are radiosensitive.

Radioisotopes that are produced by nuclear fission are distributed in the organism by the bloodstream after they enter the organism by ingestion, by inhalation, or through wounds. The amount that enters depends on the radioisotope's physical properties and its solubility in body fluids. The ultimate size of the deposit in tissue is determined by the radioisotope's chemical properties.

Various radioisotopes use different portals of entry into the individual. Some are preferentially absorbed in gastrointestinal mucosa, others gain access to the bloodstream via the respiratory tract, and still others preferentially enter the body fluids through wounds or by direct intravascular administration. The length of retention of the radioisotope in the body is determined by its effective half-life, mechanism of entry, quantity, target organ, and the processes of elimination.

Some fission products are preferentially absorbed in the intestine (calcium, strontium, iodine, cesium, etc.), whereas others (e.g., actinides) are hardly absorbed by ingestion and are primarily incorporated by inhalation or through contaminated wounds. Most of the fission products are rapidly eliminated from the body after the initial fallout from a nuclear detonation. The main biomedical hazard is due to the radioisotopes of high organospecificity and long half-life (such as ^{137}Cs, ^{90}Sr, ^{90}Y, ^{14}C, ^{3}H, ^{131}I, and transuranic elements), which invariably produce pathologic changes (including malignant tumors, spontaneous and induced mutations) as their genetic effects in the contaminated individual.

I. ROUTES OF ENTRY AND DISTRIBUTION

The four main routes of internal contamination are (1) ingestion and gastrointestinal absorption, (2) inhalation and transalveolar transfer to the bloodstream, (3) percutaneous absorption, and (4) through wounds or by direct injection into the bloodstream.

A. Ingestion

Gastrointestinal absorption of the nuclear fission products differs for the various radionuclides. Some of the ingested radioactive isotopes preferentially enter the bloodstream via the intestinal mucosa, whereas other isotopes are not absorbed in any significant amount. Of those isotopes whose principal route of entry is gastrointestinal absorption, the most significant are the isotopes of cesium (^{137}Cs), strontium (^{90}Sr), cobalt (^{60}Co), iodine (^{131}I), phosphorus (^{32}P), mercury (^{197}Hg and ^{203}Hg), radium (^{226}Ra), and tritium (^{3}H).

Gastrointestinal absorption is an important route of entry of the osteotropic alkaline earth isotopes such as ^{90}Sr. Gastrointestinal absorption is particularly important as a consequence of the delayed fallout hazards because of the contaminated biosphere and the food contaminated by nuclear fission products (farm produce and dairy products). However, the homeostatic mechanisms that govern the transfer of radioactive isotopes across the intestinal mucosa can discriminate against some of the radioisotopes that are foreign to the organism, thus favoring absorption of their homologs, which are involved in the normal homeostasis.

Over 90% of the entire process of discrimination of strontium takes place in the gastrointestinal tract, where calcium is preferentially absorbed. This phenomenon constitutes one of the methods of therapeutic removal of radioactive strontium via the intestinal tract.

Other sites where discrimination processes against radioactive strontium occur include the renal tubules, the mammary gland, and the placenta, where calcium reabsorption is favored. These biological membranes represent the sites of homeostatic protection against potentially hazardous radionuclides.

The mechanism of preferential absorption of calcium in relation to strontium in the intestinal mucosa was partly addressed by the processes of diffusion and active transport for calcium, whereas the transfer of strontium from the intestinal lumen to the circulation is mainly via diffusion (2). The ingestion of ^{137}Cs results in its rapid entry into the bloodstream. Numerous cases have been reported of accidental contamination with ^{137}Cs in humans (3, 4).

Intestinal absorption of radioactive iodine (^{131}I) is an important route of accidental contamination because the transfer of contamination from the biosphere to the human body takes place via the food chain (from pasture to dairy product to man). Numerous reports in the literature (5, 6) describe protective measures against the accidental ingestion of ^{131}I (including disposal of contaminated cattle feed and dairy products). In all cases of accidental ingestion of ^{131}I, a thyroid bioassay should be made, and therapeutic management of the contaminated patients should begin immediately. Periodic monitoring for the evidence of hypothyroidism should be performed for several years (7).

The intestinal absorption of radium (^{226}Ra) is an important cause of inducing skeletal malignancies. Over 30% of ^{226}Ra is absorbed in the intestine after accidental ingestion, and it is almost entirely deposited in the skeleton (8–10). Ingestion of ^{226}Ra has been reported in the classic work on internal contamination in dial painters who ingested luminous paints containing ^{226}Ra (11, 12). Various pathological consequences followed the ingestion of ^{226}Ra, including osteogenic sarcoma, fibrosarcoma, paranasal and mastoid carcinoma, aplastic anemia, and leukemia (13, 14).

Other radionuclides that enter the circulation via the gastrointestinal tract include tritium (^{3}H) [which penetrates intestinal mucosa in the form of tritiated water (15)] and uranium isotopes (^{234}U, ^{235}U, and ^{238}U). The uranium isotopes present a high biomedical hazard because of their long half-lives, nephrotoxicity (^{238}U), and retention in the skeletal tissue (^{234}U and ^{235}U), with a high potential of inducing malignancy in the bone and hematopoietic tissues.

B. Inhalation

The kinetics of (1) the deposition of radionuclides in the bronchial tree and alveoli and (2) the passage of radionuclides across the alveoli into the bloodstream is extremely complex, from the viewpoints of physiology and radiation toxicology (16). Inhaled radioactive particles are deposited in the upper bronchial tree on the alveolar surfaces, or, if soluble, they are absorbed into the systemic circulation.

Classic reports on the quantitative data concerning deposition of the radioactive particles in the bronchoalveolar tree were reported over 35 years ago (17, 18). Since that time, many reports have been published concerning the pathways of various radioisotopes in the respiratory system. To evaluate the radiation hazard of inhaled radioactive particles, a general model of their metabolic behavior in the respiratory system was adopted by the International Commission on Radiation Protection in 1955 (19). This model was later revised (20), with emphasis on the significance of different variables that determine the metabolic fate of inhaled radioactive particles.

According to that model, about 75% of inhaled radioactive particles are deposited in the respiratory tree, and 25% are immediately exhaled. About 50% of the inhaled particles are deposited in the upper bronchial tree; then they are moved by the ciliary epithelium to the nasopharynx. From there they are swallowed and handled according to the mechanisms of their gastrointestinal kinetics.

This is an important factor in contamination with actinides. Their intestinal absorption is negligible, but their deposition in the lung is a major radiotoxicologic hazard. To move them from the respiratory system to the gastrointestinal system is one of the aims of therapeutic management of accidentally inhaled actinides. About 25% of these inhaled particles are deposited on the alveolar surfaces; at this site, the metabolic behavior of the particles largely depends on their solubility. In general, about 10% of particles reaching alveolar surfaces are transferred into the systemic circulation. The remaining 15% ascend the bronchial tree and are ultimately eliminated by expectoration or by transport to the gastrointestinal tract.

Inhalation of radioactive particles is the main route of internal contamination with actinides (americium, plutonium, uranium, curium, polonium, radium, thorium), cobalt, cerium, iodine, and tritium. Quantitative differences exist in the kinetics of different radioisotopes that gain access to the body via the respiratory tract, but their ultimate distribution after reaching the systemic circulation depends on their metabolic pathways and organospecificity. Differences in distribution occur as a consequence of the radionuclide's portal of entry, solubility, particle size, and chemical form.

The transport of americium to extrapulmonary tissues after inhalation will be greater if the isotope is in a citrate form. Less body burden and accumulation in the target organ (liver and bone) occur if the radionuclide is present in the form of a nitrate. If americium-241 (^{241}Am) is inhaled as an oxide, the target organs will be the tracheobronchial lymph nodes, liver, lung, bone, and thyroid, in descending order of importance. When humans were accidentally contaminated by inhalation of an undetermined chemical form of ^{241}Am, the main target organs were bone and liver (21). Americium is eliminated from lung tissue by its absorption in the blood, by endobronchial ciliary mechanisms, and by expectoration or ingestion after reaching the nasopharynx (22). ^{241}Am that gains access to the systemic circulation from the lung is distributed equally in bone (45%) and liver (45%) for all of its compounds.

Internal contamination with plutonium (^{239}Pu) via the respiratory tract is the major route of accidental contamination. It accounts for over 75% of all industrial exposures to plutonium (23). Absorption from the respiratory tract depends on the compound's solubility. Soluble compounds (nitrate, citrate, and fluoride) are absorbed into the systemic circulation and deposited in the liver and bone within a few weeks. Retention of plutonium compounds (oxides) in the lung is much longer, with slow translocation into the pulmonary and tracheobronchial lymph nodes, followed by liver uptake many years after the inhalation exposure (24).

Uranium isotopes are a considerable hazard for accidental exposure through inhalation. The absorption and retention of a uranium isotope depend on its chemical form and particle size. Its biological half-life in the lung is estimated to be 120 days, with considerably longer half-life (1470 days) in the case of inhalation of uranium oxides. Soluble uranium salts are primarily absorbed by the respiratory route. Fatal cases have been reported of accidental inhalation in humans, which caused nephrotoxic changes including glomerular and tubular damage, azotemia, albuminuria, and tubular necrosis. These changes may be reversible; tolerance has been reported after subsequent exposure to soluble uranium compounds. Renal damage is caused by chemical rather than radiation injury. The less soluble uranium compounds are less readily absorbed in the lung (25).

Accidental internal contamination with the isotopes of iodine occurs mostly with ^{131}I, although about ten radioactive isotopes of iodine are produced in nuclear fission. Inhalation is not a major route of entry for iodine, but iodine is a significant radiation hazard because of its volatility. Inhaled iodine reaches equilibrium with body fluids in less than 1 hr, and it selectively accumulates in the thyroid gland. A thyroid bioassay should be performed in each case of suspected internal contamination with ^{131}I. As in other routes of internal contamination with ^{131}I, follow-up studies should be performed for many years. Some patients have developed hypothyroidism as late as 17 years after exposure (7).

Tritium presents a radiation hazard when inhaled. But the radiation–toxicology consequences for inhalation are less significant than for ingestion of elemental tritium as tritiated water.

Internal contamination with medically significant radioisotopes via inhalation has been described in humans, related to exposures from nuclear weapons and from industrial accidents. But a need exists for analyzing the various parameters of metabolic behavior and the consequences of internal contamination by various radionuclides via the respiratory route of exposure. To date, compartmental analysis, kinetics, and autopsy data have not been sufficiently well defined for human exposure. Further insight into the metabolic fate of inhaled radioisotopes is being gained from animal experiments and from excretion data in humans after pulmonary exposure.

C. Percutaneous Absorption

Normal skin is an effective mechanical barrier to internal contamination from most radionuclides. This route of entry is the least important in the transfer of radioisotopes from the contaminated biosphere to the internal environment of the human body, but still is of potential concern for internal contamination.

Studies on the percutaneous absorption of transuranic elements have been described in laboratory animals, with the absorption of 2% of plutonium through intact skin (26). Transcutaneous absorption in these exposure studies was facilitated by the high acidity and by complexing the plutonium with tributylphosphate. The amount of radionuclide absorbed also depended on the quantity of applied radionuclide and on the anatomic site of the skin to which applied.

The main pathway of a radioisotope from the skin to the systemic circulation is through hair follicles. The hair bulbs below their keratogenous zone are supplied by a highly vascularized connective tissue, part of a normal hair papilla. This rich network of blood vessels is the principal site of transcutaneous migration of the radioisotope from the contaminated skin into the systemic circulation.

The surface epithelium (epidermis), with its primary function of protecting the internal environment of the body, is less important as a route of entry for radioisotopes into the body. This is mainly because of its thick structure of many layers and because the keratinized stratified squamous epithelium of the outermost layer provides an effective mechanical barrier to the insults of the external environment.

However, it is not possible to consider the events in a nuclear accident as separate phenomena, because the possibility of combined injury produced by a nuclear weapon results in multiple and complex effects on the human body. It is possible that the protective capacity of the skin will be deranged in both primary and secondary thermal injury, which results in significant alterations of the skin and permits easier entry of the externally deposited contaminants into the body. Burned, desquamated, and necrotic skin loses its integrity and provides an open route of entry for radioactive and infectious insults to reach the internal organs. The main concern in preventing internal contamination through this route is to maintain the integrity of the skin.

II. INTERNAL CONTAMINATION THROUGH WOUNDS AND INJECTION INTO THE SYSTEMIC CIRCULATION

Radionuclides may have direct access to the internal environment of the body as a result of thermal or traumatic injury after atomic bomb exposure, industrial or laboratory accidents, or misadministration of radiopharmaceuticals in the diagnostic and therapeutic use of radioisotopes in hospitals.

Primary injuries by the blast component of the nuclear weapon explosion usually occur near the hypocenter. These take the form of vascular and visceral damage, without apparent damage to the skin. This type of primary traumatic injury is of no consequence for internal contamination. However, secondary blast injuries are a considerable concern for internal contamination through bruised, lacerated, or cut wounds; open fractures of bones; or multiple wounds caused by fragments of building material, glass splinters, wood, or any other contaminated projectile. These lesions pose a complex problem because of the complications of infection and internal contamination.

Intradermal or subcutaneous deposition of the fission products has been widely studied because of the therapeutic need to eliminate radioactive isotopes from the contamination site without interrupting the integrity of the normal integument. The quantity of absorbed contaminant will largely depend on the depth of the deposition, anatomic site of the skin, and size of the contaminated area. Some isotopes will translocate relatively rapidly from the intradermal or subdermal site of deposition (iodine, strontium, cesium, tritium), whereas others will absorb less avidly from the shallow dermal wounds (transuranic elements).

The fate of the isotopes at the site of superficial deposition will largely depend on the healing processes or the complications of the superficial lesions (e.g., eschar, fibrous tissue, infection, draining ulcers). Translocation from the intradermal sites of contamination is mainly via the lymphatic system. The ultimate deposition will depend on the physical and chemical natures of the isotope, solubility, particle size, and organotropism. Management of the contaminated intradermal and subdermal wounds is still an area in which further investigation is needed, preferably by professionals with experience in the medical and surgical management of contaminated wounds (24).

The intramuscular deposition of radioactive isotopes has been widely studied and documented in animal experiments and in accidental exposures of humans. Some radionuclides are completely and rapidly absorbed into the systemic circulation (e.g., strontium, iodine, and tritium), while others have a slower rate of translocation (e.g., transuranic elements).

Retention of radionuclides in the various organic systems can be affected by the site of initial deposition. That is, intramuscular deposition of the actinides will result mainly in final incorporation in the skeleton, with relatively low deposition in the liver, compared with intravenous injection. The other radioisotopes, such as iodine or osteotropic alkaline earths (calcium and strontium), will be much less affected by the site of primary incorporation, and will be ultimately deposited in the organs of their biologically specific avidity (thyroid and bone).

Radioisotopes that are normally widely distributed in the body fluids (cesium and tritium) will be largely unaffected by the site of their initial incorporation. The only effect of their intramuscular administration (versus intravenous) will be on the kinetics in the various compartments of the body.

The intravenous route of internal contamination results in the rapid incorporation of different radioisotopes in their respective target organs, as well as in their rapid removal through the renal, hepatobiliary, and other endogenous systems of elimination. Similar to that observed for absorption, the retention and elimination of various radionuclides depend on their chemical form in plasma. For example, strontium ions are present in plasma in the protein-bound, complexed, and free (hydrated) forms; strontium's elimination and reabsorption in the renal tubules will be determined by its chemical form. The osteotropic alkaline earth ions will be eliminated faster if they are in the ionized form.

Other radioisotopes such as actinides, which are preferentially incorporated in the liver and bone, will be largely affected in their deposition and elimination after intravenous administration (versus other parenteral routes). Actinides injected intravenously will be deposited in the liver in higher quantities than when injected intramuscularly, with a smaller percentage depositing in bone.

Over 30% of intravenously injected plutonium will be rapidly eliminated, mostly via the gastrointestinal tract by the processes of hepatobiliary and endogenous elimination. After intravenous injection, the rate and amount of deposition of transuranic elements in the liver and bone will depend on the elements' polymerized form, acidity, the presence of complexing agents, and their valence.

The intraperitoneal route of contamination occurs in radiation accidents of nuclear weapon origin or industrial origin, as well as in the misadministration of isotopes used in colloidal form to treat metastatic deposits in the peritoneal cavity (^{32}P).

III. PHYSICAL AND METABOLIC CHARACTERISTICS OF RADIOISOTOPES COMMON IN INTERNAL CONTAMINATION

Radioisotopes of medical concern can be classified according to their physical or chemical properties, their metabolic behavior, and the pathogenesis induced in the target organs of their final incorporation. Classification is extremely complex because of the many factors governing the metabolic pathways of each radioisotope. Significant differences may exist in the metabolic behavior of similar radioisotopes, and metabolic similarities may exist for radioisotopes that have dissimilar physical or chemical characteristics. Furthermore, radioactive isotopes of the same element can differ greatly in their behaviors in the living organism. So classifying radioisotopes is a complex and as-yet-unsolved problem. The problem can be addressed by considering each radioactive isotope as a separate entity with a variety of parameters and by individually considering each parameter.

Internal contamination by the fission products released in the explosion of nuclear weapons or after accidents in nuclear industry frequently occurs as simultaneous contamination by multiple isotopes and their products of radioactive decay. These mixed-fission products will make the diagnostic assessment of a contaminated patient a challenging task. Assessment requires diagnostically identifying the principal radioisotopes involved in the internal contamination so that proper therapeutic management can begin.

A. Americium

Two isotopes of americium are important in internal contamination: ^{241}Am and ^{243}Am. ^{241}Am ($t_{1/2}$ Ph = 458 years, $t_{1/2}$ Eff = 140 years) is a daughter product of plutonium, which decays to neptunium (^{237}Np) by the emission of high-energy alpha particles. It also decays by a low-energy photon emission (60 keV). ^{243}Am ($t_{1/2}$ Ph = 7950 years, $t_{1/2}$ Eff = 195 years) decays by emission of alpha particles. Both isotopes most commonly occur in the trivalent state, but they may be present in oxidation states from II to VII.

Internal contamination with americium most commonly occurs by the respiratory route or through contaminated wounds. Gastrointestinal absorption is negligible, but is higher in the young organism (27). Absorption through the skin is low, but increases if the isotopes are present in a solution of high acidity. The intramuscular route of contamination results in 10–60% absorption from the site of incorporation, depending on the chemical form of americium.

Target organs of americium are liver and bone (50–70% versus 20–30% of the retained dose, respectively) after parenteral administration. The skeleton is the primary target organ, followed by the liver. Reports exist of a high incidence of malignant changes in hematopoietic tissue, bone, and gonads after the intraperitoneal injection of americium in experimental animals (28).

Inhaled americium results in preferential deposition in the lung, tracheobronchial lymph nodes, liver, bone, and thyroid, with resulting tissue degeneration, fibrotic changes, and malignant changes. Human data on the metabolic fate of americium indicate that all americium compounds result in similar distribution in the liver (45%) and skeleton (45%), with the remainder (10%) distributed in other tissues and excreta.

B. Californium

Among 13 isotopes of californium, only one is a potential hazard of internal contamination, ^{252}Cf. It is an alpha emitter, with a $t_{1/2}$ of 2.6 years and a photon emission of 43, 100, and 160 keV. This isotope is used in radiation oncology as a neutron source for intracavitary use (28).

^{252}Cf is a serious hazard of external and internal radiation, with metabolic properties similar to other transuranic elements. It is absorbed into the systemic circulation mainly through the respiratory tract or contaminated wounds. Inhaled ^{252}Cf is retained mainly in the liver and bone, with other significant retention in the pulmonary and tracheobronchial lymph nodes. Intraveneous or intramuscular administration of ^{252}Cf results in 60% deposition in the skeleton and about 15% in the liver.

Over 90% of ^{252}Cf initially deposits in the liver; it is then eliminated by hepatobiliary secretion into the small intestine. Human exposure to ^{252}Cf has been reported after the inhalation of ^{252}Cf particles (29). The main data on biodistribution, internal dosimetry, pathology, and treatment are derived from the work on experimental animals.

C. Cerium

Two radioactive isotopes of cerium, ^{141}Ce and ^{144}Ce, are of potential significance as a hazard of internal contamination. ^{141}Ce ($t_{1/2}$ = 32 days) decays by beta and gamma emission, and is produced by neutron irradiation of stable cerium (^{140}Ce). ^{144}Ce ($t_{1/2}$ Ph = 284 days) is a fission product of uranium, and it decays by beta and gamma emission.

The route of internal exposure is mainly by inhalation. Gastrointestinal absorption is negligible in humans and in experimental animals (30). The critical organ for ^{141}Ce is the liver, and ^{141}Ce is preferentially deposited in the skeleton. Inhaled cerium is preferentially deposited in the lung, whereas the critical organ for ingested cerium isotopes is the descending colon and rectosigmoid.

D. Cesium

Among 21 radioisotopes of cesium, only 2 are medically significant for potential risk of internal contamination: ^{137}Cs and ^{134}Cs. ^{137}Cs ($t_{1/2}$ = 30 years) decays by beta emission, and its daughter-product emission of photons (E = 662 keV) accompanies its spectrum of radioactive decay. ^{134}Cs ($t_{1/2}$ = 2.1 years) decays by both beta and gamma emissions, with multiple energy levels for each mode of decay.

^{137}Cs is a product of nuclear fission, and it has been studied extensively as a significant component of radioactive fallout. As a metabolic homolog of potassium, it is uniformly distributed in the body and is eliminated by the renal system. Cesium enters the systemic circulation through either the respiratory or the gastrointestinal system. Its average biological half-life in humans is 110 days in males, 80 days in females, and 60 days in children (31). Accidental contamination with ^{137}Cs has been declining, because of its decreasing levels in the biosphere due to reduced atmospheric testing of nuclear weapons.

E. Curium

Among 13 curium isotopes, the ^{242}Cm ($t_{1/2}$ = 152 days), ^{244}Cm ($t_{1/2}$ = 16.7 years), and ^{245}Cm ($t_{1/2}$ = 9300 years) are medically significant. The main route of entry into the body is by the respiratory system. Fifteen to forty-five percent of inhaled curium is absorbed into the circulation, and 10% is retained in the skeleton.

Initial excretion of curium is by the urine. Delayed excretion is equal between the urinary and intestinal routes, because the initial deposition in the liver is slowly eliminated via the hepatobiliary mechanisms.

Bone retention of curium isotopes predominantly occurs on the mucoproteins of endosteal surfaces rather than in the bone minerals. The retention is affected by the active growth of bone and is particularly high in the areas of enchondral ossification (32).

F. Iodine

Ten radioactive isotopes of iodine are produced in the explosion of a nuclear weapon. Of all the fission products of medical interest, the radioisotope of iodine (^{131}I) is one of the most frequent concerns for internal contamination. Other isotopes of iodine (^{132}I, ^{133}I, ^{134}I, and ^{135}I) are important in early exposure to the products of nuclear fission.

^{131}I ($t_{1/2}$ = 8 days) is a principal cause of internal contamination in any nuclear incident and in early exposure to the radioactive fallout. ^{131}I decays by beta and gamma radiations. In reactor accidents, iodine is a major cause of concern for internal hazard because of its volatility and ability to enter the body via inhalation (33). In nuclear weapon testing, it is estimated that over 30,000 Ci of ^{131}I are released for each kiloton (kt) of fission energy (34). In reactor accidents, it has been estimated that over 20,000 Ci of ^{131}I were released into the atmosphere (35).

Other routes of internal contamination are by gastrointestinal absorption and by the cutaneous route of entry (intact skin, abrasions, and wounds). Contaminated grasslands after atmospheric tests of nuclear weapons are the major hazard of internal contamination because they result in contaminated dairy products. In the Marshall Islands experience, the ingestion of radioiodine was the main hazard from the standpoint of internal contamination (36).

In any case of suspected contamination with radioiodine, it is essential to determine the amount of thyroid incorporation by using the thyroid bioassay for both gamma and beta radiations. In cases of significant external contamination, the early estimate of thyroid uptake has to be interpreted with caution, because contaminated skin contributes to the findings of the thyroid assay. Bioassay of the ^{131}I body burden includes whole-body counting and studies of urinary excretion. Continuous follow-up monitoring of the thyroid should be performed routinely on all patients who are internally contaminated with radioiodine.

G. Plutonium

First in the chain of transuranic elements, plutonium is a very toxic substance. Among 15 radioactive isotopes of plutonium, 2 have been important as a potential hazard of internal contamination.

^{239}Pu ($t_{1/2}$ = 24,400 years) is an alpha emitter with infrequent gamma decay. A plutonium mass of 16 g contains 1 Ci of radioactivity. ^{239}Pu produces a fission after exposure to slow neutrons (fuel for nuclear weapons and reactors). ^{238}Pu ($t_{1/2}$ = 86 years) is an alpha emitter whose mass of 57 mg contains 1 Ci of radioactivity. Both isotopes are retained in the bone, liver, and all other tissues in the ratio of 45 : 45 : 10% of the absorbed quantity (37).

Factors that determine the distribution and retention of plutonium include the portal of entry, the valence state, polymeric particulate or soluble compounds, and chemical form. The main route of entry is inhalation. Intestinal absorption is negligible, but plutonium does gain access to the systemic circulation through intact skin (38). Entry through contaminated wounds results in a localized deposit of plutonium at the site of entry, with the formation of reactive fibrous tissue (38) and the potential induction of malignant changes.

Most of the cases of accidental contamination are through the respiratory system (75%); from there, absorption into the circulation largely depends on the solubility of the plutonium compounds. Soluble compounds are absorbed from the alveolar site to the circulation, and are ultimately deposited in the critical organs: liver and bone. Less soluble plutonium compounds are retained in the lung tissue, with slow migration to the pulmonary or tracheobronchial lymph nodes. Lung deposits of insoluble plutonium particles can be reduced by bronchopulmonary lavage.

H. Strontium

One of the most hazardous radioisotopes for internal contamination is ^{90}Sr, which is produced with five other strontium radioisotopes in the process of nuclear fission of uranium. ^{90}Sr ($t_{1/2}$ = 28 years) decays by beta emission to ^{90}Y, which is also a beta-emitting radionuclide. ^{89}Sr ($t_{1/2}$ = 51 days) and ^{85}Sr ($t_{1/2}$ = 65 days) are medically important, but their implications have been of less concern in radiation toxicology than the effects of ^{90}Sr. ^{85}Sr has been used in tracer and nuclear medicine diagnostic studies of skeletal metabolism and bone scintigraphy.

The metabolism of radiostrontium has been widely studied in animals and humans, as a consequence of a contaminated biosphere from radioactive fallout after nuclear weapons testing. The routes of entry for strontium are predominantly ingestion and inhalation, but strontium's access to the body fluids and target organs is rapid after being absorbed through skin lesions.

After its entry into the systemic circulation, strontium is rapidly deposited in the bone: first in its exchangeable fraction and then followed by its deep incorporation into the nonexchangeable bone mineral structures, through the process of exchange with the stable calcium ions and physiochemical absorption in the crystals of hydroxyapatite. The amount of ^{90}Sr in the trabecular bone can be reduced by therapeutic management to facilitate the exchange of mineral salts between bone and plasma. However, once ^{90}Sr has been incorporated into the nonexchangeable structures of the bone minerals, its therapeutic removal is impractical, if not impossible. The consequences of its retention in bone, its beta radiation, and its long half-life include genetic changes, leukemia, and osteogenic sarcoma (39, 40).

Strontium in the body behaves similarly to its metabolic homolog calcium, but some quantitative differences exist in their kinetics and the ultimate quantities retained. Biological membranes (intestinal mucosa, renal tubular epithelium, placenta, and mammary gland) possess the ability to discriminate against strontium, and favor the transfer of calcium ions. It is still controversial whether such discriminating processes affect the transfer of strontium across the basal membrane in the bone tissue. Physiological factors (such as the growth, nutritional, hormonal, and reproductive processes) that affect the metabolism and homeostatic function of bone are important in determining the ultimate fate of this greatly hazardous product of nuclear fission.

I. Tritium

Tritium (^3H) is the only isotope of hydrogen that decays to ^3He by beta emission. Tritium ($t_{1/2}$ = 12.3 years) is a normal constituent of the atmosphere and biosphere, produced by the fission of radioactive elements in the earth's crust, as well as by cosmic ray irradiation of stable nitrogen in the atmosphere. The testing of nuclear weapons has resulted in an increased concentration of tritium in the atmosphere.

The routes of entry of ^3H into an organism include inhalation, ingestion, and penetration through the skin. Ingestion of tritium, in the form of tritiated water, results in rapid and complete absorption in the body fluids, with diffuse distribution throughout the body. The body burden is monitored by using the urinary bioassay and by using liquid scintillation counting to detect its weak beta emission (E_{max} = 18 MeV).

Accidental contamination with tritium has been reported in humans (41). A multicurie dose of tritium exposure led to clinical symptoms of nausea and exhaustion, which led to death due to panmyelocytopenia. Analysis of tissue samples from casualties contaminated internally by tritium has shown the presence of tritium in the endocellular structural elements and in the body fluids (42).

J. Uranium

Three isotopes of uranium are important in medicine as potential hazards of internal contamination. ^{238}U ($t_{1/2} = 4.5 \times 10^9$ years), ^{235}U ($t_{1/2} = 7.1 \times 10^8$), and ^{234}U ($t_{1/2} = 2.5 \times 10^5$ years) are alpha, beta, and gamma emitters, with spontaneous fission below the level of criticality. Decay products of uranium isotopes include the alpha-emitting isotope of radon (^{222}Rn), which presents a hazard of internal contamination when radioactive particles are inhaled in uranium mines.

Uranium ore (U_3O_8) is obtained from mines and then concentrated and processed to ammonium diuranate (yellow cake), which is fluorinated and enriched for use as fuel for nuclear reactors or nuclear weapons. Uranium recycling is the process of obtaining uranium from the fuel dissolved in nitric acid, resulting in the removal of fission products and transuranic elements. The handling of uranium presents a hazard because of the possibility of a chemical explosion in the process of uranium oxidation.

Uranium isotopes have different metabolic behaviors in the body, depending on their physical forms. The ingestion of uranium isotopes results in relatively low absorption (1–5%). This absorbed dose is rapidly excreted through the kidneys. Other routes of internal contamination include inhalation or direct entry into the body fluids through the skin and contaminated wounds.

The critical organ for uranyl salts (U-VI) is the bone, while uranous salts (U-IV) are retained in the skeleton in a much smaller quantity. Soluble uranium (^{238}U) is rapidly eliminated through renal excretion. Less soluble compounds of uranium, particularly when enriched with ^{234}U and ^{235}U, are primarily retained in the bone or in the lung if inhaled. Soluble uranium compounds cause mainly chemical damage to the proximal convoluted tubules of the kidneys (43), with resulting albuminuria, hematuria, hyaline and granular casts, azotemia, and tubular necrosis.

Renal recovery even after exposure to high levels of uranium is quite common, and additional exposures seem to cause less damage to the kidney after its initial recovery. Urine bioassay should be routinely performed in any case of exposure to uranium compounds.

IV. THERAPEUTIC MANAGEMENT OF INTERNAL CONTAMINATION

The principal goals in reducing a radiation dose and the pathologic effects of internally deposited radionuclides are (1) to prevent the absorption of contamination from the treatment site(s) and (2) to eliminate absorbed radionuclides already in the bloodstream or in their respective target organs. It is of utmost importance to initiate therapy of the contaminated patient very early after exposure. Therapy includes the use of diluting and blocking agents to prevent and reduce gastrointestinal absorption, use of agents to decorporate radionuclides from the sites of internal deposition and mobilize them into the bloodstream, and measures to facilitate excretion through the urinary, gastrointestinal, or respiratory system.

Finally, the medical management of internal contamination includes administering chemical agents to facilitate the elimination of radioisotopes from the body by binding inorganic ions to nonionized complexes, which then can be eliminated through the kidney when present in soluble forms.

V. PREVENTION OF GASTROINTESTINAL ABSORPTION

After the ingestion of various products of nuclear fission, a high number of these products are rapidly absorbed into the systemic circulation and then deposited in their target organs. Reduction of the intestinal absorption of alkaline earth ions (calcium and strontium), cesium, cobalt, iodine, iron, gold, tritium, uranium, and radium is of special importance in this therapeutic approach. The most important methods for reducing the intestinal absorption of medically significant radioisotopes and facilitating their elimination via the fecal route are gastric lavage and the administration of emetics, ion-exchange agents, and antacids containing aluminum salts as well as guluronic and manuronic acid salts of alginates, barium sulfate, and sodium phytate.

Gastric lavage is a method of high merit in treating early exposure by ingestion. It is performed by inserting a nasogastric or orogastric tube into the stomach and repeatedly washing the stomach (by introducing water or physiological saline into the gastric lumen and then removing it by aspiration) until the aspirate is free of contaminating substance. All necessary precautions should be applied, including proper positioning of the patient during the procedure, so that the gastric lavage will be complete and will prevent the aspiration of contaminated gastric contents into the respiratory system.

Emetics may be used to complement gastric lavage, although the two methods are frequently used alone. An emetic should be used only after careful diagnosis of the contaminated patient, because it is contraindicated for a patient in shock, in altered consciousness, or after ingestion of petroleum or corrosive substances. The most commonly used emetics are apomorphine for subcutaneous administration and ipecacuana derivatives for oral administration. Sound clinical knowledge of the management of direct effects and side effects of the emetic drugs is required in each case of their use. Application is best immediately after drinking 250 ml water. Emetics act directly on the gastric mucosa and by stimulating the vomiting center in the medulla oblongata. Apomorphine acts predominantly by stimulating the vomiting center. It should be administered subcutaneously in a single dose of 5–10 mg, whereas ipecacuana derivatives can be used repeatedly (oral administration) until vomiting is induced. Both agents are readily available. The potential side effects (nausea, weakness, tachypnea, tachycardia, and hypotension) can be treated by symptomatic therapy, but frequently do not require specific treatment.

The use of laxatives has been a common therapeutic approach in reducing internal contamination. Laxatives are administered in various forms, such as (1) the rhinoleic acid-releasing drugs, which stimulate contractions of the small intestine (castor oil and cascara), and (2) saline purgatives, which inhibit the absorption of radionuclides by forming insoluble salts, by cathartic elimination from the intestine, and by their hypertonicity, which causes extraction of water from the intestinal mucosa. Detailed clinical diagnostic management is required before using laxative therapy because it is contraindicated in any case of undiagnosed abdominal pain or in an acute surgical abdominal syndrome. The use of laxatives is associated with multiple side effects (including heart dysrhythmia, tachypnea, dyspnea, intestinal irritation, exanthema, electrolyte imbalance, and syncope), which must be addressed by appropriate symptomatic therapy.

A. Alginates

In this group of ion-exchange therapeutic agents are the extracts of brown seaweeds (Pheophyceae). These compounds act by the binding of their active ingredients [alginic acids (guluronic and manuronic)] to radionuclides in the intestinal lumen. Radionuclides chelated in this way are not as well absorbed through the intestinal mucosa (44). The action of alginates has been most intensively studied in the comparative absorption of strontium and calcium through the intestinal mucosa. These cations are metabolic homologs that selectively incorporate in the skeleton. However, their metabolism is affected by the processes that control their transfer across the biological membranes, resulting in the favorable retention and transfer of calcium and also discrimination against strontium.

Alginates possess the ability to preferentially bind the strontium ion in the intestine, without much effect on the absorption of calcium. This phenomenon has been used in the therapeutic management of internal contamination by ingested strontium (45), and has resulted in significant decrease of its retention in the skeleton. Alginates are administered orally. Their main disadvantage has been high viscosity, although low-viscosity preparations (such as manucol SSLD) are available (46, 47).

Ion-exchange drugs reduce the intestinal absorption of ingested radioisotopes. These drugs include activated charcoal, sodium polystyrene sulfonate, biorex-40, and ferric ferrocyanide. They should be used with caution because of their side effects, including gastritis, anorexia, vomiting, and diarrhea. Ion-exchange resins can also interfere with the absorption of essential inorganic and organic nutrients by binding them and eliminating them from the intestinal lumen.

One of the forms of ferrocyanide used to decrease the intestinal absorption of ingested radioisotopes is Berlin blue (Prussian blue), which is particularly useful in binding and removal. This compound is commercially available in Europe. Its use in the United States is restricted to emergency situations in which FDA investigational-drug approval is required.

Aluminum-containing antacids have been effectively used to therapeutically remove strontium, with a highly significant decrease of ^{90}Sr absorption by the intestine. Aluminum phosphate administered orally reduces the absorption of strontium by over 80%. Aluminum hydroxide reduces the uptake of strontium by 50%. No side effects are associated with their therapeutic use.

Other drugs to eliminate ingested radionuclides from the digestive tract include barium sulfate, which is highly effective in reducing absorption of strontium and radium, and phytates, which reduce the absorption of calcium, iron, magnesium, and zinc ions.

B. Isotopic Dilution, Blocking Agents, and Displacement Therapy

The use of water to reduce tritium in the body fluids is a common therapeutic method, applied by the oral or the intravenous route of administration. Clinical assessment of each patient is essential to avoid possible side effects from fluid overload in patients with cardiovascular or renal disease.

In therapy using blocking agents, the uptake of radioactive iodine is inhibited by the immediate administration of stable iodide after an accidental exposure (KI and NaI). This therapy should be continued for 2 weeks to allow the elimination of the radioactive iodine and to prevent its reuptake. The FDA-recommended dose is 130 mg KI for adults daily and 65 mg daily for children.

The uptake of radioactive strontium can be reduced by administering stable strontium compounds (lactate and gluconate). The intestinal absorption of radiostrontium can be significantly reduced by oral administration of phosphates, which reduce over 60% of the strontium absorption. This effect is sometimes counterbalanced by increased tubular reabsorption of strontium, if the phosphate content is elevated in the extracellular fluid.

Tubular reabsorption of strontium will increase after the intravenous administration of phosphate. This factor reduces the net effect of diminished skeletal retention of strontium by the high phosphate content in the digestive system (48). Parenteral administration of phosphate can be used to treat internal contamination (soluble radioactive phosphorus, ^{32}P).

Calcium salts have been used to reduce the intestinal absorption of radioactive strontium (Ca-lactate and Ca-gluconate). Other stable cations (potassium and zinc) are rarely used as potential agents in managing internal contamination by ^{65}Zn or ^{42}K.

Therapeutic agents for decorporating and mobilizing the organotropic radioisotopes include hormonal preparations (PTH, corticosteroids, and calcitonin), propylthiouracic (PTU), methimazole (MMI), diuretics, expectorants, perchlorate, and ammonium chloride. Parathormone has been used in different species of experimental animals to enhance bone resorption, with the subsequent release of incorporated osteotropic radionuclides (calcium, strontium, phosphorus, and radium).

It has been demonstrated that physiological processes that result in increased catabolic processes in the skeleton produce significant reduction in the amount of incorporated bone-seeking radioisotopes. These effects have been observed in lactating animals, whose skeletal uptake of calcium and strontium was reduced by over 50% after catabolic processes of the skeleton were induced by lactation.

This reduction of bone mass and the demineralization of both the exchangeable and nonexchangeable fractions of the skeleton were observed, regardless of hyperphagia in the lactating animals.

The influence of corticosteroid hormones (prednisone, cortisone, dexamethasone, and methylprednisolone) has been studied in various experimental models in an attempt to evaluate their use in mobilizing the incorporated bone-seeking radioisotopes. No significant effect of corticosteroids was seen in the metabolic behavior of transuranium or alkaline earth isotopes in the bone, regardless of the catabolic processes induced in the skeleton by the long-term use of corticosteroids.

Propylthiouracil and methimazole decrease the synthesis of thyroid hormones (T3 and T4) by their inhibitory effect on the iodide oxidation. These antithyroid drugs are not widely accepted for use in antagonizing radioiodine uptake by the thyroid, because of their complex metabolic effects on the radioiodine in the kidney and liver, as well as numerous toxic side effects. Other antithyroid drugs (e.g., thiocyanate) are not of practical use for radioiodine elimination because of questionable effects and toxic reactions. Of all the compounds used to inhibit thyroid uptake of radioactive iodine, stable iodide is the drug of choice for the competitive inhibition of ^{131}I incorporation.

For mobilizing radiostrontium from the body, ammonium chloride was found to be of certain benefit in reducing the body burden of ^{90}Sr. However, the toxic effects (gastritis and hepatitis) of ammonium chloride make it less than an ideal drug for strontium decorporation.

Diuretic therapy with various conventional agents has been used in various studies on the excretion of internally deposited radioisotopes. Because of the complex metabolic effects of diuretic drugs, with the need for meticulous monitoring of the electrolyte and ECF metabolism, diuretic therapy of internal contamination is still an unexplored area. Ethacrinic acid is the only diuretic agent now recommended for excretion of the alkaline earth isotopes.

Treatment of patients exposed to radioactive particles via the respiratory route of contamination includes administering (1) drugs that reduce the viscosity of endobronchial mucus and (2) various mucolytic drugs that act on mucopolysaccharides and nucleoproteins in the respiratory tree, thus mobilizing its contents by expectoration. The results of testing these agents (pancreatic dornase, triton, Tween-80, ^{68}F, etc.) have been unsatisfactory in reducing the uptake of inhaled radioisotopes from the lung.

C. Treatment of Internal Contamination with Chelating Agents

Complexing agents have been used to treat internal contamination in experimental animals and in accidentally exposed humans, with more success than other therapeutic modalities. The elimination of radioactive isotopes by chelation therapy is based on the ligand's ability to form nonionized ring complexes with inorganic ions, which are then excreted by the kidney.

Treatment with chelating agents should be instituted as soon as possible after internal contamination, before the radionuclides are retained in their target organs. The hydrophilic nature of these agents makes them ineffective in reaching the isotopes that are incorporated in the endocellular environment. Therefore, many studies are concentrating on the synthesis and production of lipophilic chelating agents, for their potential use in mobilizing radionuclides from the cells for excretion by the kidney.

The effect of chelation therapy with various complexing agents has been an area of extensive experimental and clinical research. Among many chelating agents tested in the experimental and clinical trials, only a few are of practical use at the present time.

Ethylenediamine tetraacetic acid (EDTA) has been used in animal experiments and to treat poisoning in humans from various inorganic compounds. EDTA has been beneficial in the treatment of lead poisoning and in the treatment of internal contamination with zinc, copper, cadmium, chromium, manganese, nickel, and transuranic elements (41).

The parenteral administration of EDTA results in its binding of stable calcium, resulting in hypocalcemia (tetany) and toxic side effects. Among them, nephrotoxicity is the primary complication, with a potentially fatal outcome. EDTA can be used as Ca-EDTA or as Na-EDTA. The intravenous administration dose for Na-EDTA is 75 mg/kg bid, not exceeding a total dose of 550 mg/kg in the entire therapeutic regimen.

Intramuscular administration (75 mg/kg tid) of EDTA should be used with a local anesthetic because of tissue irritation and pain at the injection site. The intravenous route is the preferred method of administration, by infusion in physiological saline or 5% glucose in water.

Renal function tests and urinalysis should be performed before treatment, because EDTA therapy is contraindicated in patients with renal disease. Na-EDTA is used in a lower dose (50 mg/kg) as physiological saline or 5% glucose, not exceeding 300 mg/6-day treatment period. Oral or intramuscular administration is not used, being contraindicated in renal and hepatic disease.

Diethylenetriamine pentacetic acid (DTPA) is more effective than EDTA in the therapeutic removal of radioisotopes that are common in internal contamination. DTPA is used as Ca-DTPA or Zn-DTPA. Ca-DTPA is administered as intravenous infusion (1000 mg in 250 ml of physiological saline or 5% glucose) for a maximum of 5 consecutive days. DTPA can be obtained in the United States as an investigational new drug from the United States Department of Energy, Office of Health and Environmental Research, Human Health and Assessment Division, Washington, D.C., or the Radiation Emergency Assistance Center/Training Site, Oak Ridge Associated Universities, Oak Ridge, Tennessee.

Administration of DTPA is contraindicated in leukopenia or thrombocytopenia, renal disease, hypertension, or pulmonary disease (if used as inhalation therapy). Zn-DTPA can be used in the same dose as Ca-DTPA by the intravenous or inhalation routes, and it is less toxic than Ca-DTPA. Na-DTPA is not used because it chelates calcium, with resulting hypocalcemia and tetany.

DTPA is now the most effective agent in treating internal contamination with transuranic elements, particularly plutonium and americium. DTPA does not produce toxic symptoms if used in recommended doses administered either intravenously or by inhalation.

The treatment of internal contamination is currently limited to a few therapeutic agents, and considerable problems are associated with their use. The present therapeutic modalities are still unsatisfactory, particularly in the removal of radionuclides that are already incorporated in their respective critical organs.

In removing the most hazardous radionuclides of the transuranium series, DTPA is clearly superior to other chelating agents. However, its use is limited because it is not commercially available, its administration must be performed by qualified personnel, it is effective only in early treatment, and its strong hydrophilicity prevents it from reaching the intracellular environment. It is not practical in treating mass casualties of internal contamination, although it has distinct benefits in treating cases of sporadic contamination in a medical facility.

These factors have contributed to the continuous investigational efforts to produce new chelating agents. Derivatives of paraaminocarboxylic acid (PACA) have been studied in an attempt to synthesize adequate lipophilic agents (chelons) for the intracellular binding and removal of incorporated radioisotopes. These agents, when administered orally, rectally, or by depot, have potential use in treating mass casualties of endemic or epidemic proportions.

Other agents being studied for the potential treatment of internal contamination include synthetic polyamine catecholamides (49), various phospholipid compounds (liposomes) for encapsulation of the radiotoxic substances (50), and natural chelates isolated from the cultures of various microorganisms (51). Their place in the medical management of internal contamination is yet to be determined by experimental and clinical trials.

When live-saving measures have been instituted and the patient stabilized, diagnostic monitoring of contaminated wounds should be performed to establish the nature and quantity of possible contamination with the organotropic radionuclides. Mechanical removal is instituted by cleansing, chemical therapy, and surgical procedures. Tissue samples from the contaminated wounds, obtained in the process of debridement, are placed in a counting vial and analyzed by radioimmunoassay methods. Monitoring of the body surface, wound assessment, and tissue counts of radiation from the surgical debridement should be compared, and procedures of internal decontamination should be instituted if it has been determined that a radionuclide is present in the internal environment of the body.

Therapeutic decisions in the radionuclide decontamination of combined-injury patients are assisted considerably by available human data on the early assessment of radionuclide excretion from contaminated wounds. These data provide significant clinical aid in the determination of an optimal dosage of chelating agents that are used in internal decontamination therapy (52). Early therapeutic decisions in combined injury are of the utmost importance because the effectiveness of chelating agents is significantly reduced by delay in treatment. This is particularly important in the case of open wounds, where therapy with complexing agents is clearly indicated for all soluble compounds. Follow-up diagnostic procedures by bioassay methods should be part of the routine management for patients with combined injury, because these patients have the potential for a larger fraction of organotropic radionuclides in the extracellular fluid and parenchymal organs than patients without wounds.

It can be expected that internal contamination with organotropic radionuclides will be of particular concern in combined injury because of additional diagnostic and therapeutic requirements in monitoring the radiation type and quantity in a contaminated wound. A patient population subjected to traumatic, thermal, and infectious complications in addition to internal contamination will require particular clinical skills for the maintenance of homeostasis and the determination of clinical priorities.

REFERENCES

1. National Council on Radiation Protection and Measurements. *Radiological Factors Affecting Decision Making in a Nuclear Attack,* Rep. No. 42, p. 8. NCRPM, Washington, D.C., 1974.
2. Schachter, D., and Rosen, S. M. Active transport of Ca-45 by small intestine and its dependence on vitamin D-2. *Am. J. Physiol.* **196,** 357–362 (1959).
3. Hesp, R. *The Retention and Excretion of Caesium-137 by Two Male Subjects,* Rep. No. STI/PUB 84, p. 61. IAEA, Vienna, 1964.
4. Miller, C. E. Retention and distribution of Cs-137 after accidental inhalation. *Health Phys.* **10,** 1065 (1964).
5. Bernhardt, D. E., Carter, M. W., and Buck, F. N. Protective actions for radioiodine in milk. *Health Phys.* **21,** 401 (1971).
6. White, M. M., and Moghissi, A. A. Transfer of I-131 from milk to cheese. *Health Phys.* **21,** 116 (1971).
7. Glennon, J. A., Gordon, E. S., and Sawin, C. T. Hypothyroidism after low dose I-131 treatment of hyperthyroidism. *Ann. Intern. Med.* **76,** 721 (1972).
8. Neuman, W. F., Hursh, J. B., Boyd, J., and Hidge, H. C. On the mechanisms of skeletal fixation of radium. *Ann. N.Y. Acad. Sci.* **62,** 123 (1955).
9. Lloyd, E. The distribution of radium in human bone. *Br. J. Radiol.* **34,** 521 (1961).
10. Rowland, R. E. *Local Distribution and Retention of Radium in Man,* Rep. No. STI/PUB 65, p. 57. IAEA, Vienna, 1963.
11. Martland, H. S., Colon, P., and Kneff, J. P. Some unrecognized dangers in the use and handling of radioactive substances. *JAMA, J. Am. Med. Assoc.* **85,** 1769 (1925).

12. Martland, H. S., and Humphries, R. E. Osteogenic sarcoma in dial painters using luminous paint. *Arch. Pathol.* **7,** 406 (1929).
13. Finkel, A. J., Miller, C. E., and Hasterlik, R. J. *Radium Induced Malignant Tumors in Man,* Rep. No. ANL-7461. Argonne Natl. Lab., Argonne, Illinois, 1969.
14. Hasterlik, R. J., Miller, C. E., and Finkel, A. J. Radiographic development of skeletal lesions in many many years after acquisition of radium burden. *Radiology* **93,** 599 (1969).
15. Lambert, B. E., and Vennart, J. Radiation dose received by workers using tritium in industry. *Health Phys.* **22,** 23 (1972).
16. Langham, W. H. Physiology and toxicology of plutonium-239 and its industrial medical control. *Health Phys.* **2,** 172 (1959).
17. Abrams, R., Seibert, H. C., Potts, A. M., Forker, L. L., Greenberg, D., Postel, S., and Lohr, W. Metabolism of inhaled plutonium in rats. *Health Phys.* **2,** 172 (1959).
18. Scott, K.- L, Axelrod, D. J., Crowley, J., and Hamilton, J. G. Deposition and fate of plutonium, uranium, and their fission products inhaled as aerosols in rats and man. *Arch. Pathol.* **48,** 31 (1949).
19. Recommendations of International Commission on Radiological Protection. *Br. J. Radiol., Suppl.* **6** (1955).
20. Langham, W. H. Determination of internally deposited radioactive isotopes from excretion analysis. *Am. Hyg. Assoc. Q* **17,** 305 (1956).
21. Wrenn, McD. E., Rosen, J. C., and Cohen, N. *In Vivo Measurement of Americium-241 in Man,* Rep. No. STI/PUB 290. IAEA, Vienna, 1972.
22. *The Metabolism of Compounds in Plutonium and Other Actinides,* Publ. No. 19. International Commission of Radiation Protection, Pergamon, Oxford, 1972.
23. Ross, D. M. *Diagnosis and Treatment of Deposited Radionuclides,* Rep. No. 427. U.S. At. Energy Comm., Excerpta Medica Foundation, Amsterdam, 1968.
24. Norwood, W. D. Therapeutic removal of plutonium in humans, *Health Phys.* **8,** 747 (1962).
25. West, C. M., and Scott, L. M. Uranium cases showing long chest-burden retention. *Health Phys.* **17,** 781 (1969).
26. Durbin, P. W. Metabolism and biological effects of the transplutonium elements. In *Handbook of Experimental Pharmacology—Uranium, Plutonium, Transplutonic Elements* (H. C. Hodge, J. N. Stannard, and J. B. Hursh, eds.). Springer-Verlag, Berlin and New York, 1973.
27. Nilsson, A., and Broome-Karlson, A. The pathology of americium-241. *Acta Radiol. Ther. Phys. Biol.* **15,** 49 (1976).
28. Seaborg, G. T. Medical uses: Americium-241, californium-252. In *Handbook of Experimental Pharmacology—Uranium, Plutonium, Transplutonic Elements* (H. C. Hodge, J. N. Stannard, and J. B. Hursh, eds.). Springer-Verlag, Berlin and New York, 1973.
29. Poda, G. A., and Hall, R. M. Two californium-252 inhalation cases. *Health Phys.* **29,** 407 (1975).
30. Moskalev, Y. Experiments on distribution of cerium-144. *Med. Radiol.* **4,** 52 (1959).
31. Lloyd, R. D. Cesium-137 half-times in humans. *Health Phys.* **25,** 605 (1973).
32. Lloyd, R. D., Atherton, D. R., Mays, C. W., McFarland, S. S., and Williams, J. L. Curium extraction, retention, and distribution studies in beagles. In *Research in Radiobiology,* Rep. No. C00-119-248. University of Utah, Salt Lake City, 1973.
33. Holland, J. Z. Physical origin and dispersion of radioiodine. *Health Phys.* **9,** 1095 (1963).
34. National Council on Radiation Protection and Measurements *Management of Persons Accidentally Contaminated With Radionuclides,* Rep. No. 65, p. 84. NCRPM, Washington, D.C., 1980.
35. Conard, R. A., Dobyns, B. M., and Sutow, W. W. Thyroid neoplasia as late effect of exposure to radioactive iodine in a fallout. *JAMA, J. Am. Med. Assoc.* **214,** 316 (1975).
36. Mehl, H. G., and Rundo, J. Preliminary results of a world survey of whole-body monitors. *Health Phys.* **9,** 607 (1963).

37. Vaughan, J., Bleaney, B., and Taylor, D. M. Distribution, excretion, and effects of plutonium as a bone seeker. In *Handbook of Experimental Pharmacology—Uranium, Plutonium, Transplutonic Elements* (H. C. Hodge, J. N. Stannard, and J. B. Hursh, eds.) Springer-Verlag, Berlin and New York, 1973.
38. Lushbaugh, C. C., Cloutier, R. J., and Humason, G. Histopathologic study of intradermal plutonium and metal deposits: Their conjectural fate. *Ann. N.Y. Acad. Sci.* **145,** 791 (1967).
39. Barnes, D. W. H., Carr, T. E. F., Evans, E. P., and Loutit, J. F. Sr-90 induced osteosarcomas in radiation chimaeras. *Int. J. Radiat. Biol.* **18,** 531 (1970).
40. Loutit, J. F. Strontium-90 and leukemia. *Medical Annual Review,* p. 34. Atholone Press, London, 1967.
41. Levine, W. G. Heavy metals and heavy-metal antagonists. In *The Pharmacological Basis of Therapeutics,* (L. S. Goodman and A. G. Gilman, eds), 6th ed. Macmillan, New York, 1980.
42. Synder, W. S., Fish, B. R., Bernard, S. R., Ford, M. R., and Muir, J. R. Urinary excretion of tritium following exposure of man to HTO—A two exponential model. *Phys. Med. Biol.* **13,** 547 (1968).
43. Luessentrop, A. J., Gallimore, J. C., Sweet, W. H., Struxness, E. G., and Robinson, J. The toxicity in man of hexavalent uranium following intravenous administration. *Am. J. Roentgenol. Radium Ther. Nucl. Med.* **79,** 83 (1958).
44. Carr, T. E. F., Nolan, J., and Durakovic, A. The effect of alginate on absorption and excretion of 203-Pb in rats fed milk and normal diets. *Nature (London)* **224,** 1115 (1969).
45. Sutton, A. Reduction of strontium absorption in man by the addition of alginate. *Nature (London)* **216,** 1005 (1967).
46. Sutton, A., Harrison, G. E., Carr, T. E. F., and Barltrop, D. Reduction in the absorption of dietary strontium in children by an alginate derivative. *Int. J. Radiat. Biol.* **19,** 79 (1971).
47. Kostial, K., Durakovic, A., Simonovic, I., and Zul, V. The effect of some dietary additives on calcium and strontium absorption from the intestine in new-born and lactating rats. *Int. J. Radiat. Biol.* **15,** 63 (1969).
48. Durakovic, A. Metabolism of calcium and strontium in lactation. Ph.D. Thesis, University of Zagreb, Croatia, Yugoslavia, 1968.
49. Bergeron, R. J., Kline, S. J., Stolowich, N. J., McGowern, K. A., and Burton, P. S. Flexible synthesis of polyamine catecholamides. *J. Org. Chem.* **46,** 4524–4529 (1981).
50. Gregoriadis, G. Liposomes in therapeutic and preventive medicine; the development of the drug-carrier concept. *Ann. N.Y. Acad. Sci.* **308,** 343–370 (1978).
51. Strandberg, G. W., Shumate, S. E., and Parrott, J. R. Microbial cells as biosorbents for heavy metals: Accumulation of uranium by *Saccharomyces cervisiae* and *Pseudomonas aeruginosa*. *Appl. Environ. Microbiol.* **41,** 237–245 (1981).
52. Voeltz, G. L. What we have learned about plutonium from human data. *Health Phys.* **29,** 551–561 (1975).

CHAPTER 14

Radioprotectants

LEO GIAMBARRESI AND AARON J. JACOBS*

Armed Forces Radiobiology Research Institute
Bethesda, Maryland 20814-5145

I. MILITARY APPLICATIONS OF RADIOPROTECTORS

A major goal of military radiobiology is the development of drugs to be used to provide partial protection against radiation injury. From a military standpoint, radioprotectors will have their greatest utility in a nuclear theater of operations, where their use would protect personnel and maintain effective military performance.

Current nuclear defense practices depend primarily on physical countermeasures such as shelters and shielding. These countermeasures are effective, and the use of radioprotectors does not preclude their use. This becomes especially evident when one considers that currently known radioprotectors, by definition, must be given before irradiation to effect protection. Therefore, they will be of no value at the time of a detonation unless personnel are forewarned. But situations will occur in which tactical demand will require units to deploy in or maneuver through radioactively contaminated areas where physical protection is not feasible. In those cases, radioprotective drugs could be used to reduce or delay the incapacitating and fatal effects of nuclear radiation and to enhance postattack effectiveness. In addition, radioprotective drugs would be beneficial to personnel engaged in decontamination of fallout areas, in cleanup of radioactive accident areas, and possibly in future spaceflights.

II. CHARACTERISTICS OF A FIELD-USABLE RADIOPROTECTOR

A radioprotective drug or regimen that is to be used by personnel on a nuclear battlefield should meet several criteria. It should provide sufficient protection to

*Present address: United States Army Institute for Dental Research, Washington, D.C. 20307.

reduce the effects of radiation by at least a factor of two (dose reduction factor of >2), and the drug should be easily self-administered. Although intramuscular injection would be acceptable, oral administration is much preferred, particularly for a drug that may have to be administered repeatedly. Relatively long-term protection is necessary for a practical drug. In experimental studies in mice, protection against radiation-induced lethality has been achieved for over 4 hr after a single injection of the most effective radioprotector, WR-2721 (1). Any drug intended for human use must be well tolerated. In addition, to be practical for use by the military, the drug must not interfere with performance. Thus, a major objective is the development of a drug that produces minimal side effects and no cumulative or irreversible toxicity. Compatibility with the wide range of other drugs that will be available to personnel on the integrated battlefield is another important consideration. The drug should have a shelf life of at least 2–5 years, and should be formulated and packaged in such a manner that stability is retained under a variety of adverse conditions. Finally, the drug should not be abusable.

The ideal drug would possess all of these characteristics and significantly exceed several of them. For example, it would be highly desirable to field a drug that (1) reduces by a factor of 10 the dose of radiation that produces combat ineffectiveness, (2) offers protection lasting several days, and (3) produces no side effects or toxicity. The possibility that such a drug may be found cannot be excluded. However, based on data from experiments using the most effective radioprotectors, the listed objectives are considered to be those characteristics that realistically can be achieved using current drugs and state-of-the-art drug-development technology.

III. HISTORICAL PERSPECTIVE

The first experiment that stimulated widespread interest in developing radioprotector drugs for humans was performed over 35 years ago (2). Large doses of cysteine, a sulfur-containing amino acid, were given intravenously (iv) to rats 15 min before a lethal dose of X rays. Survival of the pretreated rats was greatly increased. During those early postwar years, the memory of Hiroshima and Nagasaki was still fresh; nuclear weapon development, testing, and proliferation were beginning; and the potential use of nuclear weapons in military strategy had become a reality. Thus, after the results of the experiment were published, it was immediately recognized that the development of a drug that could protect against the effects of ionizing radiation would have significant defensive military application.

In an effort to improve effectiveness of protection, Bacq *et al.* (3) tested β-mercaptoethylamine (MEA), a close structural analog of cysteine. MEA provided protection at considerably lower drug doses. A great deal of the subsequent research in development of radioprotectors has followed this basic pharmacological approach of modifying the chemical structure of known radioprotectors slightly and testing the new compounds for their radioprotective ability.

In the early 1950s, the Atomic Energy Commission initiated a program of radioprotector development. From this program came a series of aminothiol compounds, among which 2-aminoethylisothiouronium (AET) (4) was the most effective radioprotector developed to that time. Historically, this series is important because it provided an understanding of the structural features of aminothiols, are necessary for protection (4–6).

In 1957, the United States Army initiated an Antiradiation Drug Development Program at the Walter Reed Army Institute of Research. From 1957 to 1973, approximately 4400 compounds were developed and tested. They were almost exclusively aminothiols or aminothiol derivatives. From this program, a phosphorothioate designated as WR-2721 was developed, and it has proven to be the most effective radioprotective drug developed to date (1, 7, 8).

This brief historical overview has highlighted only some of the more important milestones in radioprotective drug development. Concurrent with these drug-development programs, a great deal of research effort has gone into investigating the mechanisms of action of the various radioprotective drugs, as well as examining the role of naturally occurring compounds and enzyme systems in mediating radiation injury. Although the radioprotector WR-2721 is a major improvement over earlier compounds in terms of effectiveness, potency, tolerance, and duration of action, several pharmaceutical and pharmacological problems must be overcome before WR-2721 can be used in the field. These problems will be discussed more fully later in this chapter. Current efforts at a number of laboratories are involved with overcoming these difficulties: with developing newer, more effective radioprotective drugs; with enhancing the body's natural defense systems against radiation injury and free-radical damage; and with stimulating immune responses to allow recovery from radiation-induced hemopoietic failure.

IV. MECHANISMS OF RADIATION INJURY AND PROTECTION

A. Radiation Injury

Injury resulting from the penetration of biological tissue by ionizing radiation is brought about by the transfer of radiation energy to critical biological macromolecules (e.g., DNA, proteins, and membrane lipids). The initial chemical injury can occur in two ways: either directly from the absorption of radiation energy by the target macromolecules themselves or indirectly from diffusible ions and free radicals [i.e., highly reactive intermediates that are produced from the absorption of radiant energy by low-molecular-weight (<500) cellular components (9)]. Since water is by far the most abundant low-molecular-weight component in the cell (constituting 70–80% of the mass of cells and tissues), the radiolytic products of water are responsible in large part for this indirect damage. The relative contribution of the direct and indirect effects to overall cellular injury is difficult to assess *in vivo.* However, since water is so abundant, the amount of energy deposited in this compartment by radiation is quite high, and it is likely that the free radicals that are produced contribute significantly to radiation damage (10, 11).

When cellular water molecules are irradiated, a wide variety of very reactive molecular species are formed. These include the hydroxide radical (OH·), the hydrated electron [e^- (aq)], and the hydrogen radical (H·). Of these, the OH· and e^- (aq) are produced in the highest concentrations, and OH· is considered to be the most damaging (10, 12, 13).

In the presence of oxygen additional reactive species, including a number of oxygen and peroxide derivatives, are also produced. The concentration of these species is dependent on the degree of oxygenation of the cell or tissue being irradiated, which may explain why oxygen effectively enhances radiation damage (11, 14).

When target macromolecules are irradiated directly or react with high-energy free-radical intermediates, the targets themselves become ionized or transformed into free radicals. So, although radiation can exert its effect by two different mechanisms, the end result is the same, i.e., the disruption of molecular structure and function, leading to altered cell metabolism and injury. The types of damage produced are discussed in detail elsewhere, and so will be only briefly reviewed here. DNA is considered to be a critical biological target molecule for radiation-induced cell death. A number of chemical alterations are induced in the DNA molecule by radiation either directly or indirectly (Fig. 14-1). These in-

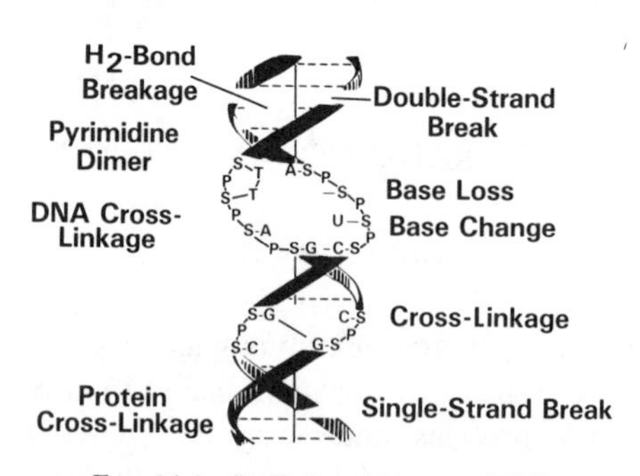

FIG. 14-1 Radiation damage to DNA.

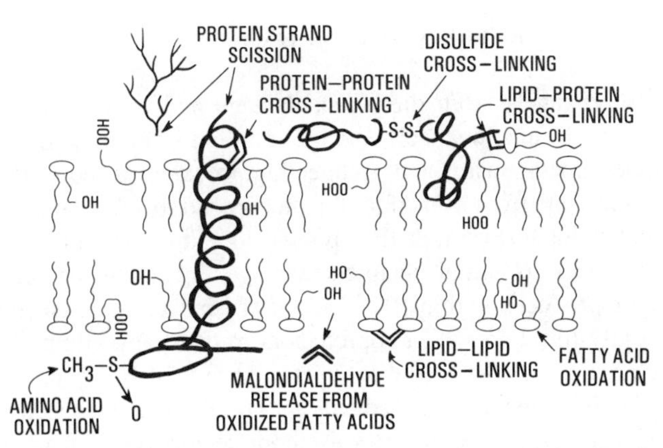

FIG. 14-2 Free-radical damage to membranes. (Reprinted with permission from B. A. Freeman and J. A. Crapo, Biology of disease: Free radicals and tissue injury. *Lab. Invest.* **47**(5), 412–426. © 1982 by U.S.–Canadian Div. of the IAP.)

clude chemical alterations in the purine and pyrimidine bases, single- and double-strand breaks, removal of bases, and cross-linking of DNA with DNA or adjacent protein molecules. These changes, depending on their type and extent, are expressed functionally in a variety of ways, including cell death.

The bulk of evidence favors DNA damage as being the most significant factor in radiation-induced cell damage. However, extensive work *in vitro* on model membrane systems, bacteria, and eukaryotic cells has shown that radiation and free radicals can produce a variety of alterations in membrane lipids and associated proteins (Fig. 14-2) (15–17). Although the characterization of radiation effects on biological membranes *in vivo* is difficult, radiation and free-radical damage to cellular membranes could contribute significantly to altered cellular function and ultimately cell death. Further, a number of the effects of radiation that lead to incapacitation and performance decrement (such as generalized weakness, hypotension, nausea, and vomiting) appear to occur too rapidly (within hours) to be due to DNA damage. It is reasonable to expect that at least some of these effects are related to radiation-induced membrane damage.

One of the major alterations that occur in membranes after irradiation or free-radical attack is lipid peroxidation (15, 16). This leads to the production of short-chain fatty acyl derivatives and other lipid by-products. Additional alterations include lipid–lipid cross-linking as well as protein–protein and lipid–protein cross-linking, oxidation of accessible amino acids, protein denaturation, and scission of protein strands. Functionally, these changes can be expressed as altered membrane fluidity and permeability, which could, for example, trigger the release of potent physiological mediators. In addition, activity of enzymes associated with these membranes may be altered by the disruption of lipid microenvironments and protein structure.

B. Time Scale for Radiation Damage and Protection

All of these changes, from the direct damage and formation of radiolytic products of water to functional cell damage resulting from disruption of cellular macromolecules, metabolism, and architecture, occur as a sequence of events traversing a time scale from a fraction of a picosecond to a few hours. Protection can be brought about at several distinct points along this chain of events (9, 10).

Radiation induces its initial chemical damage, including the production of water radiolytic products, within 10^{-12} sec after irradiation. These events occur much too rapidly for chemical radioprotectors to exert any effect. The earliest time at which radioprotectors begin to function is around 10^{-9} sec after irradiation. It is within this time that free radicals are present in high concentrations and react with critical biological molecules. Radioprotectors function at this stage by competing with these molecules for free radicals. Between 10^{-7} and 10^{-3} sec, the reactions of most water free radicals with targets are essentially completed. At this time (approximately 10^{-6} sec), radioprotectors begin to repair chemical lesions in target molecules by reducing oxidative damage induced by the free radicals. Between 10^{0} and 10^{4} sec, endogenous enzyme systems come into play to remove the more slowly reacting products of water radiolysis and to repair the chemical lesions produced in cellular macromolecules.

C. Mechanisms of Radioprotection

Several generalities regarding mechanisms of radioprotection emerge from the above consideration of the time frame for radiation damage and protection. Radioprotectors cannot prevent the extremely rapid direct absorption of energy from radiation by critical macromolecules and cellular water. The ability of radioprotectors to produce their effects, then, is a result of their capacity to inhibit indirect damage, to repair direct and indirect damage once they occur, and to facilitate the recovery of damaged cells or depleted cell populations. These are all brought about by a variety of mechanisms operating at three distinct levels of cellular organization: the molecular level, the physiological–biochemical level, and the organ level (see Table 14-1).

1. Molecular Level

At the molecular level radioprotectors may exert their effect through the fundamental physical–chemical interactions of free-radical scavenging, hydrogen atom donation, direct binding to biological molecules, and mixed disulfide formation.

TABLE 14-1

MECHANISMS OF RADIOPROTECTION

Mechanisms of radioprotection
Molecular level
Free-radical scavenging
Hydrogen donation
Binding to critical biological targets
Mixed disulfide formation
Physiological–biochemical
Hypoxia
Nonprotein sulfhydryl release
Biochemical shock
Hypothermia
Organ level
Stimulate recovery of cell populations

a. Free-Radical Scavenging. Free-radical scavenging refers to the ability of radioprotectors to remove (scavenge) the highly reactive products of water radiolysis before they can react with and damage the molecules of biological importance (18, 19). In essence, this process is a competitive reaction between radioprotectors and biological molecules. The radioprotector incurs the "damage," thereby reducing the concentration of free radicals in solution and sparing the biological target. Many effective radioprotectors are very efficient scavengers of water-derived free radicals. Thus, to the extent that indirect damage contributes to cellular injury, free-radical scavenging is likely to be an important component in the overall mechanism of radioprotection (12, 20).

b. Hydrogen Transfer. A second fundamental phenomenon that may contribute, at the molecular level, to radioprotection is hydrogen transfer or donation (12, 20). This is a rapid repair process whereby hydrogen atom loss in a biologically important molecule (R—H), brought about by direct absorption of radiant energy ($h\nu$),

$$R\text{—}H \xrightarrow{h\nu} R + H,$$

or via indirect reaction with free radicals,

$$OH\cdot + R\text{—}H \rightarrow R\cdot + H_2O$$

is compensated for by donation of a hydrogen atom by a sulfhydryl protector (P—H):

$$R\cdot + P\text{—}H \rightarrow R\text{—}H + P\cdot$$

In the absence of the protector, the biological free radical (R·) would become permanently altered by reactions with other free radicals or oxygen. Thus, in contrast to free-radical scavenging, which prevents molecular damage via mediation of only indirect effects, hydrogen donation repairs the molecular lesions after they occur, and can diminish cellular injury induced by both the direct and indirect effects of ionizing radiation.

c. Direct Binding. Although thiols are very efficient free-radical scavengers and may take part in hydrogen transfer, the general applicability of these basic mechanisms to overall protection *in vivo* has been questioned (18). The major criticism has been the realization that the cellular concentrations of radioprotectors, after administration of protective doses, may be much lower than would be required for free-radical mechanisms to play any but a minor role. However, the chemical structure of a variety of radioprotectors, most notably the aminothiols, allows them to bind electrostatically to the sugar–phosphate backbone of DNA. Thus, locally high concentrations of radioprotectors may be achieved at vulnerable sites on critical molecules to allow free-radical mechanisms to play a significant role in protection (20, 21).

The reversible binding of polybasic radioprotectors to DNA may also stabilize the secondary structure of the DNA molecule. This inhibits unwinding of the double helix in response to radiation-induced strand breaks, and reduces the rate at which DNA may be unwound for transcription and translation. In both cases, the increase in stability results in a greater possibility that lesions produced in the DNA molecule will be repaired (22).

d. Mixed Disulfide Formation. The mixed disulfide hypothesis (20, 23, 24) applies only to sulfhydryl-containing radioprotectors, and involves the reversible formation of disulfide bonds between thiol groups of proteins and radioprotectors. The formation of mixed disulfides preserves the integrity of enzymes and structural proteins that are dependent on intact sulfhydryl and disulfide groups for normal biological function. Binding of radioprotectors to proteins in this manner, in effect, alters their intrinsic radiosensitivity, and results in protection of these molecules.

2. Physiological–Biochemical Level

In addition to interacting directly with the products of radiation and target molecules, radioprotectors may function at a more complex level by inducing physiological and biochemical alterations that can mitigate the damaging effects of ionizing radiation. The hypotheses that have evolved to explain radioprotection at this level include hypoxia, release of endogenous radioprotective substances, biochemical shock, and hypothermia.

a. Hypoxia. Oxygen has long been known to be an important factor in enhancing cellular radiation sensitivity (25). Since the extent of radiation damage in a tissue is directly related to the degree of oxygenation of that tissue, it follows that drugs or treatments that can reverse this oxygen effect should result in significant radioprotection. A wide variety of agents with radioprotective abilities are known to induce localized or general hypoxia, which can be brought about in several ways: by interfering with the delivery of oxygen to irradiated tissues via induction of hemodynamic cardiovascular alterations, blocking hemoglobin function, increasing tissue oxygen utilization, inducing localized hypoxia through chemical and biochemical reactions, and depressing respiratory centers (12, 20, 26, 27). Of all the concepts of radioprotection action, hypoxia induction has the most widespread application (19). It accounts for, at least in part, the radioprotective action of a large variety of diverse chemicals, drugs, mediators, and pharmacological agents.

b. Nonprotein Sulfhydryl Release. One of the effects of at least some exogenously administered thiol or disulfide radioprotectors is to displace endogenous radioprotectant compounds that are contained in natural nonprotective mixed disulfide forms with proteins within the cell. This "nonprotein sulfhydryl (NPSH) release" hypothesis (12, 20, 28, 29) suggests that radioprotection results from the released endogenous sulfhydryl compounds (mainly glutathione) which function, in turn, to prevent radiation damage by radical scavenging and hydrogen donation. This hypothesis is based on the observation that the ability of sulfhydryl compounds to increase cellular levels of NPSH correlates well with their radioprotective effectiveness (12, 20, 29).

c. Biochemical Shock. The biochemical shock theory (18, 30, 31) is much more generalized than the theories previously described, and it takes into account the pronounced alterations that exogenous thiols induce in a cell's overall biochemical and metabolic state. The term "biochemical shock" was used to describe the many reversible biochemical changes that consistently occur in the cell's attempt to adapt to a massive infusion of thiol radioprotectors. The triggering event that initiates the sequence of events culminating in biochemical shock is the formation of mixed disulfides between radioprotectors and membrane sulfhydryls. This induces profound ultrastructural alterations in mitochondria and other organelles, which are rapidly followed by a characteristic syndrome of biochemical changes. These include disruption of cellular redox states; enhancement of glycogenolysis in the liver; and inhibition of glycolysis, protein synthesis, DNA synthesis, and cell division. Evidence is lacking as to which one or combination of these changes leads to radioresistance. However, an important component in this whole syndrome of changes is the inhibition of DNA synthesis and mitotic delay, which are known to occur in response to essentially all sulfhydryl radioprotectors, including WR-2721 (32, 33). This phenomenon may provide the cell with more time to repair the radiation damage to DNA before the next round of DNA synthesis. Although this theory fails to provide a clear mechanistic concept of how biochemical shock increases radioresistance, it does provide some suggestions, and it has been useful in elucidating some of the metabolic effects of radioprotective drugs.

d. Hypothermia. A reduction in body temperature has been noted to occur after the administration of a variety of radioprotective compounds (20). Hypothermia may mediate radioprotection by two mechanisms. The reduced metabolic activity that accompanies hypothermia may allow more complete and efficient repair of radiation damage (34, 35). Alternatively, damage-producing reactions following the absorption of radiation energy may be slower and less complete, thereby resulting in reduced sensitivity (35).

3. Organ Level

At the organ level, radioprotection can be achieved by enhancing the recovery and renewal of stem-cell populations that have been depleted by irradiation. This is discussed more fully later.

D. Summary of Mechanisms

The underlying mechanisms that explain chemical radiation protection are complex and as yet not completely understood. A number of the basic hypotheses of radioprotection have been presented, and each has its own set of strengths and deficiencies in explaining the action of radioprotectors. Although some of the factors appear to play a greater role (hypoxia induction) than others (hypothermia), no single theory can account fully for the protective effects that have been observed. The actual phenomenon of chemical radioprotection is likely to involve elements of most, if not all, of the theories discussed.

V. SCREENING AND ASSESSMENT OF RADIOPROTECTORS

Essentially any quantifiable radiobiological lesion or effect can be used to assess radioprotector effectiveness. Although all techniques may provide useful information, this discussion focuses on the most frequently used end points: animal lethality and cell lethality.

A. Mammalian Lethality

The most commonly used method for screening potential radioprotectors and determining their effectiveness is mammalian survival using mice (36). The mouse is the model of choice for a variety of reasons: it is relatively inexpensive, is housed with only moderate space requirements, is susceptible to radiation damage and protection, is easy to handle, and is available as standardized strains in large numbers. All of these factors facilitate the use of the large numbers of animals required to ensure statistical validity of the data. Other mammals (including rats, hamsters, guinea pigs, dogs, swine, and monkeys) have been used, but their utility as models for screening purposes is limited, especially when the larger animals are considered. Nevertheless, in order to determine the general applicability of a particular radioprotector and to facilitate the extrapolation of its protective action to man, it is necessary to demonstrate protection over a range of species.

1. Factors Affecting Survival Studies

Several major factors that could affect the outcome of survival studies must be considered before testing. They include choice of experimental animal; toxicity and route of administration of the prospective radioprotector; dose, dose rate, and quality of radiation; and preexposure time, i.e., time interval between drug administration and irradiation (11, 26, 27, 36). These are discussed in Appendix A.

2. Evaluation Criteria

The magnitude of chemical protection against radiation damage is most commonly assessed either by comparing percentage survival between treated and control groups at a selected lethal radiation dose or by computing a dose reduction factor (DRF) for the drug under study (26, 37). Percentage survival requires fewer animals and is more easily determined than DRF. For these reasons, percentage survival is most useful as a preliminary screening assay for potential radioprotectors. However, for critical comparison studies in which a true measure of efficacy is required, the DRF is used. In order to determine the DRF, groups of treated and control animals are exposed to several levels of radiation, and survival is monitored at 30 days. The survival curves are plotted as shown in Fig. 14-3, and the radiation LD_{50} is determined for the control group (D_0) (injected with solvent alone) and for the protected group (D_1). The DRF is computed as the ratio between D_0 and D_1:

$$\text{DRF} = \frac{LD_{50/30} \text{ days protected animals}}{LD_{50/30} \text{ days unprotected animals}}$$

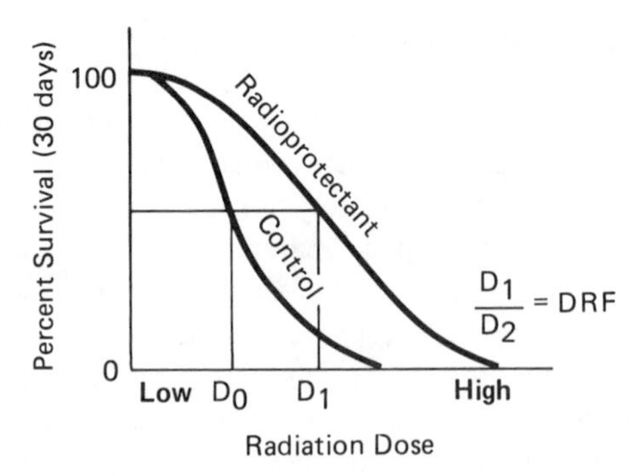

FIG. 14-3 Determination of dose reduction factor.

This means, for example, that animals treated with a radioprotector having a DRF of three can survive three times as much radiation as the unprotected animals. This is about the level that has been achieved with WR-2721 against hemopoietic death in mice. These *in vivo* screening methods using the mouse as the animal model have proven to be convenient and useful in recognizing potential radioprotectors and in quantitating their effectiveness.

B. ENDOGENOUS SPLEEN COUNTS

Recently, an alternative to 30-day mouse survival as a radioprotector-screening assay for hemopoietic protection has been suggested. This method is based on the observation that recovery from radiation injury is accompanied by formation of macroscopically identifiable nodules in the spleen, the number of which is inversely related to radiation dose (38). It has been shown that chemical protectors increase the number of nodules that appear after irradiation. This "mouse endogenous spleen count method" is performed in the same manner as the mouse survival studies except that animals are sacrificed 10 days after irradiation. Their spleens are removed and fixed histologically. Then the spleen nodules of treated and nontreated groups are counted, compared, and used to calculate a DRF. DRFs for several radioprotectors obtained by the spleen method correlate well with those obtained in survival studies, and the spleen colony method offers the advantage of being a much more rapid screening tool.

C. TISSUE CULTURE

A great deal of work on radioprotectors is being performed *in vitro* using tissue cultures of mammalian cells (39–42). Radioprotector effectiveness and DRFs are commonly determined in tissue culture by assessing the cell survival in treated and nontreated cultures based on the ability of the cell to proliferate after irradiation. This ability is quantitated by determining the number of cell colonies that develop at a specific time after irradiation by replating appropriate dilutions of treated and nontreated cells. Quantitation of the colonies is facilitated by the use of an appropriate staining material.

VI. RADIOPROTECTORS

Since the discovery in 1949 that cysteine has the ability to increase the survival of lethally irradiated mice, a staggering number of compounds have been examined for their ability to function as radioprotectors. The Army program alone developed and tested over 4000 compounds during the 1970s. Table 14-2 (43–61) is a selective list of radioprotectors (excluding aminothiols), compiled to demonstrate the diversity of compounds shown to have radioprotective properties. Several of the aminothiols are listed in Table 14-3 (62–68). The remainder of this section focuses on a discussion of some of the more effective and promising protectors highlighted in those tables. In the following discussion, radioprotector effectiveness will be described by the DRF derived from mouse $LD_{50/30}$ studies, unless specified otherwise.

A. Aminothiols

The vast majority of radioprotective agents that have been developed and tested are the aminothiols. As a group, they are the most effective class of radioprotectors. Essentially all of the compounds in this class are chemical analogs of cysteine and MEA. This is demonstrated in Fig. 14-4, which shows the structures of several aminothiols and their relationship to MEA.

1. Cysteine

The prototypical aminothiol is the sulfur-containing amino acid cysteine, which has a DRF of 1.7 when administered iv at the maximum tolerated dose of 1200 mg/kg to mice before irradiation (26). Protection is proportional to the amount of cysteine injected; however, the compound is less effective when given ip and without effect when given orally (26, 69).

2. MEA and Cystamine

The decarboxylated form of cysteine, β-mercaptoethylamine (cysteamine and MEA), is a considerably more potent radioprotective compound than cysteine

TABLE 14-2

RADIOPROTECTIVE COMPOUNDS[a]

Compound	Protective effect[b,c]	Probable mechanism of action	References
Sulfur-containing compounds		Free-radical mechanisms	
Diethyldithiocarbamate (DDC)	3	Complex[d]	27,43,44
Dimethylsulfoxide (DMSO)	2[e]		44–46
Thiourea	1		27,44
Cyanide derivatives[f]		Hypoxia	
Cyanide	2		26,27,47
Hydroxyacetonitrile	3		26,27,44
Malononitrile	2		27,44
Chelating agents		Uncertain	
EDTA	1		27,44
Metabolites		Free-radical scavenging	
Glucose	1[g]		27
Fructose	2[g]		27
α-Ketoglutarate	1		27
Hypoxia inducers		Hypoxia	
Paraminopropiophenone	3	Hemoglobin changes	25–27,37
Carbon monoxide	2		25,44
Ethanol	2[g]	Respiratory center depression	25,44
Morphine	2		25,26,44
Reserpine	2		26,44
Serotonin	3	Hemodynamic alterations	25,26,37,44
Histamine	2		25–27,44
Immunomodulators		Hemopoietic system recovery	
Glucan	1[h]		48–51
Endotoxin	1[h]		52–54
Azimexon	1[h]		55,56
Levamisole	1	Complex[d]	57
Antioxidants		Free-radical mechanisms and oxygen metabolism	
Vitamin E	1[i]		58,59
Vitamin A (β-carotene)	1		60
Superoxide dismutase	3		61
Selenium	1[i]		59

[a]Aminothiols are presented in Table 14-3.

[b]Data taken from studies using mice exposed to X or gamma radiation.

[c]Grading is according to following scale: 1, slight protection (DRF < 1.2); 2, moderate protection (DRF < 1.5); 3, good protection (DRF > 1.5).

[d]See text.

[e]Provides protection when administered topically.

[f]Highly toxic: Protective dose is very close to toxic dose.

[g]Provides protection only at extremely high doses: glucose and fructose, 13,500 mg/kg; ethanol, 6–7.5 ml/kg.

[h]Provides protection when administered after irradiation.

[i]Provides protection when given orally.

TABLE 14.3

AMINOTHIOLS[a]

Compound	Toxic[b] LD50 (mg/kg)	Protective[b] dose (mg/kg)	DRF	References
Cysteine	1700	1200	1.7	62
MEA	200	150	1.7	1,62
Cystamine	220	150	1.7	62
AET	480	400	2.1	26,62
WR-638	1120	500	2.0	8,63
WR-2721	950	500	2.7	1,8
WR-3689	1120	450	2.2	63
WR-77913	3574	2200	2.0	64
WR-151327[c]	785	315	1.9	63,65
Mercaptopropionylglycine[d]	2100	20	1.4	66,67
Glutathione	4000	4000	1.3	66,68

[a]Data taken from $LD_{50/30}$ studies using mice exposed to X or gamma radiation.
[b]By ip injection except for cysteine, which is by iv injection.
[c]Provides substantial protection against neutron irradiation.
[d]Provides significant protection when given after irradiation.

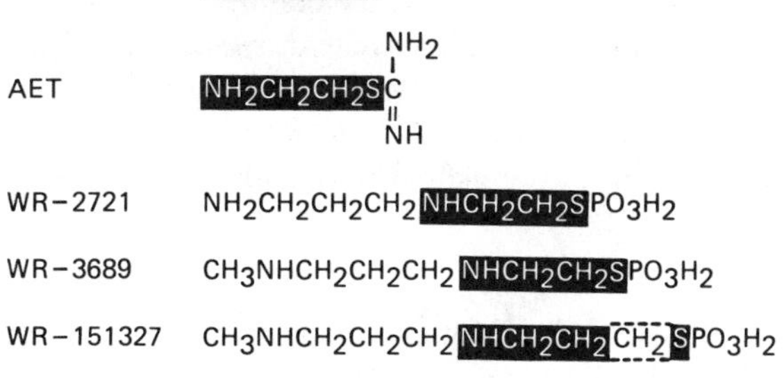

Compound	Structural Formula
Cysteine	$NH_2CH(COOH)CH_2SH$
MEA	$NH_2CH_2CH_2SH$
Cystamine	$NH_2CH_2CH_2SSCH_2CH_2NH_2$
AET	$NH_2CH_2CH_2SC(NH_2)=NH$
WR-2721	$NH_2CH_2CH_2CH_2NHCH_2CH_2SPO_3H_2$
WR-3689	$CH_3NHCH_2CH_2CH_2NHCH_2CH_2SPO_3H_2$
WR-151327	$CH_3NHCH_2CH_2CH_2NHCH_2CH_2CH_2SPO_3H_2$

FIG. 14-4 Chemical structure of selected aminothiols.

(3). MEA at a dose of 150 mg/kg provides the same degree of protection as 1200 mg/kg of cysteine when administered iv. In addition, it is more effective than cysteine when injected ip, and may confer slight protection when given orally (26). Because of its effectiveness and structural simplicity, MEA has been studied extensively, and has served as a prototype for the design of other agents. For many years, MEA was the standard against which the effectiveness of other radioprotectors was judged. Cystamine (the disulfide form of MEA) is noteworthy because, while it is equally effective as MEA on a molar basis when given parenterally, it is superior to MEA when given orally (26).

3. AET

Aminoethylisothiouronium (AET) differs in structure from cysteine and MEA in that it possesses a urea group covering the sulfur function. Its effectiveness depends on its ability to rearrange at physiological pH to form the free thiol compound mercaptoethylguanidine (MEG) (6). In mice, AET has been shown to provide DRFs of up to 2.1 at a dose of 400 mg/kg ip, and it provides modest but relatively long-lasting protection when given orally. A single oral dose of AET will confer protection to mice for at least 6 hr (26).

4. Structure–Activity Relationships

Although the protection conferred by these early drugs was significant, the major drawback precluding their use in humans is their narrow therapeutic index. These early drugs protected only at doses that were toxic or that were unacceptably close to their toxic level (Table 14-3). Nevertheless, this early work contributed significantly to further drug development by providing an understanding of structural features that were necessary for protection by aminothiols (4–6). It was determined that for optimal radioprotection, the essential structural features were a free, or "freeable," sulfhydryl group that was separated by no more than three carbon atoms from a strongly basic amino group. Additions of substituents to this nitrogen were found also to modify activity. Blocking of the thiol group in such a way that it could not be made available metabolically abolished radioprotection. However, covering the sulfur with a functional group that could be removed or altered metabolically to produce the free thiol has resulted, in many cases, in improved radioprotector effectiveness, reduced toxicity, and/or modified pharmacologic properties. A wide variety of sulfur-covering functions have been examined, and several were found to be very effective. An extensive discussion of structure–activity relationships can be found in Ref. 44 and 70.

5. WR-2721

The effect of sulfur-covering functions and nitrogen group additions is most notably demonstrated by a radioprotector developed by the Army program and designated as WR-2721. Comparison of the structure of this compound and MEA (Fig. 14-4) indicates that WR-2721 is analogous to MEA with a phosphate group on the sulfur (hence its classification as a phosphorothioate) and a propylamino group on the nitrogen function. WR-2721 is *S*-2-(3-aminopropylamino)ethylphosphorothioic acid. Its official generic name is ethiofos in the United States and gammaphos in the Soviet Union.

WR-2721 represents a significant improvement in effectiveness, potency, and reduction of toxicity over the early radioprotectors, and it has supplanted MEA as the standard against which all radioprotectors are compared. Although WR-2721 is minimally effective when given orally, it is capable of producing DRFs as high as 2.7 for gamma radiation in mice after ip administration (71). This is the highest DRF reported for any single compound with mouse lethality as the end point. The duration of protection by WR-2721 (>3 hr) is much longer than that by MEA, and WR-2721 has been shown to be effective over a broad range of species, from mice to nonhuman primates (1). Because of its increased potency and reduced toxicity, the therapeutic index for WR-2721 is much higher than that for MEA (1).

a. Pharmacokinetics. The initial steps in the metabolism of WR-2721 *in vivo* are shown in Fig. 14-5 (72–74). The first step is removal of the phosphate group covering the sulfur by the action of phosphatase enzymes. This dephosphorylation produces the free thiol, designated WR-1065, which is thought to be the active form of the drug (42, 75). The free thiol is further metabolized to its symmetrical disulfide, WR-33278. However, the overall picture of the metabolic fate and active form of exogenously administered WR-2721 is far from complete (72, 76). This is due partly to the lack of suitable methods to identify and quantitate metabolites. Methods to accomplish this have been developed recently (77), and a more complete picture of the metabolism and pharmacokinetics of WR-2721 should be forthcoming. Nevertheless, general pharmacokinetic data derived from biochemical and radiochemical studies do exist.

Plasma clearance of radiolabeled WR-2721 and its metabolites is biphasic and consists of an early rapid distribution phase with a $t_{1/2}$ of less than 10 min and a slower elimination phase with a $t_{1/2}$ of greater than 1 hr (78, 79). Tissue distribution studies have been done in mice using radiolabeled WR-2721, and localiza-

	Formula
WR-2721	$NH_2CH_2CH_2CH_2NHCH_2CH_2SPO_3H_2$
	↓
WR-1065	$NH_2CH_2CH_2CH_2NHCH_2CH_2CH_2SH$
	↓
WR-33278	$[NH_2CH_2CH_2CH_2NHCH_2CH_2S\text{-}]_2$

FIG. 14-5 Conversion of WR-2721 to free thiol and disulfide.

tion of label has been assessed by liquid scintillation counting (79–82) or whole-body autoradiography (83). In general, all of the studies agree that the liver, kidney, bladder, and salivary gland accumulate high levels of the drug; the small intestine and spleen accumulate moderate levels; and the brain, spinal cord, and most solid tumors accumulate very little, if any.

Time course studies indicate that maximum levels of drug for most tissues (the kidney, bladder, brain, and spinal cord are exceptions) are reached within 15 min after ip injection, and remain relatively constant over a 60- to 90-min period (79, 84). It is significant that the concentration of drug in those tissues increases while serum levels decline. This suggests that those tissues actively concentrate WR-2721 or its metabolites. Absorption of the drug appears to be a passive process, possibly occurring by a facilitated diffusion mechanism (79).

The tissue distribution studies reveal that some tissues accumulate large amounts of drug whereas others accumulate very little. Likewise, a difference is seen in tissue responsiveness to radioprotection by WR-2721 (84). Table 14-4 lists tissues for which WR-2721 does and does not afford protection. They are listed in order of the degree of protection provided, with the immune system and bone marrow being the most highly protected (DRF ~3). The brain and spinal cord are not protected by WR-2721.

Comparison of tissue drug distribution and tissue protection indicates that absolute tissue levels do not always correlate with protection. For example, the kidney (which absorbs high levels of WR-2721) is only weakly protected, while the bone marrow and immune system (the most effectively protected tissues) do not display the highest concentrations of the drug. Nevertheless, presence of the

TABLE 14-4

TISSUE PROTECTION BY WR-2721

Tissues protected	Tissues not protected
Immune system	Brain
Bone marrow	Spinal cord
Liver	
Skin	
Testes	
Salivary gland	
Small intestine	
Colon	
Lung	
Kidney	
Esophagus	
Oral mucosa	

drug is a prerequisite for protection. The central nervous system, in which absorption of drug is marginal to absent, is not protected. The inability of the brain and spinal cord to accumulate WR-2721 indicates that this drug does not cross the blood–brain barrier.

Several factors contribute to the discrepancies between drug levels and protection (76, 85). Essentially all of the distribution studies were performed by quantitating radioisotope levels in the various tissues after injection of animals with radiolabeled WR-2721. These isotope studies indicate total drug concentration (parent compound plus metabolites) in tissues, and their limitation lies in their inability to provide information on the precise chemical form of the drug present in those tissues. Thus, since WR-2721 is a "pro-drug," the amount of radiolabel may not reflect the actual concentrations of the active form. Other factors that would contribute are differential absorption of drug by critical target cells within an organ, and differences in the ability of the various tissues to metabolize the drug and/or concentrate it in active form in the vicinity of critical target molecules.

In spite of their limitations, the distribution studies have provided useful information on total drug partitioning. They have also shown that tissues with high levels of WR-2721 will be at least moderately protected and that absence of the drug from a tissue results in no protection.

b. Tumor Protection. As discussed above, WR-2721 provides significant radioprotection to all normal tissues examined, with the exception of the central nervous system. In contrast, this drug provides little, if any, protection to a variety of experimentally induced solid tumors in mice and rats (79, 84, 86).

Deficient absorption of WR-2721 by tumor appears to be an important factor underlying the differential protection observed between normal and tumor tissues (79, 84). Comparing the kinetics of WR-2721 absorption among tissues reveals several important differences. Figure 14-6 demonstrates the uptake of radiolabeled WR-2721 in serum, tumor tissue, and normal tissue of mice as a function of time after ip injection of WR-2721. The concentration of WR-2721 in normal tissue (the curve for liver is representative) rapidly exceeds serum levels and reaches a maximum within 15 min. This high level is maintained for an extended period even in the face of declining serum levels. On the other hand, the concentration of WR-2721 in tumor rises only gradually, never exceeds serum values over the 90-min period, and is considerably lower at all time points than in normal tissues. The difference in drug concentrations for normal and tumor tissues is most pronounced at the time of maximum radioprotection, 15–30 min after injection.

The mechanisms underlying this differential absorption and consequently differential protection of normal and tumor tissue by WR-2721 are still incompletely understood. Although drug concentrations in tumors can exceed serum levels and

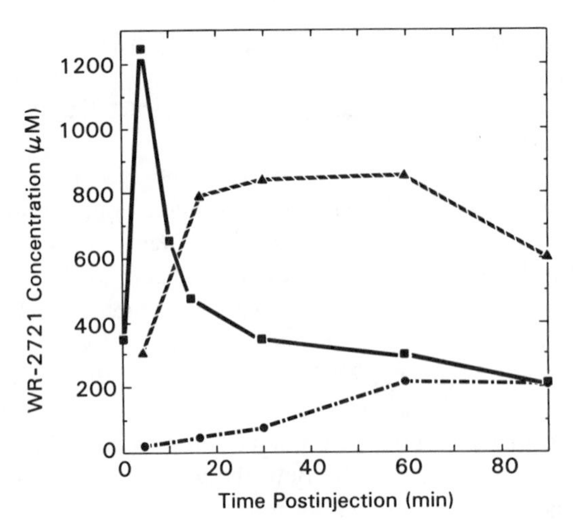

FIG. 14-6 Serum, tissue, and tumor concentration of radiolabeled WR-2721 in mice after i.p. administration. (▲) Liver, (■) serum, and (●) tumor.

approach normal tissue levels if given sufficient time (16–24 hr), the uptake of drug by tumor is accomplished primarily by a passive diffusion process (84). In contrast, normal tissue transports WR-2721 against an apparent concentration gradient. Other factors that may contribute to poor drug absorption by tumors include structural and functional alterations that are known to occur in tumor cell membranes (87), deficient tumor vascularity (86), and drug hydrophilicity (88, 89).

In spite of our inability to answer fully the questions of why differential absorption and protection occur with WR-2721 and why differential protection is not universal (90–92), the fact remains that differential protection does occur and can be quite marked. As a consequence, WR-2721 is undergoing clinical trials in the United States, Canada, and Japan as an adjunct to tumor radiotherapy. The rationale for the use of WR-2721 in these trials lies in its potential to permit more efficient killing of tumor tissue while reducing radiation damage to the adjacent protected normal tissues. The phase I trials are designed to determine maximum tolerated drug doses for single and multiple injections. These studies involve patients who require palliative radiotherapy for tumors located in or adjacent to tissues known to be protected. An acceptable tolerated dose of 740 mg/m^2 infused over a 15-min period has been established, and doses as high as 1330 mg/m^2 have been given without reaching the limits of tolerance (93). Phase II trials are designed to determine the maximum acceptable radiation dose at various sites with WR-2721 protection, and phase III trials will examine therapeutic gains provided by WR-2721 (94).

TABLE 14-5

SIDE EFFECTS OF WR-2721 IN MAN[a]

Nausea	Hypotension
Vomiting	Salivation
Diarrhea (oral)	Serum creatinine increase (oral)
Fever	Hypocalcemia

[a]See text for references.

c. Side Effects. Preclinical toxicological and pharmacological studies of WR-2721 have been conducted, and no cumulative or irreversible toxicity has been observed, even at doses substantially above the expected useful doses. However, a variety of side effects have been observed in man (Table 14-5) and in experimental animals after the administration of WR-2721 (93–95). During the course of the cancer clinical trials, over 300 patients have received single or multiple doses of WR-2721. The most frequent side effects observed were nausea, vomiting, and hypotension, which was occasionally severe. While these side effects are undesirable in any setting, they are at least manageable in a controlled clinical environment. Thus, in a clinical setting, they are acceptable to a degree, especially if the treatment results in substantial therapeutic gain. However, on the battlefield, the production of nausea, vomiting, and hypotension will severely degrade performance, and they are the major obstacles to overcome before fielding a militarily useful radioprotector. Another side effect worth noting as a potential limitation for a field-usable drug is hypocalcemia. Hypocalcemia, which has been noted to occur in man (95), could exacerbate the derangements of calcium metabolism and anemia that would occur during long-term spaceflights.

B. NEUTRON PROTECTION

The vast majority of information obtained in the field of radioprotection has been derived from studies using low-LET (linear energy transfer) X and gamma radiations. However, on the nuclear battlefield, radiation casualties also can be expected to occur from high-LET particles, primarily via external exposure to neutrons or internal exposure to ingested alpha particles. Exposure to high-LET radiation can also be expected to occur during long-term spaceflights and in tumor radiotherapy where the use of high-LET radiation therapy modalities are being investigated. Thus, protection against this type of radiation would be of direct benefit. Unfortunately, the effectiveness of radioprotectors decreases progressively as radiation becomes more densely ionizing (14, 17, 26).

To explain this inverse relationship, it is necessary to consider very briefly the nature of damage induction by high-LET radiation. Damage produced by high-LET radiation is due primarily to the direct action of highly energetic particles on critical biological molecules. In addition, the yields of water radiolytic products (e^- (aq) and OH·) as well as the ability of oxygen to enhance radiation damage decrease with increasing LET (14, 17, 96). These factors serve to limit radioprotector effectiveness because, as discussed previously, radioprotectors are unable to prevent the occurrence of direct damage, and they act (perhaps in large part) by scavenging free radicals and by inducing hypoxia. Also, the density of ionizing events induced by high-LET radiation in critical targets would require extremely high localized concentrations of radioprotector at specific sites on the target molecule to provide protection.

Although the factors limiting radioprotection against high-LET radiation are more stringent than those for low-LET radiation, substantial (albeit relatively modest) protection can be achieved. The decreased effectiveness of radioprotectors in response to high-LET radiation was noted as early as 1953, when cysteine (which has a DRF of 1.7 for X or gamma radiation) was shown to have a DRF of only 1.1 in mice against neutrons (97). Likewise, AET was shown to be only slightly effective against neutrons (27). Of the early radioprotectors, nitriles (specifically hydroxyacetonitrile) have been mentioned as being more effective than thiols in protecting mice against neutrons (26).

More recent investigations into protection against high-LET radiation have centered around the use of phosphorothioates. As in the case for low-LET radiation, the phosphorothioates have proven to be the most effective class of radioprotectors against high-LET radiation (98). Administration of WR-2721 to mice before neutron irradiation has resulted in DRFs of 1.41 for an $LD_{50/30}$ and 1.32 for an $LD_{50/7}$ (96). While WR-2721 is easily the most effective radioprotector against low-LET X and gamma radiations, it is not exceptionally better than several other drugs of this class in protection from high-LET radiation (99). However, one of the more recently developed phosphorothioates, WR-151327, shows promise as being substantially more effective than the other compounds tested. DRFs as high as 2.3 have been achieved with this compound using an $LD_{50/6}$ as an end point (65). The DRF for WR-2721 in this same assay is only 1.4 (65).

C. Immunomodulators

All organ systems are potentially vulnerable to radiation-induced injury, but the hemopoietic and lymphoid systems are especially sensitive. Whole-body exposure to doses of 200–1000 rads (depending on the species) is sufficient to produce a severe peripheral blood pancytopenia, which results from depletion of the highly radiosensitive hemopoietic stem and progenitor cells. This leads to infection, hemorrhage, anemia, and ultimately death within 30 days (19). Survival in this hemopoietic-syndrome dose range can be increased by agents that stimulate the function and recovery of the hemopoietic system. A wide variety of agents, both naturally occurring and synthetic, show promise in this regard as potential adjuncts to therapeutic and radioprotector regimens. Several are highlighted below.

1. Naturally Occurring Immunomodulators

Interferon, *Bacillus* Calmette–Guérin (BCG), *Corynebacterium parvum,* glucan, bacterial endotoxin, interleukin 1, and thymic peptides are several of the naturally occurring immunomodulators that have been shown to provide protection or therapeutic benefit from midlethal doses of radiation (200–1000 rads). Of these, bacterial endotoxin and glucan are perhaps the best characterized.

a. Endotoxin. Endotoxins (i.e., lipopolysaccharides of molecular weight of about 1,000,000 isolated from bacterial cells walls) are immunostimulants and have long been known to increase the survival of rodents (DRFs of about 1.2) after irradiation (100). These studies have been extended to include dogs and sheep, which appear to be somewhat better protected, with DRFs of about 1.4 (98). As a general rule, endotoxins are effective when administered before irradiation. However, it is of interest that in at least one study, endotoxin was found to provide some protection when given to mice as late as 24 hr after irradiation with 600 rads (52). One of the factors that limit the general applicability of endotoxin in radioprotection is the toxicity that is produced at therapeutic doses. Thus, studies demonstrating radioprotection with "detoxified" endotoxin preparations are encouraging (53, 54).

b. Glucan. Glucan, a β-1,3-linked polyglycan isolated from the inner cell wall of the yeast *Saccharomyces cerevisiae,* is a potent immune modulator that is capable of enhancing a variety of immunopoietic and hemopoietic responses (48, 49). It has been demonstrated recently that both the soluble (glucan F) and particulate (glucan P) forms of this polymer can significantly enhance survival in mice when given before a lethal dose of radiation (50). The intravenous administration of glucan P produces several severe side effects. However, glucan F is well tolerated, even at extremely high iv doses in mice, dogs, and primates. Glucan F has the added benefit of conferring radioprotection when administered after irradiation (51).

2. Synthetic Immunomodulators

a. Azimexon. Azimexon is a recently developed synthetic compound that has been shown to have substantial immunostimulatory effects. The administration of azimexon to mice before irradiation (500 rads) increased the percentage survival from 56% in the control, unprotected group to 100% in the group treated with azimexon, and prolonged the mean survival time from 17.3 to 29 days (56). This compound also protects when given after irradiation (55).

b. Levamisole and DDC. Levamisole and diethyldithiocarbamate (DDC) are two other compounds that deserve mention. As with the other agents discussed in this section, both of these compounds are immunomodulators that provide significant but modest protection (43, 57, 101). However, both contain sulfur, and therefore have the potential to function via free-radical mechanisms. In addition, levamisole has antioxidant properties and may also contribute to antioxidant defense systems (see below) (57). Related to the dual function of levamisole as an immunostimulant and antioxidant is the recent finding that vitamin E, an integral part of the antioxidant defense system, has immunomodulating activity (58). Thus, the radioprotective mechanism of these compounds is complex. This is particularly true for DDC, which is also an effective chelating agent and, in certain circumstances, can act as a radiosensitizer (43, 101). This latter activity is seen primarily in tissue culture, and is ascribed to the ability of DDC to inhibit the antioxidant enzyme superoxide dismutase.

Although the protection afforded by immunomodulators is modest compared to that provided by the phosphorothioates, they are a potentially useful class of radioprotective or radiotherapeutic compounds. As our understanding of bone marrow stem-cell physiology, kinetics of stem-cell proliferation and differentiation, and function of cellular and humoral factors increases, the potential for development of more effective immunomodulators also increases. However, at present, these compounds will have their greatest utility in combination with other, more potent radioprotectors. One factor that contributes to the potential utility of this class of compounds is the ability of many of them to enhance survival when given after irradiation.

D. Endogenous Protection (Antioxidant Defense)

Free radicals are generated *in vivo* as by-products or intermediates of normal metabolic processes, including aerobic respiration, inflammation, and xenobiotic drug metabolism and detoxification. In addition, organisms have been exposed to ionizing radiation from natural radioactivity in soil and food as well as from extraterrestrial sources since the dawn of time. Thus, in order for life to have survived on this planet, it was necessary to evolve a system of biochemical defenses to protect from free-radical damage. Several excellent and comprehensive reviews dealing with antioxidant mechanisms and the role of free radicals in tissue injury are available (13, 15, 102).

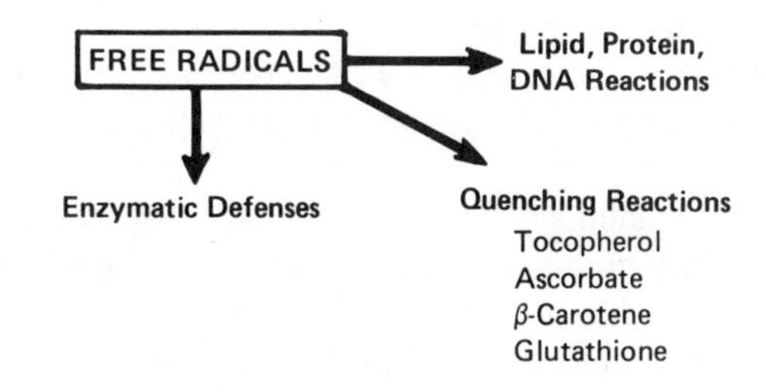

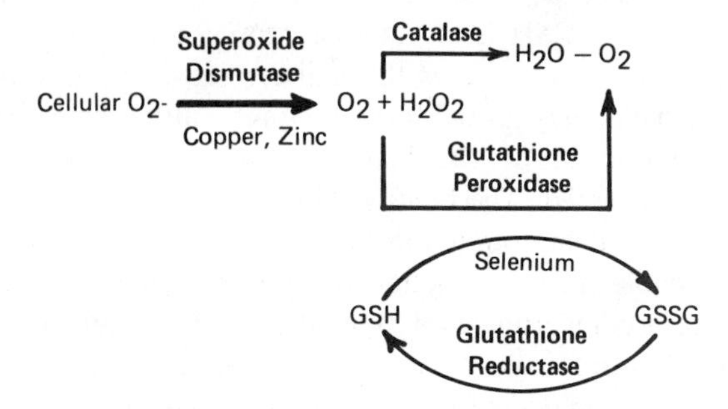

FIG. 14-7 Scheme of free-radical defense mechanisms. (Reprinted with permission from B. A. Freeman and J. A. Crapo, Biology of disease: Free radicals and tissue injury. *Lab. Invest.* **47**(5), 412–426. © 1982 by U.S.–Canadian Div. of the IAP.)

Free radicals can undergo three major reactions in cellular systems (Fig. 14-7). First, as discussed previously, free radicals can react with critical biological macromolecules and thus lead to cytotoxicity. To minimize this, a biochemical antioxidant defense system has evolved whose net effect is to lower the cellular concentrations of free radicals and thereby inhibit excessive damage to cellular components. This defense is essentially a two-component system consisting of (1) low-molecular-weight compounds that quench (i.e., scavenge) free radicals and (2) enzymes that metabolize free radicals to less reactive species.

1. Low-Molecular-Weight Compounds

The principal low-molecular-weight compounds that function as endogenous free-radical scavengers are vitamins E, C, and A and glutathione.

a. Vitamins. Vitamin E is lipid soluble and therefore can insert into cellular membranes. *In vitro* studies have shown that vitamin E can scavenge a variety of free-radical species and can protect membrane lipids from peroxidation. Similarly, vitamin A (more specifically, β-carotene) is lipophilic and is known to be an efficient oxygen radical scavenger that can also inhibit lipid peroxidation. In contrast, vitamin C is water-soluble and scavenges free radicals in aqueous compartments. Evidence from *in vitro* studies suggests that vitamins E and C may work in concert to terminate free-radical reactions. Vitamin C has been postulated to function in this regard as a secondary antioxidant by reacting with vitamin E radicals to regenerate vitamin E to its active reduced form.

b. Glutathione. Glutathione (GSH) is an important factor in this defense system. As shown in Fig. 14-7, it functions in both components of the system. GSH is a tripeptide made up of the amino acids glutamate, cysteine (the prototype aminothiol radioprotector), and glycine. It is the most abundant nonprotein sulfhydryl compound, being found in most animal tissues at high concentrations (millimolar range). GSH is an integral part of a wide variety of normal metabolic functions (103–105). It is the major storage and transport form of cysteine. The cysteine moiety provides GSH with its sulfhydryl group, which (as with all thiol radioprotectors) is the "business end" of the molecule as far as radioprotection is concerned. The relationship of this compound to free-radical defense mechanisms is demonstrated in Fig. 14-7.

GSH is water-soluble and may function as a low-molecular-weight free-radical scavenger in aqueous compartments. Essentially all intracellular GSH is present as the reduced thiol with less than 5% present in the oxidized form, glutathione disulfide (GSSG). The products of the reduction of peroxides, disulfides, and free radicals by GSH are GSH adducts of lipids or proteins, or GSSG. In order to prevent the accumulation of GSSG (which is toxic) and to maintain steady-state levels of GSH, cells can regenerate GSH from GSSG via glutathione reductase utilizing NADPH generated from the hexose monophosphate shunt for reducing equivalents.

2. Enzymes

GSH also participates in enzymatic free-radical defenses by functioning as a cofactor for glutathione peroxidase. The activity of this enzyme is dependent on selenium [a selenium-independent form has also been characterized (106)] and is tightly coupled to intracellular concentrations of GSH, glutathione reductase, and NADPH. Glutathione peroxidase converts H_2O_2 to water via oxidation of GSH, and can also metabolize lipid hydroperoxides to relatively unreactive hydroxy fatty acids.

Catalase, another enzyme that removes H_2O_2, catalyzes the divalent reduction of H_2O_2 to water. Thus, glutathione peroxidase in concert with catalase lowers the steady-state concentration of H_2O_2 which, if not degraded, would produce more potent radical species.

Both of these enzymes, in turn, act in concert with superoxide dismutase (SOD), and this is the primary means by which superoxide anions ($O_2^{\cdot-}$) are cleared from biological systems (107). SOD catalyzes the conversion (dismutation) of $O_2^{\cdot-}$ to H_2O_2 and oxygen by converting two molecules of $O_2^{\cdot-}$ to one molecule each of H_2O_2 and oxygen. At least three separate forms of SOD have been characterized. One contains copper and zinc and is present in the cytosol of eukaryotic cells. Another contains manganese and is present in eukaryotic and prokaryotic cells. The third contains iron and is exclusively prokaryotic.

3. Role in Radioprotection

This network of biochemical defenses is quite capable of handling normal metabolic loads of free radicals and reactive oxygen species. However, radiation (as well as disease states, certain drugs, and other stresses) can overwhelm this system, allowing free radicals to reach vital targets and cause serious damage. This serves as the basis for research efforts whose aim is to enhance radioprotection by augmenting cellular defenses.

One of the theories of radioprotector action discussed previously proposes that exogenously administered thiols may augment natural defenses by displacing endogenous nonprotein thiols, principally GSH, from their mixed disulfide forms. This would increase the amount of intracellular GSH available to serve as a cofactor for GSH peroxidase and to take part in free-radical scavenging, hydrogen donation, and mixed disulfide formation.

Alternatively, the exogenous administration of various components of the defense system may augment defenses and result in radioprotection. The administration of SOD to rats and mice has been shown to prolong survival after lethal doses of radiation. DRFs as high as 1.56 have been achieved in mice treated with SOD (61). Vitamin E fed to mice at only three times the minimal dietary level has increased survival (58). Similarly, selenium (4 ppm) administered to mice in drinking water has also increased 30-day survival after irradiation. Of significance is the finding that a combination of dietary vitamin E supplementation with selenium administration improved survival to a greater degree than did either treatment alone (59). These studies indicate that dietary supplementation with natural substances may be a convenient way to provide protection with minimal side effects.

E. Combined Treatments

In an effort to improve radioprotector effectiveness and decrease toxicity, combinations of radioprotective agents have been examined. In many cases, a synergistic effect is seen in which the combination provides greater protection than either agent alone. The synergistic effect of vitamin E and selenium has already been mentioned. A number of other combinations or "cocktails" have been tried (27, 67, 108, 109), and several will be highlighted below.

5-Methoxytryptamine (MOT), an analog of serotonin, is slightly less protective (DRF = 1.60) at a dose of 50 mg/kg than is AET (DRF = 1.65) at a dose of 300 mg/kg. The simultaneous administration of both of these compounds increased the $LD_{50/30}$ of mice from 500 to 1025 rads, corresponding to a DRF of 2.05 for the combination (67, 108). This synergistic effect becomes particularly significant when the amounts of MOT and AET injected are considered. In the combined experiment, the amount of MOT used was one-fourth of that required to produce a DRF of 1.6; for AET, the dosage used was one-half of that required to produce a DRF of 1.65. Similarly, a mixture of AET and cysteine improved the survival of irradiated mice more effectively than either agent alone. Cysteine in this experiment had a DRF of 1.35, and AET had a DRF of 1.65; however, the combination achieved a DRF of 2.15 (67).

In another study, various combinations of AET, glutathione, serotonin, MEA, and cysteine were examined (109). AET, MEA, or serotonin used alone provided similar protection (DRF = 1.7), cysteine was less effective (DRF = 1.12), and glutathione was marginally protective (DRF = 1.05). The most effective regimen found was a combination of all five agents, which produced a DRF of 2.8. In this combination, the doses of AET and MEA used were two-thirds and one-half, respectively, of those used individually. When this treatment was supplemented with transplants of syngeneic bone marrow, the DRF was increased to a remarkable 3.7. Although bone marrow transplants for every, or perhaps any, battlefield nuclear casualty would be impractical given the current state of the art, this experiment does demonstrate that dramatic increases in protection can be achieved by using combinations of different radioprotective agents and treatments.

2-Mercaptopropionylglycine (MPG) is marketed in Japan as a detoxifying agent under the trade name Thiola. Although this compound is only moderately protective when given before irradiation, it is of considerable interest as a radioprotector because of its low toxicity (Table 14-3) and because of its action as a radiotherapeutic agent (66, 67). Administration of MPG to mice 4 hr after exposure to 900 rads (a dose lethal to 100% of the control animals) resulted in 58% survival, and MPG given as late as 24 hr after irradiation was still able to enhance the 30-day survival (16.7%). Of relevance to the present discussion is the demonstration that the concurrent administration of MPG and AET before irradiation results in a synergistic radioprotective effect. Interestingly, when AET was given before irradiation and MPG after, no increase in effectiveness was found (67).

Some of the results from these studies and others have been impressive, and they appear to offer real potential for providing effective protection with decreased toxicity. In spite of this, remarkably little combined modality work has been done since the bulk of these studies were carried out in the 1950s and 1960s. With the development of the more effective phosphorothioates and the demonstration of significant (although modest) protection by vitamins, minerals, and immunomodulators, the revival of this approach is warranted and is being pursued (110).

VII. SUMMARY

The beginning of the search for a drug that would minimize the effects of ionizing radiation commenced in the 1940s with *in vitro* studies (Ref. 25, pp. 8–10) and culminated in 1949 with the demonstration of *in vivo* protection by cysteine. Since that time, a great deal of effort has been expended to develop a drug suitable for use by military personnel on the nuclear battlefield. An equally intense basic research effort, which has proceeded concurrently with and contributed to the developmental research, has gone into studying mechanisms of radiation injury and radioprotection. That our understanding of these processes is still incomplete and no entirely suitable radioprotective regimen yet exists underscores the complexity of the problem.

Nevertheless, substantial progress has been made on both fronts. On the developmental front, the early radioprotectors provide valuable clues to the molecular structural features required for protective activity. From these studies, newer radioprotectors (typified by WR-2721) were developed, which were vast improvements in effectiveness and toxicity over the earlier compounds. Experimentation with different formulations of WR-2721 using state-of-the-art drug-formulation technology may alleviate the major problems associated with this drug (comparatively poor oral effectiveness, nausea, vomiting, hypotension), and may lead to its use as a first-generation antiradiation drug. As a direct result of the drug-development program, the long-sought but heretofore elusive goal (Ref. 25, pp. 165–167) of using radioprotectors as adjuncts to tumor radiation therapy is closer to becoming a reality. WR-2721 is presently undergoing phase I and II clinical trials, based on the observation that this drug affords differential protection to normal and tumor tissue.

On the basic research front, our understanding of radiation protection mechanisms, radioprotector drug pharmacokinetics, free-radical defense mechanisms, and cellular responses to radiation has expanded. All of these factors contribute to the development of more effective drugs or treatments, and either open up new approaches or expand old approaches to achieve radioprotection.

A number of promising new compounds have been developed in the Army's program. Two noteworthy candidates for further investigation are WR-3689 and WR-151327. WR-3689 is a close analog of WR-2721 and possesses very similar radioprotective properties. However, it is considerably less toxic. The pharmacology, toxicology, and physicochemical properties of this drug are currently under investigation. WR-151327, another phosphorothioate, appears to have superior effectiveness against high-LET radiation.

In addition to the development of new compounds, other approaches to radioprotection are being investigated. These include methods to enhance recovery of the hemopoietic system after irradiation through the use of immunomodulators and methods to enhance the effectiveness of natural antioxidant defense mechanisms either endogenously by altering levels of glutathione, or exogenously by administration of relevant vitamins, minerals, and enzymes. These latter approaches, when used alone, provide only modest protection. But it is reasonable to expect that combining their use with less toxic doses of radioprotective drugs would result in an effective and less toxic or nontoxic radioprotective regimen.

If a nuclear exchange should occur tomorrow, the only defenses available are physical shielding, prudent adherence to established military nuclear-defense protocols, and use of conventional medical therapeutic modalities. The only other potentially radioprotective practice that can be instituted today is the provision of good nutrition with adequate supplements of vitamins and minerals that possess antioxidant properties.

For the future, it is not overly optimistic to expect that a first-generation radioprotective regimen will become available. Current evidence suggests that optimum radioprotection in the future might be achieved through a combined treatment regimen, including subtoxic doses of one or several radioprotectors in combination with good nutrition, vitamin supplementation, and immunomodulators.

APPENDIX A: FACTORS AFFECTING SURVIVAL STUDIES

A. Animal Model

Wide variations in drug tolerance, drug effectiveness, and radiation response are seen for different animal species. As discussed previously, the mouse is the model most frequently used. However, selection of a species is only the first step. Other factors that will influence the results are the strain, sex, age, and general health status of the animal. These factors must be considered and standardized.

B. Drug Toxicity and Administration

Before a radioprotective agent is tested, it is necessary to have some indication of its toxicity. Toxicity studies are performed using mice identical to those that will be used in the radioprotection studies. From these, the toxic LD_{10} or LD_{50} is found, and generally one-half to two-thirds of that dose is used for initial screening.

The route of administration can markedly affect the protective action and toxicity of a compound. In small rodents, the most convenient and most commonly used method is ip injection. Intravenous injection is also used frequently. This route results in more rapid distribution of the drug, but toxic reactions are more common. The effectiveness of radioprotectors given orally is, as a general rule, much lower than that found with ip or iv injection. In fact, a large number of the most effective radioprotectors provide comparatively poor protection when given orally due to poor absorption and breakdown of the drug by stomach acid, digestive enzymes, intestinal flora, or first-pass metabolism in the liver.

C. Radiation

The dose, dose rate, and quality of radiation must also be considered. Most commonly, *in vivo* radioprotection studies are performed at radiation doses in the range of the $LD_{50/30}$ or $LD_{100/30}$, i.e., those doses of radiation lethal to 50 or 100% of the animals within 30 days. The dose of radiation as well as the postirradiation observation period depends on the type of information desired. They are discussed more fully below. In general, radiation dose rates vary from between 40 and 200 rad/min. Radiation quality is also a factor. Most testing of radioprotectors has used low-LET X or gamma rays from external sources. High-LET radiation is much less frequently used.

D. Preexposure Time

The time interval between radioprotector administration and irradiation that will result in maximum protection varies markedly from one compound to another. Optimum preexposure times must be determined for each compound. For most radioprotectors, maximum protection is achieved when they are given between 15 and 60 min before irradiation.

E. Postirradiation Observation Time

The duration of the postirradiation observation period also varies. The 30-day survival test is most commonly chosen, on the basis that relatively few animals succumb to acute radiation damage after this time. An increase in survival over a 30-day period is an indication of the ability of a radioprotector to protect against death from the hemopoietic syndrome, which results from exposures of <1000 rads. Longer periods of observation and lower doses of radiation are required in order to assess the effectiveness of drugs in preventing or inhibiting the delayed effects of radiation, such as increased tumor incidence and cataracts. Two additional observation periods that are used, 6 and 2 days, provide information on the ability of a drug to protect against gastrointestinal syndrome death (1000–3000 rads) and central nervous system syndrome death (>5000 rads), respectively.

REFERENCES

1. Davidson, D. E., Grenan, M. M., and Seeney, T. R. Biological characteristics of some improved radioprotectors. In *Radiation Sensitizers: Their Use in the Clinical Management of Cancer* (L. W. Brady, ed.), pp. 309–320. Masson, New York, 1980.
2. Patt, H. M., Tyree, E. B., Straube, R. L., and Smith D. E. Cysteine protection against X-irradiation. *Science* **110,** 213 (1949).
3. Bacq, Z. M., Hervé, A., Lecomte, J., Fischer, P., Blavier, J., Dechamps, G., Le Bihan, H., and Rayet, P. Protection contre le rayonnement X par la β-mercaptoethylamine. *Arch. Int. Physiol.* **59,** 442 (1951).
4. Doherty, D. G., and Burnett, W. T., Jr. Protective effect of S, β-aminoethylisothiuronium Br HBr and related compounds against X-radiation death in mice. *Proc. Soc. Exp. Biol. Med.* **89,** 312–314 (1955).
5. Doherty, D. G., Burnett, W. T., Jr., and Shapira, R. Chemical protection against ionizing radiation. II. Mercaptoalkylamines and related compounds with protective activity. *Radiat. Res.* **7,** 13–21 (1957).
6. Shapira, R., Doherty, D. G., and Burnett, W. T., Jr. Chemical protection against ionizing radiation. III. Mercaptoalkylguanidines and related isothiuronium compounds with protective activity. *Radiat. Res.* **7,** 22–34 (1957).
7. Piper, J. R., Stringfellow, C. R., Jr., Elliot, R. D., and Johnston, T. P. S-2-(ω-aminoalkylamino)ethyl dihydrogen phosphorothioates and related compounds as potential antiradiation agents. *J. Med. Chem.* **12,** 236–243 (1969).
8. Yuhas, J. M., and Storer, J. B. Chemoprotection against three modes of radiation death in the mouse. *Int. J. Radiat. Biol.* **15,** 233–237 (1969).
9. Singh, A., and Singh, H. Time-scale and nature of radiation-biological damage: Approaches to radiation protection and post-irradiation therapy. *Prog. Biophys. Mol. Biol.* **39,** 69–107 (1982).
10. Chapman, J. D., and Reuvers, A. P. The time-scale of radioprotection in mammalian cells. *Experientia, Suppl.* **27,** 9–18 (1977).
11. Selman, J. *Elements of Radiobiology*. Thomas, Springfield, Illinois, 1983.
12. Klayman, D. L., and Copeland, E. S. Radioprotective Agents. In *Kirk-Othmer Encyclopedia of Chemical Technology,* (M. Grayson and D. Eckroth, eds.), 3rd ed., Vol. 19, pp. 801–832. Wiley, New York, 1982.
13. Greenstock, C. L. Redox processes in radiation biology and cancer. *Radiat. Res.* **86,** 196–211 (1981).
14. Prasad, K. N. *CRC Handbook of Radiobiology*. CRC Press, Boca Raton, Florida, 1984.
15. Freeman, B. A., and Crapo, J. D. Biology of disease: Free radicals and tissue injury. *Lab. Invest.* **47,** 412–426 (1982).

16. Edwards, J. C., Chapman, D., Cramp, W. A., and Yatvin, M. B. The effects of ionizing radiation on biomembrane structure and function. *Prog. Biophys. Mol. Biol.* **43,** 71–93 (1984).
17. Alper, T. *Cellular Radiobiology.* Cambridge Univ. Press, London and New York, 1979.
18. Bacq, Z. M., and Goutier, R. Mechanisms of action of sulfur-containing radioprotectors. *Brookhaven Symp. Biol.* **20,** 241–262 (1967).
19. Fabrikant, J. I. *Radiobiology.* Year Book Med. Publ., Chicago, Illinois, 1972.
20. Copeland, E. S. Mechanisms of radioprotection—A review. *Photochem. Photobiol.* **28,** 839–844 (1978).
21. Pihl, A., and Sanner, T. Chemical protection against ionizing radiation by sulphur-containing agents. In *Radiation Protection and Sensitization* (H. Moroson and M. Quintiliani, eds.), pp. 43–55. Taylor & Francis, London, 1970.
22. Brown, P. E. Mechanism of action of aminothiol radioprotectors. *Nature (London)* **213,** 363–364 (1967).
23. Eldjarn, L., and Pihl, A. On the mode of action of x-ray protective agents. I. The fixation *in vivo* of cystamine and cysteamine to proteins. *J. Biol. Chem.* **225,** 499–510 (1956).
24. Eldjarn, L., and Pihl, A. *Progress in Radiobiology,* p. 249. Oliver & Boyd, London, 1956.
25. Pizzarello, D. J.,and Colombetti, L. G., eds. *Radiation Biology.* CRC Press, Boca Raton, Florida, 1982.
26. Thomson, J. F. *Radiation Protection in Mammals.* Reinhold, New York, 1962.
27. Bacq, Z. M. *Chemical Protection against Ionizing Radiation.* Thomas, Springfield, Illinois, 1965.
28. Revesz, L., and Modig, H. Cystamine-induced increase of cellular glutathione level: A new hypothesis of the radioprotective mechanism. *Nature (London)* **207,** 430–431 (1965).
29. Modig, H. G., and Revesz, L. Non-protein sulfhydryl and glutathione content of Ehrlich ascites tumor cells after treatment with the radioprotectors AET, cysteamine and glutathione. *Int. J. Radiat. Biol.* **13,** 469–477 (1967).
30. Bacq, Z. M., and Alexander, P. Importance for radioprotection of the reaction of cells to sulphydryl and disulphide compounds. *Nature (London)* **203,** 162–164 (1964).
31. Bacq, Z. M., and Van Canegham, P. The shock produced by large doses of radioprotective SH or SS substances. In *Radiation Damage Panel Proceedings,* PL-311/14, pp. 141–147. IAEA, Vienna, 1968.
32. Goutier, R. Effects on cell growth processes (mitosis, synthesis of nucleic acids and of proteins). In *Sulfur-Containing Radioprotective Agents* (Z. M. Bacq, ed.), Int. Encycl. Pharmacol. Ther., Sect. 79, pp. 283–301. Pergamon, Oxford, 1975.
33. Giambarresi, L. I., Murray, W. E., Jr., and Catravas, G. WR 2721 inhibition of DNA synthesis in rats following partial hepatectomy. *Fed. Proc., Fed. Am. Soc. Exp. Biol.* **3,** 681 (1985).
34. Klayman, D. L., and Copeland, E. S. The design of antiradiation agents. In *Drug Design* (E. J. Ariens, ed.), Vol. 6, pp. 82–142. Academic Press, New York, 1975.
35. Belli, J. A., and Bone, F. J. Influence of temperature on the radiation response on mammalian cells in tissue culture. *Radiat. Res.* **18,** 272–276 (1963).
36. Griffith, W. H., and Dyer, H. M. *A Study of Research Methodology for Use in the Development of Anti-radiation Agents,* Comm. Rep. Life Sci. Res. Off., Fed. Am. Soc. Exp. Biol., Washington, D.C., 1966.
37. Carr, C. J., Huff, J. E., Fisher, K. D., and Huber, T. E. Protective agents modifying biological effects of radiation. *Arch. Environ. Health* **21,** 88–98 (1970).
38. Kinnamon, K. E., Ketterling, L. L., Stampfli, H. F., and Grenan, M. M. Mouse endogenous spleen counts as a means of screening for anti-radiation drugs. *Proc. Soc. Exp. Biol. Med.* **164,** 370–373 (1980).
39. Antoku, S. Chemical protection against radiation-induced DNA single-strand breaks in cultured mammalian cells. *Radiat. Res.* **65,** 130–138 (1976).

40. Vergroesen, A. J., Budke, L., and Vos, O. Protection against X-irradiation by sulphydryl compounds. II. Studies on the relation between chemical structure and protective activity for tissue culture cells. *Int. J. Radiat. Biol.* **13,** 77–92 (1967).
41. Vos, O., Budke, L., and Grant, G. A. *In vitro* evaluation of some latent radioprotective compounds. *Int. J. Radiat. Biol.* **30,** 433–448 (1976).
42. Mori, T., Watanabe, M., Horikawa, M., Nikaido, P., Kimura, H., Aoyama, T., and Sugahara, T. WR-2721, its derivatives and their radioprotective effects on mammalian cells in culture. *Int. J. Radiat. Biol.* **44,** 41–53 (1983).
43. Milas, L., Hunter, H., Ito, H., and Peters, L. J. *In vivo* radioprotective activities of diethyldithiocarbamate (DDC). *Int. J. Radiat. Oncol. Biol. Phys.* **10,** 2335–2343 (1984).
44. Foye, W. O. Radiation-protective agents in mammals. *J. Pharm. Sci.* **58,** 283–300 (1969).
45. Ashwood-Smith, M. J. The radioprotective action of dimethylsulfoxide and various other sulfoxides. *Int. J. Radiat. Biol.* **3,** 41–48 (1961).
46. Kim, S. E., and Moos, W. S. Radiation protection by topical DMSO application. *Health Phys.* **13,** 601–606 (1967).
47. van der Meer, C., Zaalberg, O. B., Vos, O., Vergroesen, A. J., and van Bekkum, D. W. On the mechanism of the radioprotective action of cyanide. *Int. J. Radiat. Biol.* **4,** 311–319 (1961).
48. Patchen, M. L. Immunomodulators and hemopoiesis. *Surv. Immunol. Res.* **2,** 237–242 (1983).
49. Patchen, M. L., DiLuzio, N. R., Jacques, P., and MacVittie, T. J. Soluble polyglycans enhance recovery from cobalt-60-induced hemopoietic injury. *J. Biol. Response Modif.* **3,** 627–633 (1984).
50. Patchen, M. L., and MacVittie, T. J. Stimulated hemopoiesis and enhanced survival following glucan treatment in sublethally and lethally irradiated mice. *Int. J. Immunopharmacol.* **7,** 923–932 (1985).
51. Patchen, M. L., and MacVittie, T. J. Comparative effects of soluble and particulate glucan on survival in irradiated mice. *J. Biol. Resp. Mod.* **5,** 45–60 (1986).
52. Ainsworth, E. J., and Hatch, M. H. The effect of *Proteus morganii* endotoxin on radiation mortality in mice. *Radiat. Res.* **13,** 632–638 (1960).
53. Snyder, S. L., Walker, R. I., MacVittie, T. J., and Sheil, J. M. Biologic properties of bacterial lipopolysaccharides treated with chromium chloride. *Can. J. Microbiol.* **24,** 495–501 (1978).
54. Bertok, L. Radio-detoxified endotoxin as a potent stimulator of nonspecific resistance. *Perspect. Biol. Med.* **24,** 61–66 (1980).
55. Bicker, U., Friedberg, K. D., Hebold, G., and Mengel, K. Reduction of acute toxicity of cyclophosphamide and X-rays by the new immunomodulating compound BM 12.531. *Experientia* **35,** 1361–1363 (1979).
56. Stylos, W. A., Chirigos, M. A., Papademetriou, V., and Lauer, L. The immunomodulatory effects of BM 12.531 (azimexon) on normal or tumored mice: *In vitro* and *in vivo* studies. *J. Immunopharmacol.* **2,** 113–132 (1980).
57. Dobbs, C. R., Weiss, J. F., Kumar, K. S., and Chirigos, M. A. Antioxidant and radioprotective properties of levamisole. In *Oxygen and Oxy-Radicals in Chemistry and Biology* (M. A. J. Rodgers and E. L. Powers, eds.), pp. 622–624. Academic Press, New York, 1981.
58. Srinivasan, V., Jacobs, A. J., Simpson, S. A., and Weiss, J. F. Radioprotection by vitamin E: Effects on hepatic enzymes, delayed type hypersensitivity, and postirradiation survival of mice. In *Modulation and Mediation of Cancer by Vitamins,* pp. 119–131. Karger, Basel, 1983.
59. Jacobs, A. J., Rankin, W. A., Srinivasan, V., and Weiss, J. F. Effects of vitamin E and selenium on glutathione peroxidase activity and survival of irradiated mice. In *Proceedings of the 7th International Congress on Radiation Research* (J. J. Broerse, G. W. Barendsen, H. B. Kal, and A. J. van der Kogel, eds.), pp. D5–15. Martinus Nijhoff, Amsterdam, 1983.
60. Seifter, E., Rettura, G., Padawer, J., Stratford, F., Goodwin, P., and Levenson, S. M.

Supplemental vitamin A and β-carotene reduce morbidity and mortality in mice subjected to partial or whole body irradiation. *First Conference on Radioprotectors and Anticarcinogens,* Abstract, pp. 62–63 (1982).
61. Petkau, A. Radiation protection by superoxide dismutase. *Photochem. Photobiol.* **28,** 765–774 (1978).
62. Straube, R. L., and Patt, H. M. Chemical protection against ionizing radiation. *Annu. Rev. Pharmacol.* **3,** 293–306 (1963).
63. Brown, D. Q., Pittock, J. W., and Rubinstein, J. S. Early results of the screening program for radioprotectors. *Int. J. Radiat. Oncol. Biol. Phys.* **8,** 565–570 (1982).
64. Mendiondo, O. A., Connor, A. M., and Grigsby, P. Sodium hydrogen-*S*-(3-amino-2-hydroxypropyl) phosphorothioate (WR-77913): Toxicity and bone marrow radioprotection. *Int. J. Radiat. Oncol. Biol. Phys.* **8,** 553–555 (1982).
65. Sigdestad, C. P., Gidina, D. J., Connor, A. M., and Hanson, W. R. A comparison of radioprotection from three neutron sources and cobalt-60 by WR-2721 and WR-151327. *Radiat. Res.,* in press.
66. Sugahara, T., and Srivastava, P. N. MPG (2-mercaptopro-pionylglycine): A review on its protective action against ionizing radiations. In *Modification of Radiosensitivity of Biological Systems,* Proc. Advisory Group, pp. 77–87. IAEA, Vienna, 1976.
67. Sztanyik, L. B., and Santha, A. Synergistic effect of radioprotective substances having different mechanisms of action. In *Modification of Radiosensitivity of Biological Systems,* Proc. Advisory Group, pp. 47–59. IAEA, Vienna, 1976.
68. Chapman, W. H., and Cronkite, E. P. Further studies of the beneficial effect of glutathione on X-irradiated mice. *Proc. Soc. Exp. Biol. Med.* **75,** 318–322 (1950).
69. Patt, H. M., Mayer, S. H., Straube, R. L., and Jackson, E. M. Radiation dose reduction by cysteine. *J. Cell. Comp. Physiol.* **42,** 327–341 (1953).
70. Sweeney, T. R. A survey of compounds from the antiradiation drug development program of the U.S. Army Medical Research and Development Command. September 1979.
71. Yuhas, J. M. Biological factors affecting the radioprotective efficiency of S-2-(3-aminopropylamino)ethyl phosphorothioic acid (WR-2721). LD50(30) doses. *Radiat. Res.* **44,** 621–628 (1970).
72. Anderson, K. W., Krohn, K. A., Grunbaum, Z., Phillips, R. B., Mahler, P. A., Menard, T. W., Spence, A. M., and Rasey, J. S. Analysis of S-35 labelled WR-2721 and its metabolites in biological fluids. *Int. J. Radiat. Oncol. Biol. Phys.* **10,** 1511–1515 (1984).
73. Purdie, J. W. Dephosphorylation of WR-2721 to WR-1065 *in vitro* and effect of WR-1065 and misonidazole in combination in irradiated cells. In *Radiation Sensitizers: Their Use in the Clinical Management of Cancer* (L. W. Brady, ed.), pp. 330–333. Masson, New York, 1980.
74. Purdie, J. W. A comparative study of the radioprotective effects of cysteamine, WR-2721, and WR-1065 in cultured human cells. *Radiat. Res.* **77,** 303–311 (1979).
75. Mori, T., Nikaido, O., and Sugahara, T. Dephosphorylation of WR 2721 with mouse tissue homogenates. *Int. J. Radiat. Oncol. Biol. Phys.* **10,** 1529–1531 (1984).
76. Utley, J. F., Seaver, N., Newton, G. L., and Fahey, R. C. Pharmacokinetics of WR-1065 in mouse tissue following treatment with WR-2721. *Int. J. Radiat. Oncol. Biol. Phys.* **10,** 1525–1528 (1984).
77. Swynnerton, N. F., McGovern, E. P., Mangold, D. J., and Nino, J. A. HPLC assay for S-2-(3-aminopropylamino)ethyl phosphorothioate (WR 2721) in plasma. *J. Liq. Chromatogr.* **6,** 1523–1534 (1983).
78. Millar, J. L., McElwain, T. J., Clutterbuck, R. D., and Wist, E. A. The modification of melphalan toxicity in tumor bearing mice by *S*-2-(3-aminopropylamino)-ethylphosphorothioic acid (WR 2721). *Am. J. Clin. Oncol.* **5,** 321–328 (1982).
79. Yuhas, J. M. Active versus passive absorption kinetics as the basis for selective protection of

normal tissues by S-2-(3-aminopropylamino)-ethylphosphorothioic acid. *Cancer Res.* **40,** 1519–1524 (1980).
80. Washburn, L. C., Carlton, J. E., and Hayes, R. L. Distribution of WR-2721 in normal and malignant tissues of mice and rats bearing solid tumors: Dependence on tumor type, drug dose and species. *Radiat. Res.* **59,** 475–483 (1974).
81. Washburn, L. C., Rafter, J. J., and Hayes, R. L. Prediction of the effective radioprotective dose of WR-2721 in humans through an interspecies tissue distribution study. *Radiat. Res.* **66,** 100–105 (1976).
82. Utley, J. F., and Kane, L. J. Differential absorption of ^{35}S-WR-2721 in normal and malignant tissue of animals. In *Radiation Sensitizers: Their Use in the Clinical Management of Cancer* (L. W. Brady, ed.), pp. 516–518. Masson, New York, 1980.
83. Utley, J. F., Marlowe, C., and Waddell, W. J. Distribution of ^{35}S-labeled WR 2721 in normal and malignant tissues of the mouse. *Radiat. Res.* **68,** 284–291 (1976).
84. Yuhas, J. M., Spellman, J. M., and Culo, F. The role of WR 2721 in radiotherapy and/or chemotherapy. In *Radiation Sensitizers: Their Use in the Clinical Management of Cancer* (L. W. Brady, ed.), pp. 303–308. Masson, New York, 1980.
85. Yuhas, J. M., and Phillips, T. L. Pharmacokinetics and mechanisms of action of WR-2721 and other protective agents. In *Radioprotectors and Anticarcinogens* (O. F. Nygaard and M. G. Simic, eds.), pp. 639–653. Academic Press, New York, 1982.
86. Yuhas, J. M., and Storer, J. B. Differential chemoprotection of normal and malignant tissues. *J. Natl. Cancer Inst. (U.S.)* **42,** 331–335 (1969).
87. Nicolson, G. L., and Poste, G. The cancer cell: Dynamic aspects and modifications in cell-surface organization (Part 2). *N. Engl. J. Med.* **295,** 253–258 (1976).
88. Yuhas, J. M., Davis, M. E., Glover, D., Brown, D. Q., and Ritter, M. Circumvention of the tumor membrane barrier to WR-2721 absorption by reduction of drug hydrophilicity. *Int. J. Radiat. Oncol. Biol. Phys.* **8,** 519–522 (1982).
89. Brown, D. Q., Yuhas, J. M., MacKenzie, L. J., Graham, W. J., and Pittock, J. W., III. Differential radioprotection of normal tissues by hydrophilic chemical protectors. *Int. J. Radiat. Oncol. Biol. Phys.* **10,** 1581–1584 (1984).
90. Denekamp, J., Stewart, F. A., and Rojas, A. Is the outlook grey for WR-2721 as a clinical radioprotector? *Int. J. Radiat. Oncol. Biol. Phys.* **9,** 1247–1249 (1983).
91. Yuhas, J. M. Efficacy testing of WR 2721 in Great Britain or everything is black and white at the gray lab. *Int. J. Radiat. Oncol. Biol. Phys.* **9,** 595–598 (1983).
92. Stewart, F. A., Rojas, A., and Denekamp, J. Radioprotection of two mouse tumors by WR-2721 in single and fractionated treatments. *Int. J. Radiat. Oncol. Biol. Phys.* **9,** 507–513 (1983).
93. Kligerman, M. M., Glover, D. J., Turrisi, A. T., Norfleet, A. L., Yuhas, J. M., Coia, L. R., Simone, C., Glick, J. H., and Goodman, R. L. Toxicity of WR-2721 administered in single and multiple doses. *Int. J. Radiat. Oncol. Biol. Phys.* **10,** 1773–1776 (1984).
94. Phillips, T. L. Rationale for initial clinical trials and future development of radioprotectors. In *Radiation Sensitizers: Their Use in the Clinical Management of Cancer* (L. W. Brady, ed.), pp. 321–329. Masson, New York, 1980.
95. Glover, D., Riley, L., Carmichael, K., Spar, B., Glick, J., Kligerman, M. M., Agus, Z. S., Slatopolsky, E., Attie, M., and Goldfarb, S. Hypocalcemia and inhibition of parathyroid hormone secretion after administration of WR-2721 (A radioprotective and chemoprotective agent). *N. Engl. J. Med.* **309,** 1137–1141 (1983).
96. Rasey, J. S., Nelson, N. J., Hamler, P., Anderson, K., Krohn, K. A., and Menard, T. Radioprotection of normal tissues against gamma rays and cyclotron neutrons with WR-2721: LD50 studies and ^{35}S-WR-2721 biodistribution. *Radiat. Res.* **97,** 598–607 (1984).
97. Patt, H. M., Clark, J. W., and Vogel, H. H., Jr. Comparative protective effect of cysteine

against fast neutron and gamma irradiation in mice. *Proc. Soc. Exp. Biol. Med.* **84,** 189–193 (1953).

98. Sigdestad, C. P., Connor, A. M., and Scott, R. M. Chemical radiation protection of the intestinal epithelium by mercaptoethylamine and its thiophosphate derivative. *Int. J. Radiat. Oncol. Biol. Phys.* **1,** 53–60 (1975).
99. Connor, A. M., and Sigdestad, C. P. Chemical protection against gastrointestinal radiation injury in mice by WR 2822, WR 2823, or WR 109342 after 4 MeV X ray of fission neutron irradiation. *Int. J. Radiat. Oncol. Biol. Phys.* **8,** 547–551 (1982).
100. Ainsworth, E. J., Larsen, R. M., Mitchell, F. A., and Taylor, J. F. Survival-promoting effects of endotoxin in mice, dogs, and sheep. In *Radiation Protection and Sensitization* (H. L. Moroson and M. Quintiliani, eds.), pp. 381–388. Taylor & Francis, London, 1970.
101. Weiss, J. F., Jacobs, A. J., and Rankin, W. A. Effects of diethyldithiocarbamate and WR 2721 on delayed-type hypersensitivity and survival of irradiated mice. In *Proceedings of the 7th International Congress on Radiation Research* (J. J. Broerse, G. W. Barendsen, H. B. Kal, and A. J. van der Kogel, eds.), pp. C1–37. Martinus Nijhoff, Amsterdam, 1983.
102. Fantone, J. C., and Ward, P. A. *Oxygen-Derived Radicals and Their Metabolites: Relationship to Tissue Injury.* Upjohn Company, Kalamazoo, Michigan, 1985.
103. Meister, A. Metabolism and functions of glutathione. *Trends Biochem. Sci.* **6,** 231–234 (1981).
104. Meister, A., and Tate, S. S. Glutathione and related γ-glutamyl compounds: Biosynthesis and utilization. *Annu. Rev. Biochem.* **45,** 559–604 (1976).
105. Meister, A. Glutathione metabolism and transport. In *Radioprotectors and Anticarcinogens* (O. F. Nygaard and M. G. Simic, eds.), pp. 121–151. Academic Press, New York, 1983.
106. Lawrence, R. A., and Burk, R. F. Species, tissue, and subcellular distribution of non-selenium dependent glutathione peroxidase activity. *J. Nutr.* **108,** 211–215 (1978).
107. Brawn, K., and Fridovich, I. Superoxide radical and superoxide dismutases: threat and defense. *Acta Physiol. Scand., Suppl.* **492,** 9–18 (1980).
108. Sztanyik, L. B., and Varteresz, V. Radioprotective effect of a mixture of AET and 5-methoxytryptamine in X-irradiated mice. In *Radiation Protection and Sensitization* (H. L. Moroson and M. Quintiliani, eds.), pp. 363–367. Taylor & Francis, London, 1970.
109. Maisin, J. R., Mattelin, G., Fridman, Manduzio, A., and van der Parren, J. Reduction of short- and long-term radiation lethality by mixtures of chemical protectors. *Radiat. Res.* **35,** 26–44 (1968).
110. Armed Forces Radiobiology Research Institute. *Department of Defense 5-Year Plan FY 1986–1990 for Ionizing Radiation Biomedical Research.* Armed Forces Radiobiol. Res. Inst., Bethesda, Maryland, 1985.

CHAPTER 15

Psychological Effects of Nuclear Warfare

G. ANDREW MICKLEY

Armed Forces Radiobiology Research Institute
Bethesda, Maryland 20814-5145

History teaches that "battle" has devastating effects on the psychology of its participants. If severe enough and long enough, armed conflict will ultimately break all of those committed to it. Neuropsychiatric casualty counts of World War II suggest that, depending on the battle, from 18 to 48% of all casualties were psychiatric. The Arab–Israeli Yom Kippur War of 1973 lasted only 3 weeks, yet 10% of the Israeli casualties were psychiatric. Clearly, psychological factors play an important role in determining the final outcome of any battle (1).

No definitive data speak directly to the issue of human psychological responses to the use of nuclear weapons. For example, most experiences of nuclear accident victims have been poorly documented as to mental alterations. In addition, patients exposed to ionizing radiation as part of cancer treatments also frequently receive various drugs to suppress side effects of the radiation treatments and to enhance the effectiveness of the radiation. Because these patients are usually quite sick even before the radiation therapy, it is difficult to assess the effects of the radiation itself. Finally, the data derived from the World War II atomic bombings may not be relevant to military planners because the information is derived from a mainly civilian population. Civilians may or may not react to the use of nuclear weapons in a manner similar to that expected from a military force. The radiation doses received by the people of Hiroshima and Nagasaki are not well described.

Given that the directly relevant human radiation data are imperfect, some investigators (2, 3) have chosen to emphasize models of the psychological effects of nuclear weapons. In this context, the psychological reactions to nuclear battle are assumed to be similar to those after an intense conventional battle or a natural disaster such as a flood or earthquake. Although this approach has merit, it also has a number of problems (4). In particular, it does not account for the unique stress of ionizing radiation exposure, with its variety of idiosyncrasies and the implications of those exposed. Persons exposed to radiation on a nuclear battlefield may have little or no initial knowledge of the severity of their radiogenic injuries. This uncertainty and each individual's interpretation of the uncertainty may impact his/her performance. A knowledgeable person may realize that radiation exposure may have long-term effects, such as leukemia and genetic damage. Radiation also produces direct changes in the central nervous system (CNS), which may in turn affect psychological variables. Moreover, the sheer magnitude of destruction provided by an atomic blast may result in a correlated increase in psychological trauma. In conventional warfare, only a relatively small number of persons are involved in the battle, but in a nuclear confrontation, most combatants will feel personally engaged (5). Because the bombings of Hiroshima and Nagasaki provide the only available data that reflect the combined results of blast, thermal insults, and radiation insults on a large human population, this author relied heavily on these experiences. In addition, the military members who were present in both cities during the bombings provide a few examples of military actions in a nuclear environment (6). These accounts may have value for present-day military planners and students.

This report is divided into five parts. (1) Discussion of the psychological milieu before a nuclear confrontation. (2) Acute psychological reactions to nuclear warfare (some of which may reflect, in part, direct radiogenic alteration of nervous system functions). (In this context I will present animal data that suggest how radiation directly influences the brain and "psychological variables" in experimental animals.) (3) Chronic psychological effects of a nuclear confrontation. (4) Issues concerning treatment of these psychological changes. (5) Prevention of adverse psychological reactions to nuclear warfare.

I. PSYCHOLOGICAL MILIEU BEFORE CONFLICT

A number of articles in the literature address the impact of nuclear weapons on the psychology of modern man. When faced with the ultimate horror of warfare, the responses are fairly predictable: fear, dread, and, ultimately, denial. Some have suggested that the anxiety level of the world's population has been raised substantially by the prospect of a nuclear confrontation. But this anxiety does not stimulate discussion and thought on the questionable efficacy of nuclear conflict and how to deal with the situation if it occurs. Rather, the present psychological milieu is best characterized by a numbing denial of both (1) the possibility of nuclear conflict itself and (2) the usefulness of preparation for such a conflict. If this trend exists in the military, it could significantly affect the way one prepares for and functions in such a conflict (7–9).

II. ACUTE PSYCHOLOGICAL REACTIONS

One aspect of the psychological effects of nuclear weapons that has not received much attention is that component derived from the direct interaction of nervous tissue and radiation. Neurons were once thought to be relatively resistant to irradiation. However, the substantiating data came from studies that measured cell death rather than disruptions of cell function. Actually, increasing evidence suggests that ionizing radiation may alter neural physiology and function at doses substantially below those required to produce morphological changes and death. For example, changes in the amplitude and frequency of electroencephalograph (EEG) tracings occur after 100–400 rads of X rays (10). Changes in metabolism of the neurotransmitter dopamine have also been reported in the brain after 1000 rads of cobalt-60 (11). Similarly, levels of the putative neuromodulator β-endorphin are altered in irradiated mice and monkeys at doses that do not kill neurons (12, 13). It has also been reported recently that neuronal sodium channels may lose their ability to respond properly to stimulation after only 100 rads of high-energy electrons (14). Behavioral changes have been reported after only 30 rads of ionizing radiation (10). These data suggest that alteration in CNS functioning is likely to occur after relatively low doses of radiation. It would not be surprising if psychological correlates were associated with these changes in brain function.

Since ionizing radiation alters brain function in a direct way, it may be possible to study this component of the psychological effects of irradiation by reviewing some of the work with experimental animals. This approach has the disadvantage of ignoring (for the moment) the psychogenic aspects of a reaction to nuclear confrontation, but it has the advantage of being able to control radiation dose and behavioral testing. In an attempt to validate this approach, the animal data will be correlated with some of the accounts of survivors of the World War II nuclear bombings.

A. Motivation

A variety of data derived from animal experiments suggests that motivation is altered after exposure to ionizing radiation. The individual's tendency to perform is governed by a number of factors, including capacity of the animal, rewards and/or punishments present, and perception of these reinforcements. If it can be determined that the subject in an experiment has the capacity to perform in the presence of previously motivating stimuli but does not do so, then it may be inferred that some change in the motivation of the individual has occurred. Some data collected at the Armed Forces Radiobiology Research Institute address this

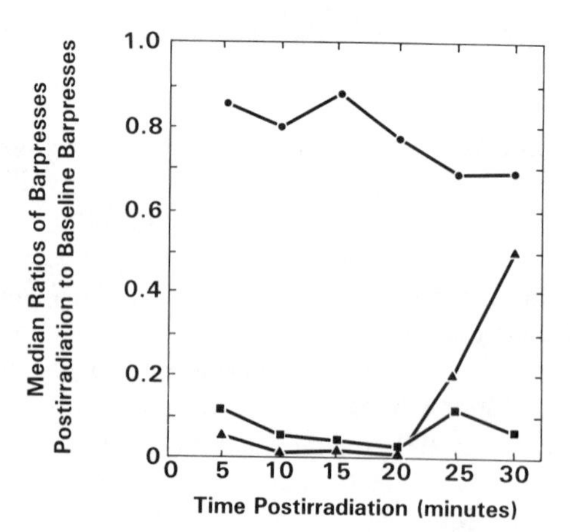

FIG. 15-1 Vigorous self-stimulation of lateral hypothalamus persisting after radiation exposure even though similar behaviors mediated by other subcortical structures were strongly attenuated. (●), Lateral hypothalamus; (▲), septum; (■), substantia nigra.

issue. For example, Burghardt and Hunt (15) have demonstrated that, after irradiation, rats will decrease the number of times they press a bar which, when activated, gives them information concerning *when* shock will occur. However, they significantly *increase* the number of times they press a bar to delay footshock. These data suggest that the animal is fully capable of performing the actual task of barpressing, but "chooses" to do so only under certain conditions.

Another study from the author's laboratory (16) also supports this concept. Rats will work in order to receive mild electrical stimulation of particular brain areas. In one rat, we implanted electrodes into two of these brain areas (lateral hypothalamus and lateral septum). Before irradiation, the subject barpressed at the same rate for activation of both sites. However, after irradiation, the subject worked for stimulation of only one of the brain areas (see Fig. 15-1). Clearly, the animal had the capacity to perform the task (it was still barpressing for electrical activation of one of the sites) and was sufficiently motivated to press both bars before irradiation. However, after irradiation, the animal's motivations had been altered, thus producing a decrease in responding.

The data from Hiroshima and Nagasaki suggest that a similar change in motivation might be expected to occur in humans exposed to a nuclear weapon detonation. Hachiya describes a scene in his classic book *Hiroshima Diary* as follows.

> Those who were able walked silently toward the suburbs in the distant hills, their spirits broken, their initiative gone. When asked whence they came, they pointed toward the city and said "that way"; and when asked where they were going, pointed away from the city

> and said, "this way." They were so broken and confused that they moved and behaved like automatons.
>
> Their reactions astonished outsiders who reported with amazement the spectacle of long files of people holding stolidly to a narrow, rough path when close by was a smooth, easy road going in the same direction. The outsiders could not grasp the fact that they were witnessing the exodus of a people who walked in the realm of dreams (17, p. 54).

These descriptions are consistent with others from the Hiroshima experience in which "fatigue," "mental weakness," "spiritual desolation," or "closing off" have been reported. Certainly, in the case of the atom bomb survivors, this change in motivation cannot be attributed solely to a dose of radiation. These people were often physically injured and, in many cases, had just witnessed an incredible destruction of their homes as well as deaths of family members. Thus, there is very likely a psychogenic component in these behavioral and psychological changes, which may compound the radiation-induced alterations just described in experimental animals.

Despite the emotional deadening and "mental weakness" reported by almost everyone influenced by the bombing of Hiroshima, it is remarkable how much activity (physical movement) was shown by some of the survivors. However, some of this activity seemed ill directed and uncoordinated.

> There was no organized activity. The people seemed stunned by the catastrophe and rushed about as jungle animals suddenly released from a cage. Some few apparently attempted to help others from the wreckage, particularly members of their family. . . . However, many injured were left trapped beneath collapsed buildings as people fled by them in the streets (18, p. 29).

This, of course, raises the issue of panic. Was there mass panic in the population of Hiroshima after the dropping of the atomic bombs? The answer is, probably not. Although several isolated instances of aimless and hysterical behavior (mainly frantic flight from the cities) have been reported, they do not seem to be typical behaviors. Reports from the "Morale Division" of the United States Army (which interviewed the survivors of the bombing) (19) do not support the claim that a sizable portion of the population behaved in an ineffective or distraught way. However, the report also indicated that although they did not *exhibit* panic, many people *felt* terrified or fearful. In only a few cases can one surmise from the interviews that individuals might have shown uncontrolled emotional behavior. Instead, more prominent were the examples of compliant, subdued behaviors, which may have resulted, in part, from some radiogenic CNS effects or other injuries: "Many, although injured themselves, supported relatives who were worse off. Almost all had their heads bowed, looked straight ahead, were silent, and showed no expression whatever" (6). Therefore, it seems that instead of panic, depression and lethargy were more typical reactions to the disaster.

Clearly, either a chaotic or apathetic response to a bombing is not adaptive in a military environment. Some evidence exists that inhabitants of Hiroshima who had a specific job to perform or a goal to meet tried valiantly to do so after the bombing. John Hersey, in his book *Hiroshima* (6), describes a group of wounded soldiers who were attempting to struggle out of the disaster area in military formation: "At Misasa Bridge, they encountered a long line of soldiers making a bizarre forced march away from the Chugoku Regional Army Headquarters in the center of the town. All were grotesquely burned, and they supported themselves with staves or leaned on one another." One account of a young Japanese soldier is particularly relevant here:

> We were under military order to return to our unit immediately in case of any attack or emergency, so I returned almost without thinking. . . . At first I couldn't get through . . . so in the evening I started out again. This time I didn't try to help anyone but just walked through them. I was worried about the Army camp because according to what . . . people told me, it had simply gone up in flames and disappeared. I was also a bit ashamed about having taken such a long time to return. But when I finally got back to the camp, just about everyone was dead—so there was no one to scold me. . . . Next thing I did was to look for the ashes of the military code book—since we had a military order to look for this book even if it were burned, as it was a secret code which had to be protected. Finally I located the ashes of the book, and wrapped them in a furoshiki and carried this around with me. I wanted to take it to the military headquarters as soon as possible, but when I finally did take it there in the morning, the officer scolded me for doing such a stupid thing. . . . I was fresh from the Military Academy and my head was full of such regulations (20, p. 28).

Thus, although a generalized motivational decrement may have characterized much of the population of Hiroshima, behaviors directed toward a well-defined goal seemed to persist after the catastrophe.

B. Social Relations

It is of psychological and social importance that, in the extremely traumatic situation after the atomic explosions, most people behaved in a manner compatible with social norms.

> To Father Kleinsorge, an Occidental, the silence in the grove by a river, where hundreds of gruesomely wounded suffered together, was one of the most dreadful and awesome events of his whole experience: The hurt ones were quiet; no one wept, much less screamed in pain; no one complained; none of the many who died did so noisily; not even the children cried; very few people even spoke. And when Father Kleinsorge gave water to some whose faces had been almost blotted out by flash burns, they took their share and then raised themselves a little and bowed to him in thanks (6, p. 32).

One might speculate that this reaction may be more characteristic of an eastern philosophy. It is difficult to predict if similar restraint would exist in a Western military population.

C. Attentional Factors

With individual motives disrupted, persons seemed most likely to pursue goals defined by others. For example, a victim of the Hiroshima bombing recounts, "All the people were going in that direction and so I suppose I was taken into this movement and went with them. . . . I couldn't make any clear decision in a specific way . . . so I followed the other people. . . . I lost myself and was carried away" (20). A component of this phenomenon may be explained by attentional focusing. That is, these people tended to focus on a particular aspect of their environment and pursue it, often to an illogical or inappropriate end. The soldier mentioned above persisted in the pursuit of his assigned task, ignoring the fact that his world had totally changed by the dropping of the atomic bomb (his behavior would not necessarily be discouraged by military commanders). A similar phenomenon has been reported in irradiated animals (21). Irradiated rhesus monkeys (<700 rads neutron–gamma) performed better than did controls on a task that placed a premium on attention to a particular portion of their environment. These subjects were not easily distracted by peripherally placed stimuli. Other investigators (10) found that after low doses of radiation (<1 rad/sec), the cortical EEG of experimental animals became desynchronized and resembled that of a fully alert and attentive animal. These experimental data suggest that low or intermediate doses of radiation may produce arousal and greater focus of attention, which may then be directed toward a well-learned task. Perhaps the performance of routine military tasks might actually be enhanced after irradiation. Caution should be taken in interpreting these data since they may reflect a phenomenon that is dependent on a particular task and a narrow range of radiation doses.

D. Learning and Memory

Some evidence suggests that, at some doses of radiation, functions of learning and memory may be altered. For example, Meyerson (10) has demonstrated that rabbits can learn to associate a tone and a light with apnea produced by the inhalation of ammonia vapor. He found that once this classically conditioned response was established, the tone and light presentation alone produced apnea. However, after irradiation (1.5 krads), this conditioned response was absent or considerably reduced in duration. In contrast with this effect, the apnea produced by the ammonia itself was enhanced after exposure (suggesting that the animal was still capable of this response). The animal literature suggests that it may take these relatively high doses of radiation to produce deficits in learning. Other studies (22) have reported *enhanced* learning capacities and retention after exposure to low doses of ionizing radiation. Perhaps the most conservative conclusions from the animal literature are that one can expect radiogenic changes in learning and memory, and that these alterations may be dose dependent.

Interviews with persons irradiated at Hiroshima indicated only rare instances of acute retrograde amnesia in the population (18). But 5 years after the attack, deficits in memory and intellectual capacity were noted in individuals experiencing radiation sickness (23). More acute human radiogenic impairments of memory have been reported in the Soviet literature (22).

III. CHRONIC PSYCHOLOGICAL REACTIONS

The initial reactions to a nuclear weapon detonation may be quite different from those that occur after a few hours, days, or weeks. While it is probable that both acute and chronic psychological reactions are partially physiological, psychogenic changes in emotionality, personality, and somatic effect are usually seen after a period of time. Changes in emotionality and temperament have been noted in irradiated experimental animals, but this literature may be less useful in describing human reactions over the long term, as unique human psychosocial factors are brought into play (10).

A. Fear and Anxiety

Some of the psychological changes experienced by persons exposed to nuclear weapons detonation are the same as those experienced by persons involved in other disasters. However, the magnitude and type of destruction imposed by a nuclear weapon will probably, at a minimum, intensify almost any psychological reaction. Although the people of Japan were accustomed to the destructive nature of conventional bombing, the effects of the atomic bomb were so much more horrible that the people's reactions differed somewhat (19).

The perception of large numbers of burned, cut, and maimed persons was a major source of emotional trauma after the bombing of Hiroshima. Many survivors located only a short distance from the center of the explosion appeared to have undergone a double emotional shock: the first from the explosion's physical impact and the second when they ran into the streets and saw large numbers of casualties. For those persons at the periphery who escaped the full physical violence of the explosion, the first emotional impact seems to have occurred when they saw the streams of injured victims pouring out of the destroyed areas. Apparently, it was not only the large number of casualties but also the specific character of the injuries (particularly the grossly altered physical appearance of those with severe burns) that produced emotional disturbances in the people who saw them. For example,

> I walked past Hiroshima station . . . and saw people with their bowels and brains coming out . . . I saw an old lady badly burned and carrying a suckling infant in her arms . . . I saw many children . . . with dead mothers . . . I just cannot put into words the horror I felt (17, p. 50).

> I had to cross the river to reach the station. As I came to the river and went down the bank to the water, I found that the stream was filled with dead bodies. I started to cross by crawling over the corpses, on my hands and knees. As I got about a third of the way across, a dead body began to sink under my weight and I went into the water, wetting my burned skin. It pained severely. I could go no further, as there was a break in the bridge of corpses, so I turned back to the shore, and started to walk upstream, hoping to come upon another way across (24, p. 76).

B. Phobias

Given these horrors, it is not surprising that various sources indicate that severe anxiety persisted for many days and, in some cases, for weeks and months after the bombings. One of the most frequent types of sustained emotional disturbances appears to have been a phobia-like fear of exposure to another disaster. This reaction consisted of strong feelings of anxiety accompanied by exaggerated efforts to ward off new threats. A vivid description of anxiety states evoked by minimal signs of potential danger has been given by T. Hagashi (18), a physician in Hiroshima, who was one of the special informants on postdisaster reactions interviewed by the United States Strategic Bombing Survey investigators: "Whenever a plane was seen after that, people would rush into their shelters. They went in and out so much that they did not have time to eat. They were so nervous they could not work. . . ." Hersey (6) describes incidents such as the following: "It began to rain. . . . The drops grew abnormally large, and someone [in the evacuation area] shouted, 'The Americans are dropping gasoline. They're going to set fire to us!' "

Further indication (18) of sustained apprehensiveness among the populace of Hiroshima comes from the anxiety-laden rumors that were reported to have been circulated during the postdisaster period. For example, one woman reports: "I heard that people who had not been wounded and seemed to be all right would begin feeling out of sorts and all of a sudden drop dead. It made me panicky.

Here I was bustling around now, but I might go off myself.'' To some extent, ''fear rumors'' may have been touched off or reinforced by the unexpected appearance of many cases of radiation sickness. During the weeks following the atomic explosion, many unusual signs of organic pathology began to appear among survivors: loss of hair, high fever, excessive fatigue, hemorrhagic spots under the skin, and other severe symptoms of radiation sickness. Witnessing the agonizing deaths of children and relatives probably intensified the sustained fear reactions created by the disaster (18).

C. Survivor Guilt

Although social customs seemed amazingly resilient after the atomic bombings (see Section II), not everyone acted in a completely altruistic fashion at all times. It was simply impossible to do so, given the sheer number of casualties. Some people fought fires and fed the hungry at large, but most people (outside the service professions) restricted their assistance, when they could give it, to people they knew: ''Under many houses, people screamed for help, but no one helped; in general, survivors . . . assisted only their relatives or immediate neighbors, for they could not comprehend or tolerate a wider circle of misery'' (5). As one survivor summarized, ''The idea of 'love thy neighbor as thyself' that I always believed in, had disappeared some place. I guess it was too much for any of us'' (25). Nagai has written that he believes that persistent ''survivor-guilt'' is an inevitable consequence of atomic bombing, because most survivors could not avoid behaving negligently in one way or another. People who were in the heart of the city were able to survive only by running away from the fires without stopping to rescue others. People who were in a position to give aid could not simultaneously perform all the duties and obligations of rescuing the wounded, rushing to their own families, assisting neighbors, carrying out their civil defense assignments, saving valuable materials at the offices or factories where they worked, preserving treasured household articles, etc. Although independent observations (18) indicate that some survivors experienced temporary guilt reactions following the A-bombings, no satisfactory evidence supports the claim that such reactions persisted in large numbers of survivors or for very long periods of time (i.e., years).

D. Depression

Although acute depression (evidenced by weakness and lethargy) characterized much of the population of Hiroshima for the first few days after the bombing, it is difficult to say if significant numbers of people experienced chronic depression. Although individual questionnaire responses from residents of Hiroshima seemed, in many cases, to characterize a depressive reaction, statistical analyses revealed no greater incidence of depression than in other Japanese cities (18). Yet this may be misleading since postwar apathy seemed to characterize most of the population of Japan. Long-term depressive reactions have been reported to occur after catastrophic natural disasters (e.g., floods) (26) and are by no means exclusively characteristic of nuclear disasters.

E. Psychosomatic Symptoms

Thus far I have described isolated psychiatric symptoms that may have been evident in a significant number of A-bomb survivors. But what about aggregates of symptoms in the form of neurotic reactions? T. Nishikawa and S. Tsuiki (23) discovered 533 patients with neurosis-like symptoms among 7297 patients exposed to ionizing radiation during the Japanese bombings. The patients were divided into two groups: those with symptoms of atomic bomb radiation illness and those without symptoms. Neurosis-like symptoms predominated in the former group, where they were about twice as common as in the latter. These Japanese researchers pointed out that some of the cases were recognizable as "pure neuroses" caused by psychogenic factors (other than the bombings), but that other cases could have been caused by functional disorders of brain or body due to the radiation. Not surprisingly, others have noted that the more severe the symptoms of atomic bomb radiation illness that the victims suffered, the stronger were their neuropsychiatric aftereffects. The latter symptoms included weariness, lack of "spirit," a tendency toward introversion, and bad memory.

Lifton (4) has suggested that there may exist in some patients what he calls a psychosomatic-like "atomic bomb neurosis." Here the survivor's identification with the dead and the maimed initiates a vicious circle on the psychosomatic plane of existence. The survivor is likely to associate the mildest everyday injury or sickness with possible radiation effects, and anything that he/she relates to radiation effects becomes associated with death. An example is this quote from such an individual:

> Frankly speaking, even now I have fear. . . . Even today people die in the hospitals from A-bomb disease, and [when I hear about this] I worry that I too might sooner or later have the same thing happen to me. . . . I have a special feeling that I am different from ordinary people . . . that I have a mark of wounds—as if I were a cripple. . . . It is not a matter of lacking something externally, but rather something like a handicap—something mental that does not show—the feeling that I am mentally different from ordinary people . . . so when I hear about people who die from A-bomb disease or who have operations because of this illness, then I feel that I am the same kind of person as they . . . (20, p. 108).

Physicians may be caught in a conflict between the humanitarian provision for medical need and the danger of encouraging the development of hypochondria, general weakness, and dependency in survivors.

It should be noted that serious permanent psychological derangements and loss of touch with reality were rare after the atomic bomb attacks, just as they were after large-scale conventional bombings (5). A few scattered cases of psychosis were reported, along with several cases of serious neurosis with decided fatigue, organic symptoms, and extended terror compulsions. But it is difficult to distinguish the numbers of these ailments from those expected in other populations.

IV. SUMMARY OF PSYCHOLOGICAL EFFECTS

Although the Japanese A-bomb experience is the best available model, it is difficult to determine how much information this model and correlated animal data can provide concerning psychological changes in a military nuclear confrontation. The question of a good model is complicated even more by the fact that all psychological effects (like all physiological effects) are dependent on the dose of radiation received, distance from ground zero (and correlated blast and thermal effects), and all of the indefinable personal psychological and social baggage that a potential nuclear victim brings with him/her to the battle. However, if we can assume a certain degree of congruity between the psychological response of the Japanese and the expected response of our military, then the summary in the following paragraph may apply.

With ionizing radiation exposure will come alteration of CNS physiology. These initial changes may have acute behavioral and psychological correlates such as generalized reduction of motivation. This may produce symptoms of lethargy and fatigue and may also inhibit the likelihood of generalized panic. Most persons who are strongly compelled to do a particular well-learned task will probably be able to do so as long as other effective systems are intact. Persons will still be able to take direction from others; in fact, they may be more likely to do so if radiogenic focusing of attention is a factor. Social order (military protocol) will probably remain intact. The capacity to learn and remember may be changed, especially in the long term. Individual experience with, and the observation of, horrible wounding and destruction produced by a nuclear weapon could be expected to have immediate psychological effects on military members. If they react like the citizens of Hiroshima, they will be fearful and anxious, perhaps even more so than during a conventional conflict. These symptoms may be intensified by rumor and by any misinformation the individuals receive concerning the threat. Group cohesion will contribute to the existence of "helping" behaviors, but self-preservation may be more compelling for many. Longer term psychological reactions may include phobias and a variety of psychosomatic symptoms. Guilt concerning questions of personal survival and inadequacies in performance could contribute to the development of neurotic symptoms, as will the severity of physical wounding. Psychotic reactions are probably less likely to occur.

V. REQUIREMENTS AND PROBLEMS OF CARE

Ross (7) has postulated that the psychological effects of a nuclear disaster will be distributed consistent with a normal curve. Here, as in other disasters, the majority of survivors (about 75%) would manifest a few of the symptoms described above. About half of the remaining persons would be almost totally unaffected, while the rest would manifest many, or a high degree of, acute and chronic psychological changes.

This text does not predict the degree of impact that the psychological changes produced by nuclear weapons may have on the performance of particular military units or on the outcome of battles. Glass (2) and Vineberg (3) have addressed these issues in their extensive works. These psychological effects may or may not be a problem for the provider of military health care. Some of the individuals exhibiting these symptoms in a minor degree will never be seen medically. However, the literature suggests that those who do find their way to psychological treatment should be handled in conventional ways. Sessions (27), in his review of the literature, points out, "Those reactions which persist and are severe enough to warrant treatment . . . generally respond favorably to practical therapy techniques developed through trial and error efforts during World War I, World War II, and the Korean campaign." These techniques involve the principles of "proximity, recency, and expectancy." Individuals respond better if they receive therapy as soon as possible and as near as possible to the scene of the disaster. The principle of expectancy requires the provider to calmly accept the person's problems and regard his/her manifestations as a temporary incapacity from which, after a brief rest, recovery is expected. Persons with situationally induced, acute psychological disorders worsen or improve, depending on what is expected from them by those responsible for their treatment. Sessions (27) continues as follows:

> Such simplified treatment techniques are designed to allow the individual a respite from the traumatic situation, rest, and emotional support. This aids in establishing communication with the individual, who can then receive information and be motivated to participate in rescue and defensive efforts. Resumption of purposeful activities and reorientation to group and social obligations is seen as important to rapid improvement.

Although this conventional approach to treatment may be quite useful, there are several unique characteristics of a nuclear conflict that will compound problems for the health care provider. One is the uncertainty of personal injury. Most people now realize that radiation exposure can be quite lethal even though initial effects may be minimal. This uncertainty concerning one's health after irradiation will increase the load of medical treatment. It has been shown in previous studies of disasters that threats or dangers that cannot be reliably perceived by the senses can cause considerable psychological disturbance. For example, a mass poisoning of bootleg whiskey in Georgia led to a large number of people attending casualty departments. When tested, about 40% were unaffected by lethal alcohol; some of these confessed that they did not know if they were affected, but they wanted to be checked. After a nuclear attack, many people might wish to be reassured that they have not been exposed to appreciable levels of radiation (28). This could add to the already chaotic situation surrounding treatment centers. Military plans presently call for personal radiation dosimetry that is not "self-readable." Instead, planned radiation dose assessment may be made on all combatants at medical aid stations.

What about the situation at treatment centers? Knowing that medical care is available has always provided comfort to combatants. If the medical load becomes too extensive and reasonable care cannot be given to casualties, morale will suffer. Burn cases, for example, place a great strain on medical personnel. The British Army Operational Research Group, using evidence from the English experiences of World War II, estimated an average time of 52 min for three persons to simply dress a burned hand. Extrapolations from their data suggest that the requirement for treating 1000 serious burn cases would be 5000 health professionals and 235 tons of supplies. Based on a case in which a 38-year-old man was accidently exposed to 200 rads of cobalt-60, others have estimated conservatively that the cost of treating such an individual would be $22,000 (in 1982) (29). It is doubtful that such extensive care could be guaranteed to mass numbers of battlefield casualties. The detrimental effect of inadequate medical care on morale was confirmed in the Hiroshima experience, in which many medical facilities were destroyed. Care was so limited that this, in and of itself, may have been a factor in some acute depressive reactions to the bombing (6, 9).

In addition, the concept of removing a combatant from the field for psychological treatment and then returning him better prepared to deal with the stresses of combat may be less useful in a nuclear conflict. Removal from the conventional battle allows psychological and physical healing. On the other hand, although removal from the nuclear battle provides a psychological respite, the progressing physical radiation effects may constantly erode the individual's ability to perform a task that is necessary for success of a military mission. It is questionable whether psychologically impaired, irradiated persons should be removed from the battlefield with any expectation of their return.

VI. PREVENTION OF ADVERSE PSYCHOLOGICAL REACTIONS TO NUCLEAR WARFARE

Can anything be done to prevent psychological changes after a nuclear confrontation? Proper training and preparedness seem to provide some degree of protection. This is confirmed by the remarkable experiences of nine persons who survived the Hiroshima bombing and then fled to Nagasaki in time for the second A-bomb (24). These survivors of Hiroshima remembered very well what they had done that allowed them to live. They drew heavily on this experience and related it to others in Nagasaki:

> Yamaguchi's lecture on A-bomb precautions, he pointed out later, was not lost upon his colleagues. With the young designer's words still fresh in their minds [at the time of the second bombing] they leaped for the cover of desks and tables. "As a result," said Yamaguchi, "my section staff suffered the least in that building. In other sections there was a heavy toll of serious injuries from flying glass" (24, p. 109).

Vineberg (3) extensively analyzed the type of training most beneficial in this regard and concluded that emphasis should be on (1) realism in order to reduce the shock aftermath of a nuclear confrontation, (2) accurate information concerning the threat, and (3) information limited to that readily comprehended and assimilated by the average individual and also directed toward promoting his/her personal welfare. More work needs to be done to meet these training needs and thus prepare for the expected psychological reactions to nuclear warfare.

Forces of social cohesiveness may also affect the psychological and performance variables after a nuclear weapon detonation. Powerful psychological support is offered to individuals by their immediate primary group. Various historical accounts suggest that a soldier isolated from his fellows was more likely to surrender than was another member of his group in the same tactically hopeless situation but who was still bound by the continuous ties of fighting, eating, and sleeping with his fellow soldiers. The ability of the primary group to maintain its integrity and to resist disintegration will materially affect the capacity of its members to withstand the stress of a nuclear confrontation. Nevertheless, we should recognize that disruption of the primary group by loss of personnel and leadership, breaks in communication, and deterioration of supply and medical care is more likely to occur in nuclear combat than in conventional confrontations (3).

The data presented here suggest that profound acute psychological changes may occur after a nuclear conflict. Longer term anxiety and other neurotic symptomatology may persist as well. Clearly, the author has taken a somewhat conservative stance in this report. Others, less guarded with their interpretations of the available data, would predict much more significant and long-lasting psychological changes than those mentioned here. Lifton (4), for example, summarizes as follows:

> The question so often asked, ''Would the survivors envy the dead?'' may turn out to have a simple answer. No, they would be incapable of such feelings. They would not so much envy as, inwardly and outwardly, resemble the dead (p. 292).

We hope that the future fails to provide us the definitive data to resolve such speculations as these.

REFERENCES

1. Marlow, D. H. Cohesion, anticipated breakdown, and endurance in battle: Considerations for severe and high intensity combat (unpublished paper). Walter Reed Army Institute of Research, Washington, D.C., 1982.
2. Glass, A. J. Psychological considerations in atomic warfare. *U.S. Armed Forces Med. J.* **7,** 625–639 (1956).
3. Vineberg, R. *Human Factors in Tactical Nuclear Combat,* RT 65-2, Contract DA 44-188-ARO-2. Human Resources Research Office, George Washington University, Alexandria, Virginia, 1965.
4. Lifton, R. J. Psychological effects of the atomic bombings. In *Last Aid* (E. Chivian, S. Chivian, R. J. Lifton, and J. E. Mack, eds.), pp. 48–68. Freeman, New York, 1982.
5. von Greyerz, W. *Psychology of Survival.* Am. Elsevier. New York, 1962.
6. Hersey, J. *Hiroshima.* Alfred A. Knopf Co., New York, 1981.
7. Ross, W. D. The emotional effects of an atomic incident. *Cincinnati J. Med.* **33,** 39–41 (1952).
8. Amme, C. H. Psychological effects of nuclear weapons. *U.S. Nav. Inst. Proc.* **86,** 26–36 (1960).
9. Thomas, L. *Late Night Thoughts on Listening to Mahler's Ninth Symphony.* Viking Press, New York, 1983.
10. Kimeldorf, D. J., and Hunt, E. L. *Ionizing Radiation: Neural Function and Behavior.* Academic Press, New York, 1965.
11. Hunt, W. A. Personal communication. Armed Forces Radiobiol. Res. Inst., Bethesda, Maryland.
12. Mickley, G. A., Stevens, K. E., Moore, G. H., Deere, W., White, G. A., Gibbs, G. L., and Mueller, G. L. Ionizing radiation alters beta-endorphin-like immunoreactivity in brain but not blood. *Pharmacol., Biochem. Behav.* **19,** 979–983 (1983).
13. Danquechin-Dorval, E., Mueller, G. P., Eng, R. R., Durakovic, A., Conklin, J. J., and Dubois, A. Involvement of endogenous opiates in radiation-induced suppression of gastric emptying. *Gastrointest. Motil., Proc. Int. Symp., 9th, 1983* (1984).
14. Wixon, H. N., and Hunt, W. A. Ionizing radiation decreases veratridine-stimulated uptake of sodium in rat brain synaptosomes. *Science* **220,** 1073–1074 (1983).
15. Burghardt, W. F., and Hunt, W. A. Characterization of radiation-induced performance decrement using a two-lever shock-avoidance task. *Radiat. Res.* **103,** 149–157 (1985).
16. Mickley, G. A., and Teitelbaum, H. Persistence of lateral hypothalamic-mediated behaviors after a supralethal dose of ionizing radiation. *Aviat., Space Environ. Med.* **49,** 868–873 (1978).

17. Hachiya, M. *Hiroshima Diary* (W. Well, transl.), Univ. of North Carolina Press, Chapel Hill, 1955.
18. Janis, I. L. *Air War and Emotional Stress,* McGraw-Hill, New York, 1951.
19. U.S. Strategic Bombing Survey. *The Effects of Strategic Bombing on Japanese Morale.* U.S. Govt. Printing Office, Washington, D.C., 1947.
20. Lifton, R. J. *Death in Life: Survivors of Hiroshima.* Random House, New York, 1967.
21. Brown, W. L., and McDowell, A. A. Some effects of radiation on psychologic processes in rhesus monkeys. In *Response of the Nervous System to Ionizing Radiation* (T. J. Haley and R. S. Snider, eds.), pp. 729–746. Academic Press, New York, 1962.
22. Furchtgott, E. Behavioral effects of ionizing radiations: 1955–61. *Psychol. Bull.* **60,** 175–199 (1963).
23. Committee for the Compilation of Materials on Damage Caused by the Atomic Bombs in Hiroshima and Nagasaki. Psychological trends among A-bomb victims. In *Hiroshima and Nagasaki: The Physical, Mental and Social Effects of the Atomic Bombings* (E. Ishikawa and D. Swain, transl.). Basic Books, New York, 1981.
24. Trumbull, R. *Nine Who Survived Hiroshima and Nagasaki.* E. P. Dutton & Co., New York, 1957.
25. Nagai, T. *We of Nagasaki.* Meredith Press, New York, 1958.
26. Gleser, G. C., Green, B. L., and Winget, C. *Prolonged Psychosocial Effects of Disaster.* Academic Press, New York, 1981.
27. Sessions, G. R. *The Psychological Effects of Tactical Nuclear Weapons,* DNA Intermediate Dose Program Rep. (in press).
28. British Medical Association, Board of Science and Education. *The Medical Effects of Nuclear War.* Wiley, New York, 1983.
29. Gellhorn, A. The immediate medical response. In *Last Aid* (E. Chivian, S. Chivian, R. J. Lifton, and J. E. Mack, eds.). Freeman, New York, 1982.

CHAPTER 16

Effects of Ionizing Radiation on Behavior and the Brain

WALTER A. HUNT

Armed Forces Radiobiology Research Institute
Bethesda, Maryland 20814-5145

A major concern of military planners is whether military personnel will be able to perform their duties after exposure to radiation. The degree to which irradiation affects operations may depend on the nature of the duties and the consequences of nonperformance. The more critical the role of one or a few individuals in successfully completing a mission, the more important the issue of performance decrement. For example, a jet fighter pilot might tolerate less of a performance decrement than an infantryman, due to the nature of the required performance. Consequently, it is important to know under what circumstances any deterioration of performance might be expected and the characteristics of the decrement.

It is difficult to predict expected decrements in performance based on available information from human exposures. From the clinical literature, it is well known that exposure to low doses of ionizing radiation can induce emesis. However, instances of human exposure to high, lethal doses are few. Some of these exposures, resulting from nuclear accidents, clearly indicate that irradiation can induce severe alterations in behavior. For example, in an accident at Los Alamos Scientific Laboratory in 1958, a man received a 4500-rad, whole-body dose of radiation. Shortly after irradiation he exhibited a transient deterioration of behavior, characterized by incoherence, ataxia, and general incapacitation. This behavioral decrement was relatively short-lived, lasting about 1 hr and 40 min. The man ultimately died.

Since documentation of behavioral decrement in victims of nuclear accidents is rare, an adequate description of how radiation might degrade performance can come only from controlled experiments using laboratory animals. Although extrapolations of data from such animals to human situations can be difficult, they provide at least a first approximation. They can be valuable tools in the initial formulation of casualty criteria.

Experiments reported over the last 30 years have clearly indicated that exposure to ionizing radiation alters behavior and neurobiology in a complex and multifaceted manner. Such factors as dose, quality of radiation, and the nature of the task being performed are important determinants of the effect of irradiation on behavior. In this chapter, an overview of the behavioral effects of radiation exposure is presented, along with discussion of the biological changes that are relevant to behavior.

I. BEHAVIORAL ALTERATIONS

A. Emesis

Emesis is one of the most important responses to irradiation that can impact performance. The consequences of emesis can range from minimal to almost total debilitation, depending on intensity of the irradiation and circumstances under which it occurs. As an example, a pilot vomiting into his face mask could have dire consequences. Emesis can occur at sublethal doses of radiation, and it may be the only immediate reaction to irradiation that may impair performance.

Emesis may be seen after the absorption of only a few hundred rads of radiation. It is generally dose-dependent up to 1000 rads. Animal studies have indicated that at doses above 1000 rads, the incidence of emesis decreases (1). The ED_{50} for emesis is about 500 rads in the monkey (2), compared to 180 rads in man (3). If the monkey is sitting on a moving platform, the ED_{50} is about 260 rads (2), probably reflecting the interaction of radiation and motion sickness.

The initial stimulus for emesis is thought to be irritation of the gut and then interaction of a chemical mediator with the area postrema in the brain stem. The area postrema is often referred to as the chemoreceptor trigger zone. Lesions of the area postrema will block radiation-induced emesis (4).

Another behavioral response to radiation exposure that may be related to emesis is the development of a conditioned taste aversion (CTA). A CTA will develop to a novel tasting solution that is normally preferred by the animal, when consumption is paired with irradiation (5). A CTA can be obtained with doses of radiation less than 50 rads.

The possible relationship between a CTA and emesis is based on a number of similarities. For example, both responses are biological reactions intended to prevent poisoning. Also, CTA depends on the integrity of the area postrema, as does radiation-induced emesis. Lesioning the area postrema prevents the development of CTA (6). Irradiation of the abdomen is more effective than irradiation of the head in inducing both CTA and emesis (7–9). And finally, neither CTA nor emesis is blocked by vagotomy (9,10). These observations have led to the conclusion that a humoral factor in the blood plays a role in both phenomena. Based on these similarities, the CTA paradigm may serve as a useful model for the study of emesis.

B. Behavioral and Neural Activation

Under some conditions, radiation exposure can lead to stimulation of the nervous system and concomitant changes in behavior. Animals are transiently aroused by doses of radiation of about 1000 rads (11). This was demonstrated in rats that had been trained to sleep in a radiation exposure room. When irradiated, the animals awakened for 1 min and then returned to their quiescent state. In a particular strain of mouse, exposure to 1000 or 1500 rads of radiation stimulated locomotor behavior for at least 30 min (12).

Exposure to ionizing radiation alters the seizure threshold. A single dose of 15–500 rads of X rays significantly reduces the threshold for electroshock-induced convulsions in a dose-dependent manner (13). This effect lasts up to 8 months with the highest dose of radiation. The pattern of the seizure is also dose-dependent. The durations of the seizure and of the clonic phase are less after lower doses of radiation. However, after higher doses of radiation, the duration of tonic extension increases, while that of tonic flexion decreases. Shielding experiments indicate that radiation acts primarily on the brain (14). After doses of radiation of 500–10,000 rads, the intensity of seizures is reduced.

In addition to changing the seizure threshold, exposure to radiation modifies potentials in the olfactory cortex that are evoked by stimulating the lateral olfactory tract (15). The amplitude of the evoked potentials is increased, and the latency between the stimulus and the response is shortened. This action of radiation on evoked potentials lasts at least 1 month. No studies to date have addressed whether these excitatory effects of radiation have any behavioral consequences.

C. Behavioral Decrement

A variety of studies have examined the disruptive effects of ionizing radiation on behavior in a number of species. These studies have clearly shown that irradiation with doses above the lethal level degrades behavior. However, the nature of the behavioral changes and the doses of radiation required to induce them have varied, depending on the species used. Monkeys exhibit a transient decrement in performance similar to that described in humans who were accidentally irradiated (16–19). The reactions of rats and miniature pigs are similar in quality but different in sensitivity and time course (19,20). Radiation-induced incapacitation in dogs is not transient.

Decrement of behavior after exposure to ionizing radiation has been studied extensively in the rhesus monkey. The decrement examined is early transient incapacitation (ETI), defined as the complete cessation of performance for 1 min. On a visual discrimination task, ETI develops within a few minutes after a supralethal dose of mixed-spectrum radiation (neutron/gamma = 0.4), and it lasts for about 30 min (18,21). The performance then recovers, but not completely, to preexposure levels. The ED_{50} is about 1600 rads. Under these experimental conditions, few monkeys exhibit an ETI at doses below 1000 rads, but most of them do at 5000 rads.

Although ETI is the predominant response between 1000 and 5000 rads, other qualitative differences have been reported. In some cases, monkeys perform normally without developing ETI until several days before death. In other cases, the animals are incapacitated quickly and die within a few hours. Few multiple ETIs are observed.

Using a physical activity task where monkeys are trained to avoid shock in a running wheel, monkeys irradiated in a mixed-spectrum field (neutron/gamma = 3.0) are incapacitated with an ED_{50} of about 2000 rads (16). The dose–response curve is quite steep. Few monkeys have an ETI at about 1300 rads, but most do at 2300 rads. Unlike a visual discrimination task, multiple ETIs develop during the physical activity task, and the time lost in performance increases in a dose-dependent manner.

The effectiveness of radiation in degrading performance depends, in part, on the quality of radiation being used. In rats, high-energy electrons are more effective than gamma photons in degrading performance (22,23). In monkeys, gamma radiation is more effective than neutron radiation (20).

The sensitivity of animal behavior to radiation appears to be related to the task being performed. For example, using a delayed match-to-sample task that integrates motor performance, decision making, and short-term memory, the performance of monkeys is impaired after doses of gamma radiation of 500–1000 rads (17). If the required response time on a visual discrimination task is shortened, degradation of performance is found more frequently with doses below 1000 rads (19). It has been suggested that the successful completion of a task in a radiation environment may depend on the physical and cognitive demands of the task, and that the poor performance of the animals in the latter study is due in part to a motor decrement.

The contribution of motor dysfunction to radiation-induced decrements of behavior depends on the nature of the task and its physical requirements. Using a platform avoidance task where rats are trained to jump up onto a platform to avoid shock, irradiation degrades performance in a dose-dependent manner (22).

Although irradiated animals do not jump on the platform to avoid shock, they can escape the shock once it is presented. In another type of test for performance degradation, rats can either press a lever to avoid shock or press another lever to obtain visual and auditory cues that tell the animals when the shock will occur. It was found that successful avoidance of the shock is not maintained after irradiation (24). However, once shocked, the animals respond at higher rates than control animals in attempts to avoid further shock. By so doing, they do not respond on the lever to obtain cues that would aid in avoiding shocks. Instead, the irradiated animals respond *to* the shock. In the above cases, the primary deficit that is responsible for the degradation of a task may be an inability to properly process the sensory stimuli that make possible the successful completion of the task. The ability of radiation exposure to disrupt sensory processes has not been systematically studied.

Sensitivity to the development of behavioral decrement after irradiation may be sex specific. Irradiated male rats do not perform as well as female rats on an avoidance task (25), so it may be said that female rats are somewhat resistant to radiation. It appears that the female hormone estradiol is responsible for this resistance.

II. PHYSIOLOGICAL MECHANISMS AND PERFORMANCE DECREMENT

A. General Physiological Considerations

The mechanisms underlying the behavior decrement that is observed after exposure to ionizing radiation are unknown, but several physiological changes have been suggested as causative factors. One change is an early transient hypotension that correlates with the development of early transient behavioral decrement (26,27). The reasoning centers on the idea that the sudden hypotensive event might reduce cerebral blood flow and create a hypoxic condition. However, cerebral blood flow is not clearly changed after irradiation in spite of the presence of hypotension (28,29). (See discussion of this in Chapter 8.) Also, the administration of pressor agents such as norepinephrine does not prevent incapacitation (30). Furthermore, studies reveal that radiation-induced behavioral decrement occurs in spontaneously hypertensive rats to the same extent as in normotensive rats, even though the transient hypotension that develops yields blood pressures equivalent to those of the normotensive rats (31).

Another mechanism that might underlie performance decrement is the peripheral release of histamine. Histamine is released from mast cells into the bloodstream after exposure to ionizing radiation (32). Attempts to alter the development of behavioral decrement by pretreating animals with antihistamines before exposure have been encouraging (33,34). Treated monkeys are able to perform and survive longer than controls after irradiation. Antihistamines also reduce the degree of radiation-induced hypotension. On the other hand, a CTA induced by radiation is not prevented by antihistamines (35,36). As shown in a more recent study, pretreatment with antihistamine attenuates the decrement of active avoidance behavior after irradiation (37). However, this effect can be found only in male rats, not female. As stated earlier, this sex-specific action may be related to the greater sensitivity of male animals to radiation (25).

B. Neurobiological Effects

The effects of ionizing radiation on neurobiological mechanisms have not been studied extensively. From the available reports, it is not clear whether the observed alterations have any behavioral or physiological consequences.

Recent studies have demonstrated parallel changes in neurotransmission and behavior after exposure to radiation. One area of the brain that is specifically affected is the caudate nucleus of the basal ganglia, an area involved in motor coordination. Radiation exposure significantly modifies the neurons that use dopamine as their neurotransmitter. Available evidence indicates that doses of 10,000 rads of radiation depress the transmission of dopamine (38,39). This effect correlates with the behavioral changes found under similar experimental circumstances.

Dopamine is an inhibitory transmitter. So alterations of this inhibitory input to the caudate nucleus could lead to corresponding changes of motor activity. Increases in dopaminergic activity are associated with stimulation of motor behavior and vice versa. Therefore, the radiation-induced changes in dopaminergic activity may reflect alterations in the flow of impulses along nigrostriatal fibers. In further support of this concept, stimulation of the substantia nigra (which synapses in the caudate nucleus) normally increases locomotor behavior. However, this response is depressed after high doses of radiation that reduce dopaminergic transmission (40).

Another neurochemical end point that correlates with early treatment behavioral decrement is the concentration of adenosine 3′,5′-cyclic monophosphate (cyclic AMP) in several areas of the brain. Cyclic AMP is one of the cyclic nucleotides that act as second messengers in synaptic transmission. After irradiation, concentrations of cyclic AMP are significantly reduced in rats, but they return to control values within 2–4 hr (41). These effects are not due to alterations in the activities of the enzymes that are responsible for the synthesis and degradation of cyclic nucleotides (42). Similar effects have been observed in monkeys (43).

The endogenous opioid peptides have been implicated in the actions of radiation on behavior. When mice of a particular strain are irradiated with moderate doses of cobalt-60, their locomotor activity is increased by opiates (12). If the animals are pretreated with naloxone (an opiate antagonist) or chronically stressed (which desensitizes the endogenous opioid system), the radiation-induced increase in locomotor activity is prevented (44). In addition, mice that will usually self-administer morphine, presumably for its reinforcing properties, consume less morphine after irradiation (45). Taken together, these results suggest that radiation exposure stimulates the release of endogenous opioid peptides.

Ionizing radiation can interact with the most basic processes of neural excitation. The initiation and propagation of action potentials are based on the transient inward movement of sodium ions. Nerve endings stimulated with neurotoxins take up less sodium after irradiation (46). This effect is consistently shown after doses of either high-energy electrons or gamma photons as low as 100 rads (47). These are doses well within the range used for therapeutic irradiations of brain tumors. In addition, the effect of radiation appears to be in a specific area of the sodium channel deep in the membrane bilayer, and to reflect a reduced number of functional sodium channels. The accumulation of calcium by nerve endings is not altered by radiation exposure (48). Disruption of the normal function of the sodium channel might play a role in the neural effects of radiation, and may have consequences on behavior.

III. SUMMARY

Exposure to ionizing radiation has multiple effects on the brain and behavior. These changes depend largely on the dose received. After doses below 1000 rads, the major effect of radiation is to induce emesis. Also, behavioral activation and increased susceptibility to seizures have been reported. Because high lethal doses of radiation depress the nervous system, animals are unable to properly perform tasks for which they have been trained. In some cases, this effect is transient, but in others it is permanent. The cause of these behavioral decrements appears to be related to cognitive and/or motor dysfunctions.

Several measures of CNS function correlate with some of the behavioral effects that are induced by radiation. Changes in the dopaminergic system in the caudate nucleus parallel behavioral decrements. Dopaminergic activity is reduced when performance is impaired. Also, opioid activity is enhanced after lower doses of radiation when locomotor activity is stimulated. Some of the effects of radiation may be mediated by the dysfunction of basic neural processes, such as stimulus-induced sodium influx.

REFERENCES

1. Middleton, G. R., and Young, R. W. Emesis in monkeys following exposure to ionizing radiation. *Aviat., Space Environ. Med.* **46,** 170–172 (1975).
2. Mattsson, J. L., and Yochmowitz, M. G. Radiation-induced emesis in monkeys. *Radiat. Res.* **82,** 191–199 (1980).
3. Lushbaugh, C. C. Human radiation tolerance. In *Space Radiation Biology and Related Topics* (C. A. Tobias and P. Todd, eds.), pp. 475–522. Academic Press, New York, 1974.
4. Borison, H. L. Area postrema: Chemoreceptor trigger zone for vomiting—is that all? *Life Sci.* **14,** 1807–1817 (1974).
5. Smith, J. C. Radiation: Its detection and its effects on taste preferences. *Prog. Physiol. Psychol.* **4,** 53–117 (1971).
6. Rabin, B. M., Hunt, W. A., and Lee, J. Attenuation of radiation- and drug-induced conditioned taste aversions following area postrema lesions in the rat. *Radiat. Res.* **93,** 388–394 (1983).
7. Smith, J. C., Hollander, G. R., and Spector, A. C. Taste aversions conditioned with partial body radiation exposures. *Physiol. Behav.* **27,** 887–901 (1981).
8. Rabin, B. M., Hunt, W. A., and Lee, J. Effects of dose and of partial body ionizing radiation on taste aversion learning in rats with lesions of the area postrema. *Physiol. Behav.* **32,** 119–122 (1984).
9. Borison, H. L. Site of emetic action X-radiation in the cat. *J. Comp. Neurol.* **107,** 439–453 (1957).
10. Rabin, B. M., Hunt, W. A., and Lee, J. Effects of subdiaphramatic vagotomy of the acquisition on a radiation-induced conditioned taste aversion. In preparation.
11. Hunt, E. L., and Kimeldorf, D. J. Behavioral arousal and neural activation as radiosensitive reactions. *Radiat. Res.* **21,** 91–110 (1964).
12. Mickley, G. A., Stevens, K. E., White, G. A., and Gibbs, G. L. Endogenous opiates mediate radiogenic behavioral change. *Science* **220,** 1185–1186 (1983).
13. Pollack, M., and Timiras, P. S. X-ray dose and electroconvulsive responses in adult rats. *Radiat. Res.* **21,** 111–119 (1964).
14. Sherwood, N. M., Welch, G. P., and Timiras, P. S. Changes in electroconvulsive thresholds and patterns in rats after X-ray and higher-energy proton irradiation. *Radiat. Res.* **30,** 374–390 (1967).
15. Timiras, P. S., Woolley, D. E., Silva, A. J., and Williams, B. Changes in the electrical activity of the olfactory cortex induced by radiation and drugs. *Radiat. Res.* **30,** 391–403 (1967).
16. Bruner, A., Bogo, V., and Jones, R. K. Delayed match-to-sample early performance decrement in monkeys after ^{60}Co irradiation. *Radiat. Res.* **63,** 83–96 (1975).
17. Franz, C. G. Effects of mixed neutron–gamma total-body irradiation on rhesus monkeys physical activity performance. *Radiat. Res.* **101,** 434–441 (1985).
18. Young. R. W. The relative effectiveness of high-energy neutrons in producing behavioral incapacitation, emesis, and mortality in monkeys. In preparation.
19. Bogo, V., Franz, C. G., and Young, R. W. The effects of radiation on the speed-stressed, visual discrimination performance of rhesus monkeys. In preparation.
20. George, R. E., Chaput, R. L., Verrelli, D. M., and Barron, E. L. The relative effectiveness of fission neutrons for miniature pig performance decrement. *Radiat. Res.* **48,** 332–345 (1971).
21. Thorp, J. W., and Young, R. W. *Neutron Effectiveness for Causing Incapacitation in Monkeys,* Sci. Rep. SR72-5. Armed Forces Radiobiol. Res. Inst., Bethesda, Maryland, 1972.
22. Hunt, W. A. Comparative effects of exposure to high-energy electrons and gamma radiation on active avoidance behavior. *Int. J. Radiat. Biol.* **44,** 257–260 (1983).

23. Bogo, V. Effects of bremsstrahlung and electron radiation on rat motor performance. *Radiat. Res.* **100,** 313–320 (1984).
24. Burghardt, W. B., and Hunt, W. A. Characterization of radiation-induced performance decrement using a two-lever shock-avoidance task. *Radiat. Res.* **103,** 149–157 (1985).
25. Mickley, G. A. Behavioral and physiological changes produced by a supralethal dose of ionizing radiation: Evidence for hormone-influenced sex differences in the rat. *Radiat. Res.* **81,** 48–75 (1980).
26. Phillips, R. D., and Kimeldorf, D. J. The effect of whole-body X-irradiation on blood pressure in the rat. *Radiat. Res.* **18,** 86–95 (1963).
27. Chapman, P. H., and Young, R. J. Effect of cobalt-60 gamma irradiation on blood pressure and cerebral blood flow in the "Maccaca Mulatta." *Radiat. Res.* **35,** 78–85 (1968).
28. Chapman, P. H., and Young, R. J. Effect of high energy X-irradiation of the head on cerebral blood flow and blood pressure in the Maccaca mulatta. *Aerosp. Med.* **39,** 1316–1320 (1968).
29. Turbyfill, C. L., Roudon, R. M., and Kieffer, V. A. Behavior and physiology of the monkey (Maccaca mulatta) following 2500 rads of pulsed mixed gamma–neutron radiation. *Aerosp. Med.* **43,** 41–45 (1972).
30. Turns, J. E., Doyle, T. F., and Curran, C. R. *Norepinephrine Effects on Early Postirradiation Performance Decrement in the Monkey,* Sci. Rep. SR71-16. Armed Forces Radiobiol. Res. Inst., Bethesda, Maryland, 1971.
31. Mickley, G. A., Teitelbaum, H., Parker, G. A., Vieras, F., Dennison, B. A., and Bonney, C. H. Radiogenic changes in the behavior and physiology of the spontaneously hypertensive rat: Evidence for a dissociation between acute hypotension and incapacitation. *Aviat. Space Environ. Med.* **53,** 633–638 (1982).
32. Doyle, T. F., and Strike, T. A. Radiation-released histamine in the rhesus monkey as modified by mast-cell depletion and antihistamine. *Experientia* **33,** 1047–1049 (1977).
33. Doyle, T. F., Turns, J. E., and Strike, T. A. Effect of antihistamine on early transient incapacitation of monkeys subjected to 4000 rads of mixed gamma–neutron radiation. *Aerosp. Med.* **42,** 400–403 (1971).
34. Doyle, T. F., Curran, C. R., and Turns, J. E. The prevention of radiation-induced, early, transient incapacitation of monkeys by an antihistamine. *Proc. Soc. Exp. Biol. Med.* **145,** 1018–1024 (1974).
35. Rabin, B. M., Hunt, W. A., and Lee, J. Studies on the role of central histamine in the acquisition of a radiation-induced conditioned taste aversion. *Radiat. Res.* **90,** 609–620 (1982).
36. Rabin, B. M., Hunt, W. A., and Lee J. State-dependent interactions in the antihistamine-induced disruption of a radiation-induced taste aversion. *Radiat. Res.* **90,** 621–627 (1982).
37. Mickley, G. A. Antihistamine provides sex-specific radiation protection. *Aviat., Space Environ. Med.* **52,** 247–250 (1981).
38. Hunt, W. A., Dalton, T. K., and Darden, J. H. Transient alterations in neurotransmitter activity in the caudate nucleus of rat brain after a high dose of ionizing radiation. *Radiat. Res.* **80,** 556–562 (1979).
39. Hunt, W. A., and Dalton, T. K. Reduction in dopamine metabolism in the caudate nucleus after exposure to ionizing radiation. In preparation.
40. Mickley, G. A., and Teitelbaum, H. Persistence of lateral hypothalamic-mediated behaviors after a supralethal dose of ionizing radiation. *Aviat., Space Environ. Med.* **49,** 868–873 (1978).
41. Hunt, W. A., and Dalton, T. K. Reduction in cyclic nucleotide levels in the brain after a high dose of ionizing radiation. *Radiat. Res.* **83,** 210–215 (1980).
42. Hunt, W. A., and Dalton, T. K. Synthesis and degradation of cyclic nucleotides in brain after a high dose of ionizing radiation. *Radiat. Res.* **85,** 604–608 (1981).
43. Catravas, G. N., Wright, S. J., Trocha, P. J., and Takenaga, J. Radiation effects on cyclic

AMP, cyclic GMP, and amino acid levels in the CSF of the primate. *Radiat. Res.* **87,** 198–203 (1981).
44. Mickley, G. A., Stevens, K. E., White, G. A., and Gibbs, G. L. Changes in morphine self-administration after exposure to ionizing radiation: Evidence for the involvement of endorphins. *Life Sci.* **33,** 711–718 (1983).
45. Mickley, G. A., Sessions, G. R., Bogo, V., and Chantry, K. H. Evidence for endorphin-mediated cross-tolerance between chronic stress and the behavioral effects of ionizing radiation. *Life Sci.* **33,** 749–754 (1983).
46. Wixon, H. N., and Hunt, W. A. Ionizing radiation decreases veratridine-stimulated uptake of sodium in rat brain synaptosomes. *Science* **220,** 1073–1074 (1983).
47. Mullin, M. J., Hunt, W. A., and Harris, R. A. Ionizing radiation alters the properties of sodium channels in rat brain synaptosomes. *J. Neurochem.* **47,** 489–495 (1986).
48. Ely, M. J., and Catravas, G. N. Calcium accumulation and retention by synaptosomes with high-energy electrons. *Radiat. Res.* **88,** 623–630 (1981).

CHAPTER 17

Nuclear Weapons Accident Response Procedure†

RAYMOND L. CHAPUT*

Armed Forces Radiobiology Research Institute
Bethesda, Maryland 20814-5145

Since development of the first nuclear weapon in 1945, the United States has never had a nuclear weapon accident involving a nuclear yield. However, accidents have occurred that released radioactive contamination because of fire or detonation of the high explosive. The two most recent accidents involving radioactive contamination occurred near Palomares, Spain, in January 1966, and Thule, Greenland, in January 1968. The recovery effort in these accidents involved United States resources from all Services, the Atomic Energy Commission (now the Department of Energy), the State Department, and other federal agencies. Recovery efforts lasted several months and required extensive scientific and technical operations and logistical support.

Historically, the Services have established response forces along their respective organizational lines. This resulted in nonstandard terminology and procedures, which have contributed to the response problem of past accidents. The Joint Chiefs of Staff directed the Defense Nuclear Agency to develop, in coordination with the Services, a general technical manual to summarize Department of Defense (DoD) assigned responsibilities and to provide procedural guidance for a joint response to accidents involving a nuclear weapon or its components. *The Nuclear Weapon Accident Response Procedures* (NARP) *Manual* was developed to satisfy the requirement by the Joint Chiefs of Staff.

This chapter provides an overview of the problem of response to a nuclear weapon accident, the fundamentals of response to an accident, and a summary of

†The material in this chapter was extracted from the Nuclear Weapon Accident Response Procedures (NARP) Manual, DNA Publication No. 5100.1, Defense Nuclear Agency, Washington, D.C., January 1984.

*Present address: Biomedical and CBR Defense Exploratory Technology, Office of Naval Technology, Washington, D.C. 22217-5000.

the NARP Manual. The manual provides a summary of procedural guidance, technical information, and DoD responsibilities, to assist DoD forces in preparing a response to a nuclear weapon accident.

RESPONSE TO A NUCLEAR WEAPON ACCIDENT

In a nuclear weapon accident, three primary problem areas must be addressed: public affairs, radiological concerns, and recovery of the weapon. Other areas of concern may include support requirements for the primary problem areas, e.g., communications, security, logistics, and medical and legal support. Any nuclear weapon accident can result in an immediate but temporary nonradiological threat to public safety from toxic or explosive hazards. Some nuclear weapon accidents may release contamination that could create long-term concerns for public health.

A. Public Affairs

Public affairs encompasses much more than the release of information to the public. The degree of public support for accident-response actions and the impact of that support on nuclear issues of national interest will affect the manner in which the public perceives an accident and the actions taken in response to the accident.

Public concerns that can be expected following a nuclear weapon accident include

(1) Danger to people in the vicinity.
(2) Effects on the health of persons exposed to contamination.
(3) Credibility of information being provided about the accident and its effects.
(4) General safety of the United States nuclear program.

Public needs that can be expected following a nuclear weapon accident include

(1) Treatment of casualties.
(2) Confirmation either that people were not exposed to radiological hazards or, if exposed, that the effect of such exposure on health has been adequately diagnosed.
(3) Decontamination of contaminated areas.
(4) Guidance on precautions to be taken by those in contaminated or potentially contaminated areas.
(5) Shelter, food, and clothing if the accident results in the evacuation of people from their homes.

(6) Reparations for direct and incurred damages caused by the accident.
(7) Information about the accident and about ongoing and planned response activities to meet the needs stated above.

The Department of Defense Directive 5230.16 (titled *Public Affairs*) states that, in general, although it is DoD policy to neither confirm nor deny the presence of a nuclear weapon at a specific location, the response force commander is authorized to make two exceptions. First, he must confirm the presence of a nuclear weapon when public safety is endangered. Second, he may confirm or deny the presence of a weapon as necessary to calm public alarm. No other variations from DoD policy are authorized. If the exceptions are invoked, confirmation or denial of the presence of a weapon must be made as soon as possible to prevent undue public concern and to establish credibility of the response force. Notification of the public must be made through any means available and in simultaneous coordination with state and local authorities, if possible. Once the public is notified of the involvement of a nuclear weapon, frequent updates on the situation and the response actions being taken must be given to the public.

B. Radiological Concerns

All nuclear weapon accidents will not necessarily disperse radiological material. If a weapon explodes, however, there will probably be extensive radiological contamination. Although a fire is less serious than an explosion, it may cause substantial contamination. The rapid determination of the presence or absence of a radiological problem is a critical part of the initial response to an accident. If a radiological problem exists or if it is caused during weapon-recovery operations, specialized DoD and Department of Energy (DOE) radiological response teams will be required to define the extent of contamination and to simultaneously provide a credible demonstration of preparedness and ability to protect the public and return the area to normal use.

1. Radiological Safety

Alpha particles are the primary radiological hazard associated with a nuclear weapon accident, not the better known beta or gamma radiation that results from a nuclear detonation or a reactor accident. Alpha particles are massive particles that can be blocked by a piece of paper or the outer layer of skin. Alpha particles normally are a health hazard only when introduced into the body through a wound, inhalation, or ingestion. Beta and gamma radiations, on the other hand, possess greater penetrating power and, if present in sufficient quantity, may be a health hazard to nearby people. Quantities of beta and gamma radiations that are sufficient to pose a significant health problem normally will not be present at a nuclear weapon accident. Contamination is the physical deposit of radioactive materials on or in other materials, including people. The hazard of deposition and incorporation of alpha contamination in the body is different from the hazard of beta contamination, although similar health effects may result from equivalent radiation doses. Hazards of contamination and basic principles of radiation protection are described in Appendix A.

2. Decontamination and Restoration of the Site

The process of restoring a contaminated area may include the removal, dilution, or fixing of contamination to levels that will not be detrimental to health over a lifetime of exposure and that will be as close to background levels as can be reasonably achieved (see Appendix B). This process will be the most time-consuming portion of a nuclear weapon accident response, and will be a coordinated federal, state, and local effort. One of the most difficult steps is the selection of criteria for determining when site restoration has been completed. Many, if not most, of the legal claims against the government will be related to restoration of the site.

3. The Radiological Response

If contamination has been dispersed by an accident, the first radiological response actions will be to identify the affected area, to minimize any radiological hazard to people in the area, and to determine the amount of contamination present. Information on the size of the area and the amount of contamination present is essential to prevent speculation about health effects of the accident and to permit plans for site-restoring actions. Steps to obtain this information should be initiated as soon as possible. Initial information on the affected area may come from projections based on worst-case conditions. It is the response force that must obtain the facts to define the actual problem instead of the projected problem. Preaccident plans should identify those actions that will be taken immediately to identify the affected area, to minimize the hazard to the public, and to measure actual contamination. The plan should also identify limitations on organizational resources in performing such actions and should specify how resources initially will be used when available. (See Appendix C for a checklist for actions before an accident and Appendix D for a checklist for immediate actions.) Specific actions that must take place in order to resolve a radiological problem include the following:

(1) Determination of whether or not contamination was released;

(2) Provision of precautionary measures to be taken by persons in potentially contaminated areas;

(3) Identification and monitoring of potentially contaminated people at the accident scene;
(4) Delineation of the affected area, e.g., ground survey of suitable low-level contour, such as three times background;
(5) Determination of the various levels of contamination present within the contaminated area;
(6) Establishment of a bioassay program to determine if anyone actually received a radiation dose.

C. Recovery of the Weapon

All nuclear weapon accidents will, by definition, have some form of weapon problem. Security and weapon concerns should not prevent or interfere with basic medical and humanitarian responses to accident victims. Public affairs and radiological issues must be addressed concurrently with the weapon-recovery effort. The weapon presents a technical problem for Explosive Ordnance Disposal (EOD) personnel and the DOE Accident Response Group. The weapon also presents a need to provide appropriate security for the weapon and its components. Procedures for making the weapon safe and security procedures for nuclear weapons are documented in EOD and security publications. A weapon involved in an accident may have been subjected to severe stress, so qualified EOD and/or DOE response personnel should take as much time as necessary to thoroughly assess possible damage before the weapon is moved. If the high explosives are detonated during the accident or during recovery operations, it may be necessary to search for classified and hazardous parts. Coordination between EOD and radiological-response personnel is essential to ensure that operations continue with minimum risk to personnel.

D. Associated Areas of Concern

In a major accident, the normal problems of logistical support and communications will need to be resolved. Legal and medical problems that are unique to radiological accidents will also require solution. Perhaps the most difficult problem is the coordination of actions by the various agencies and organizations that will respond to a nuclear accident. There will be few, if any, established lines of command authority between agencies and many overlapping areas of joint responsibility. To overcome the initial confusion and provide an effective coordinated response, the response force commander must be aware of other responding agencies and their basic capabilities, and be prepared to recommend, request, or direct initial actions using these capabilities. To do this effectively, interagency preaccident coordination may be required. Depending on the expertise of the response force staff, a more effective response may be possible. Through prearrangement, a nearby specialized team of another responding agency, which will be immediately available, could be given responsibility. Upon its arrival, that team will have authority for coordinating and directing some areas of the accident response.

E. Response to a Nuclear Weapon Accident

The response procedures considered here are based on two force levels. The first level is the initial force on-scene, including any regional forces that may be required if an accident occurs near a base with only capability for a humanitarian emergency response. The second level is the service response force, a force with sufficient assets to meet the unique requirements of a nuclear weapon accident. The Department of Defense Directive 3025.1 (*Use of Military Resources During Peacetime Civil Emergencies Within the United States, Its Territories, and Possessions*), DoD Instruction 5100.52 (*Radiological Assistance in the Event of an Accident Involving Radioactive Materials*), and DoD Directive 5200.8 (*Security of Military Installations and Resources*) provide authority to the response force commander to respond to a nuclear weapon accident and to assist civil authorities in restoration operations after removal of the weapon. The primary objectives of the accident response are to ensure that the weapon is safe from nuclear or conventional explosion, to recover all classified materials, and to assist in restoring the affected area to normal use. The response effort can be divided into the initial phase and the follow-on phase.

1. Initial Phase

The initial phase includes the notification of an accident and the immediate emergency measures to be taken by the nearest DoD and/or DOE installation(s) to provide a federal presence and humanitarian support. The initiation of nuclear weapon accident response actions by the National Military Command Center, Service command centers, and other agencies and organizations (that will respond or assist in the response) begins automatically after notification of an accident. So accidents must be reported immediately, using the fastest means available. Upon receiving notification of an accident, the appropriate response forces are identified and tasked, and specialized teams are alerted and prepared for immediate deployment. Figure 17-1 shows a simplified chain of notification that results from the initial accident report to the National Military Command Center and/or the Service command centers. The initial phase of response requires defining and stabilizing the situation to the extent possible by the initial

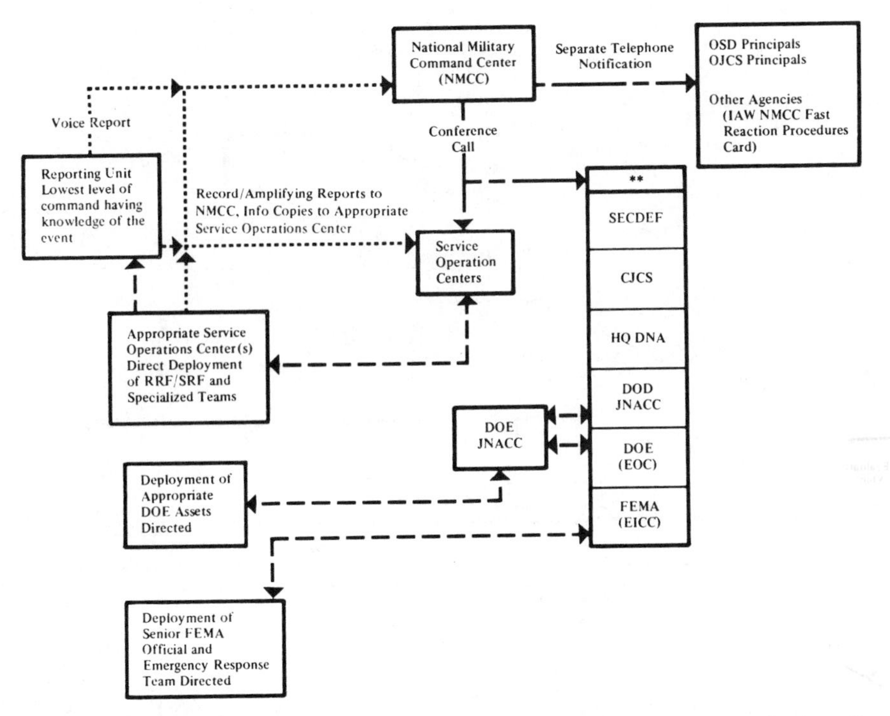

FIG. 17-1 Simplified chain of notification of a nuclear weapon accident. (**), Conference call includes reporting unit's command center and appropriate service operations center (AOC, NCC, AFOC); (——) OPREP 3 reporting procedures; (-- — -), NMCC (National Military Command Center) notification, coordinating, and command procedures; (– – – –), Service/agency notification and command procedures.

forces present. These actions include suppression of fire, reconnaissance, rescue and treatment of casualties, and assessment of hazards to public health and safety. Other actions of immediate concern include establishing communications between the accident site, the supporting military installation, and the command centers; providing information related to public affairs; and establishing security for the nuclear weapon, its associated components, and classified material. Figure 17-2, although not all-inclusive, illustrates the interrelationship of initial actions.

2. Follow-On Phase

The follow-on phase of the response effort includes the conduct of all operations that are required to recover the weapon and restore the environment to an acceptable condition, using the combined assets of the various agencies and

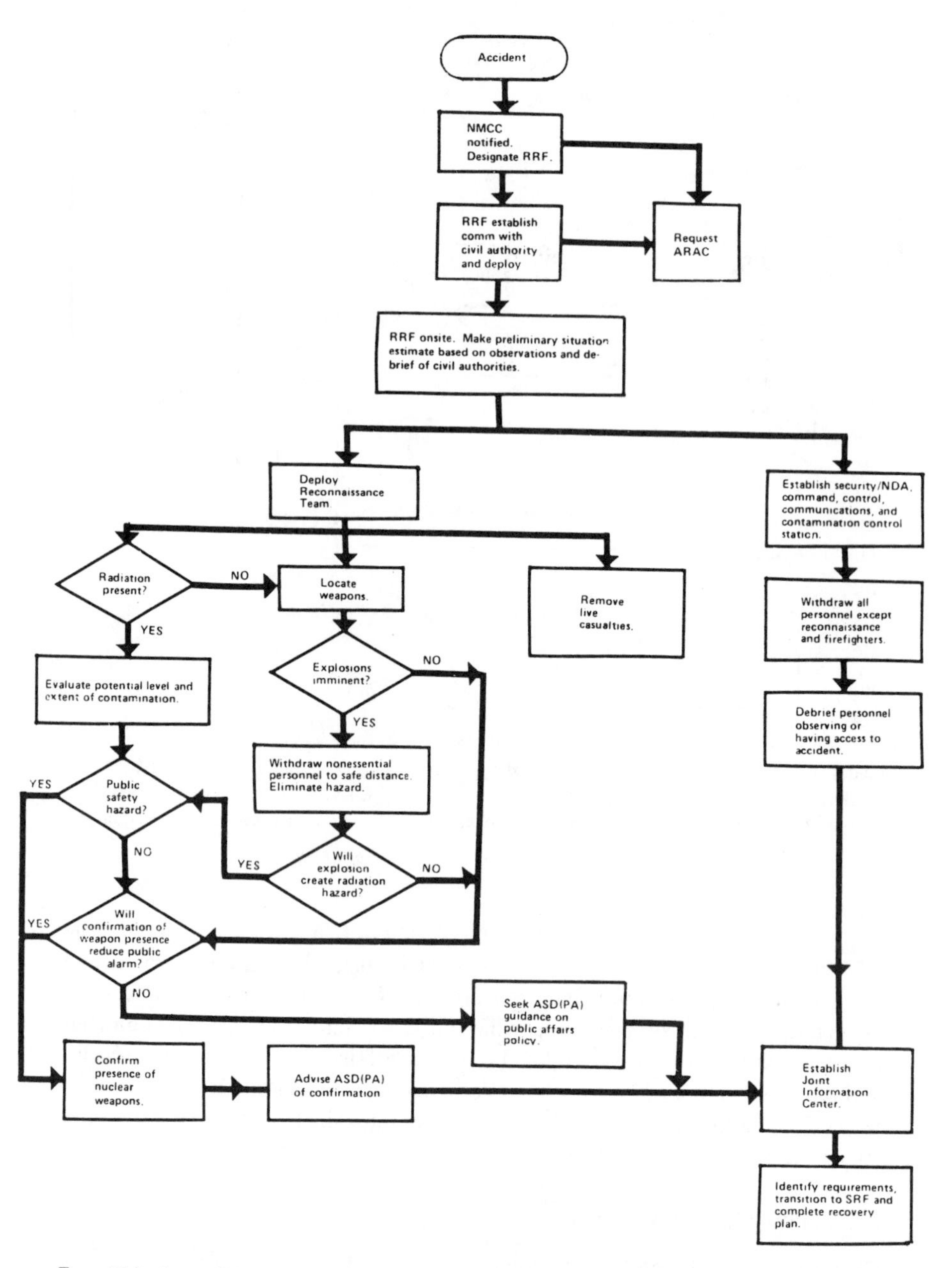

FIG. 17-2 Interrelationship of initial actions during a nuclear weapon accident response.

organizations involved in the response. Appendix E provides a checklist of actions to be taken as soon as resources and personnel permit.

F. Primary Non-DoD Agencies for Response to Nuclear Weapons Accidents

1. Department of Energy

The Department of Energy is responsible for sending appropriate DOE response elements to the scene of a DoD nuclear weapon accident or significant incident. The specific response elements, including any necessary specialized equipment, will be chosen by the HQ DOE Emergency Operations Center (in Germantown, Maryland) to best meet the accident or incident situation. The Emergency Operations Center will advise the DOE element of the Joint Nuclear Accident Coordinating Center.

The Department of Energy has established the Accident Response Group as its primary accident-response element. It consists of scientists and technical specialists with equipment that is ready to be sent on short notice to the scene of a nuclear accident. The Group will advise and assist the DoD on-scene commander in operations for weapon recovery and in evaluating, collecting, handling, and mitigating radioactive material and other hazards associated with weapons.

The Accident Response Group will provide technical advice and assistance to the on-scene commander in the following:

(1) Support of the EOD teams in their procedures for making the weapon safe and their procedures for recovery;

(2) Determination of the extent of any radioactive hazards;

(3) Minimizing hazards to the public;

(4) Collection, identification, packaging, and disposition of weapon components, weapon debris, and the resulting radioactive material;

(5) Identification and protection of information on nuclear weapon design and other restricted data;

(6) Discussions with foreign or local government officials on matters within areas of special DOE competence.

The Department of Energy will also have off-site technical responsibilities. It will coordinate off-site radiological monitoring and assessment activities of federal agencies for the Federal Emergency Management Agency.

2. Federal Emergency Management Agency (FEMA)

The Federal Emergency Management Agency is responsible for the establishment of federal policies and the coordination of all civil executive agencies. The primary role of FEMA during a nuclear weapon accident is to coordinate requests for assistance from federal agencies and to ensure that off-site actions and response activities of federal, state, and local officials are mutually supportive and coordinated with the on-site actions of DoD and DOE. This role of FEMA is carried out through a senior FEMA official.

3. State and Local Governments

Off-site authority and responsibility for a nuclear weapon accident rests with local and state officials. The state governor is responsible for the health, safety, and welfare of persons within the territorial limits of the state during periods of emergency or crisis, and should be expected to direct measures that must be taken in order to satisfy that responsibility. Land temporarily placed under federal control by the establishment of a National Defense Area to protect United States Government classified materials will revert back to state control upon disestablishment of the National Defense Area. The on-scene commander will assist the state to ensure the protection of the public.

APPENDIX A: CONTAMINATION HAZARDS AND BASIC PRINCIPLES OF RADIATION PROTECTION

A. Contamination Hazards

Contamination can be introduced into the body through wounds caused by the accident, or through wounds caused later by materials or surfaces contaminated during the accident. When responding to an accident involving injury, members of the response force must always be aware that administration of lifesaving first aid is of primary importance.

The greatest hazard from inhalation occurs immediately after an accident if contamination has been released. If a weapon's high explosives detonate, the explosion will create a cloud of contamination, which gradually dissipates and settles from the air as it moves downwind. If a weapon burns, contamination is carried into the air by the smoke and thermals from the fire, which also is dispersed by the wind. In either case, once the explosion and/or fire is over and the resulting contamination has settled or dispersed (approximately 2 hr), the remaining inhalation hazard resulting from resuspension of radioactive particles is significantly lower.

Ingestion of contamination may result from eating food or using utensils on which airborne contamination has settled. It is possible that some airborne contamination may be ingested during the passing of the cloud of contamination. The principles of decorporation therapy are discussed in Chapter 13.

B. Basic Principles of Radiation Protection

There are four basic radiation protection principles: quantity, time, distance, and shielding. These factors are interrelated.

1. Quantity

The exposure rate from radioactive material is directly related to the quantity of the material present. For the types of radioactive materials present at a nuclear weapon accident, the total quantity present does not normally present a significant external irradiation hazard. Since the primary radiation threat in a weapon accident is from inhalation or ingestion of dispersed material, the quantity is usually expressed per unit area or volume, e.g., microcurie per square meter ($\mu Ci/m^2$) for quantities on the ground and microcuries per cubic meter ($\mu Ci/m^3$) or per cubic centimeter ($\mu Ci/cm^3$) for quantities in the air. Field measurements of quantity are normally expressed in instrument-dependent units of counts per minute (cpm) or counts per second (cps). They must be converted to definitive units such as $\mu Ci/m^2$ for meaningful comparison.

2. Time

Any radioactive material will emit a known amount of activity or radiation per unit time. For the type and quantities of radiation present at a nuclear weapon accident, months or years would normally be required for external radiation exposure to be a hazard. The length of time of exposure to the radioactive materials present at a nuclear weapon accident is related to a health hazard primarily through the amount of material that may be deposited in the lungs through inhalation over a period of time. The average inhalation rate for a normal person is 30 m^3 of air for each 24-hr day. This rate provides a basis for calculating the amount of contamination that may be deposited in the lungs if the amount of airborne contamination is known. The rate at which contamination may be inhaled is highest during the initial period following the accident, when a substantial quantity of contamination is airborne. If there is no airborne contamination or if respiratory protection is being worn, exposure time is not a critical factor in a nuclear-weapon-accident response unless limited fission occurred, which produces beta–gamma emitters.

3. Distance

The protective principle of distance [in which radiation intensity varies inversely with the square of the distance (if the distance doubles, the intensity is reduced by a factor of four)] applies primarily to beta/gamma radiation (not normally a significant part of the radiological problem in a nuclear weapon accident). Alpha particles, the source of the primary radiological problem, will travel only about 1 cm in air.

4. Shielding

Shielding results from the ability of a material to reduce or stop radiation. The alpha emissions of primary concern in a nuclear weapon accident can be stopped by a piece of paper, and will not penetrate the outer layer of skin. Beta omissions can be stopped by a sheet of aluminum, whereas gamma emissions may require several inches of lead to stop them. Since alpha emitters are the predominant concern after a nuclear weapon accident, any light clothing or gloves used to prevent contamination of underclothing or the body will automatically provide protection from external radiation.

APPENDIX B: CHECKLIST FOR ACTIONS TO RESTORE A SITE

(1) Provide required administrative, medical, and logistic support, including that needed by DOE response organizations.

(2) Coordinate with Service and National Transportation Safety Board accident-investigation teams.

(3) Coordinate communications assets and radio-frequency requirements of all response organizations.

(4) Publish a Communications–Electronics Operating Instruction for use by all response organizations.

(5) Request radio-frequency clearances if required.

(6) Obtain additional communications assets as required.

(7) Monitor levels of public understanding; identify and respond to concerns about nuclear issues.

(8) Establish channels for coordination of technical legal matters with higher headquarters and the principal legal advisors of other participating federal departments and agencies.

(9) Establish a heat-injury prevention program for temperatures above 70°F.

(10) Coordinate site-restoration action with FEMA, responsible civil authorities, and military organizations.

(11) Initiate actions to establish cleanup procedures and standards.

(12) Conduct environmental assessments.

(13) Complete the decontamination, including fixing, reducing concentration, or removing the contaminant.

(14) Restore the contaminated area to conditions suitable for normal use.

(15) Develop plans for monitoring and assessing the radiation after the restoration.

(16) Protect government property.

(17) Provide necessary operational security.

(18) Counter any potential terrorist and/or radical group activities or intelligence-collecting efforts.

(19) Establish a personnel replacement/rotation program to support the operation; if appropriate, minimize radiation exposure to radiation workers.

APPENDIX C: CHECKLIST FOR ACTIONS BEFORE AN ACCIDENT

(1) Prepare and periodically review the base or response-force accident response plan. Key elements for frequent review include the following:
   (a) Roster of key response-force personnel and the assigning and training of key staff members;
   (b) Initial actions to be taken after arrival at the accident scene;
   (c) Procedures for establishing communications from off-base or remote accident sites;
   (d) Provisions and authority for release of information to the public;
   (e) Actions recommended for civil authorities to take and assistance that civil authorities may be able to provide;
   (f) How deployment is accomplished, e.g., Emergency Operations Center and key personnel/equipment as advance party, or all together;
   (g) Review of equipment (capability, availability, packaging requirements for movement, operating status, etc.);

(2) When possible, conduct periodic field and command-post exercises (including civil emergency-response agencies, such as fire, police, and DoD or DOE specialized teams, if any are nearby) to test the mutual support arrangements.

APPENDIX D: CHECKLIST FOR IMMEDIATE ACTIONS

(1) Upon arrival on-scene, identify the forces present (civil and military) and their capabilities. Determine actions that have been taken or initiate actions to treat or evacuate casualties. Cordon off the area. Reduce immediate hazards such as fires. Coordinate response activities.

(2) Take emergency actions to establish control of the accident site and to

safeguard the nuclear weapon, weapon components, and other classified material. Advise all response personnel of the explosive, radiological, toxic, and other possible hazards that may be present.

(3) Determine as quickly as possible if radioactive contamination is present. Request immediate deployment of Army RADCON and RAMT, Air Force AFRAT and ATRAP, the DNA Advisory Team, and the appropriate DOE RAT for the accident area.

(4) Seek the assistance and cooperation of state and local authorities, and advise them of possible hazards.

(5) Protect the public from hazards, and mitigate the health and safety problems by providing advice on safety measures to be taken.

(6) Advise medical facilities that receive the casualties of possible or actual contamination.

(7) Determine the status and location of all weapons involved, including whether high-explosive detonation may have occurred.

(8) Confirm the presence of a nuclear weapon(s) at the accident scene, when appropriate.

(9) Establish a positive communications link with the military communications system, e.g., by radio or an open telephone line to a command center.

(10) Keep the National Military Command Center and/or the Service command centers informed of conditions at the accident scene and the status of recovery efforts.

(11) Establish direct communications with the Office of the Assistant Secretary of Defense (Public Affairs).

(12) Request a project code number from the Joint Chiefs of Staff or from the Service for citing funds.

(13) Request appropriate communications support, e.g., HAMMER ACE or other contingency satellite system.

(14) Establish a Joint Information Center as appropriate.

APPENDIX E: CHECKLIST FOR ACTIONS AS SOON AS RESOURCES AND PERSONNEL PERMIT

(1) Initiate surveys to determine the extent of contamination.

(2) Determine and plot the contaminated areas.

(3) Control any exposure of the general public and response force personnel to contamination.

(4) Identify individuals who may have been exposed to contamination and decontaminate them.

(5) Establish, in coordination with civil authorities, a bioassay screening

program for response personnel and the public who may have internal contamination.

(6) Conduct assessment of damage by the weapon.

(7) Perform procedures to make the weapon safe as required.

(8) Initiate a systematic search for the weapon and its components until they are accounted for.

(9) Perform necessary actions to transport or ship the weapon and components to the appropriate disposal area.

(10) Establish and direct the activities of a Joint Information Center near the scene of a nuclear accident that results, or appears likely to result, in effects outside the boundaries of a DoD facility.

(11) Establish a National Defense Area, if required, and dissolve it after the nuclear weapon, nuclear weapon components, and classified materials have been recovered.

(12) Establish a Joint Radiological Control Center to coordinate radiological control and health physics matters between the various radiological response elements, both off-site and on-site.

(13) Establish liaison with state and local legal agencies and law enforcement agencies.

(14) Establish a claim-processing facility.

(15) Determine the availability of assets and facilities at or near the scene of the accident, and initiate actions to obtain support for satisfying the requirements of the response force.

(16) Inform the senior FEMA official, upon arrival, of all on-site activities that could have an impact off-site.

(17) Establish an admistrative group to ensure that actions taken and evidence retained for use by the investigation board are documented.

CHAPTER 18

Management of Radiation Accidents

JAMES J. CONKLIN AND RODNEY L. MONROY*

Armed Forces Radiobiology Research Institute
Bethesda, Maryland 20814-5145

I. ROLE OF THE MILITARY PHYSICIAN IN MANAGEMENT OF RADIATION ACCIDENTS

The previous chapter reviewed the procedures to be followed after a nuclear weapon accident. It is clear from Table 18-1 that the greatest risk of radiation exposure is from an accident. The military physician who is knowledgeable in biological hazards and the effects of radiation can do great service to the public. This service will be in assessing the risk to individuals and society from radiation accidents; and in allaying public fear of the unseen and unknown when a significant risk is not present. In the event of an accident near a military hospital, it is necessary for knowledgeable and responsible military physicians to assess the size of the risk and to protect and inform the public. Although a nontechnical authority or leader may speak out to avert or curb panic in a crisis, we still hope that the military physician will be knowledgeable in discussing radiation effects, predicting those effects, and establishing exposure criteria in non-nuclear-weapon emergency plans. Military physicians should participate in community programs to educate the public about radiation. They should emphasize the distinction between somatic risks to individuals and genetic risks to the population, and stress that one should not extrapolate back and forth between genetic and somatic risks or between individual and population risks.

II. PLANNING FOR RADIATION ACCIDENTS

A critical facet of planning for emergencies is to know the kind of emergency that requires preplanning. A radiation accident implies an unintended exposure

*Present address: Transplantation Research Program Center, Naval Medical Research Institute, Bethesda, Maryland 20814-5145.

TABLE 18-1

MAJOR RADIATION ACCIDENTS WORLDWIDE 1944 TO APRIL 1984

Type of injury[a]	Number of accidents[b]	Individuals ≥ dose criteria[c]	Fatalities
WBI ≥ 25 rems	46	150	9 (United States 2)
Local + ≥ 600 rems	40	53	9 (United States 3)
Local + ≥ 600 rems	138	157	0
Int. dose ≥ MPBB	38	63	4 (United States 2)
Mixed (Marshallese)	1	110	1
Totals		533	23 (United States 7[d])

[a]WBI, Whole-body irradiation.
[b]More than one type of injury in one accident.
[c]DOE–NRC accident dose criteria.
[d]Three SL-1 fatalities not included.

of intense radiation, an uncontrolled release of large quantities of radioactive material, or a combination of the two, which results in potential for causing serious injury or death. The probability that a radiation accident will occur is rare; however, during the period 1940–1984, approximately 250 radiation accidents were reported, involving over 600 exposed persons (1). The most important facet of radiation accident management is that the radiologic aspects seldom require emergency action. The irradiated or contaminated person must receive emergency medical care irrespective of his radiologic condition. To date, no report exists of a secondary injury to medical personnel as a result of their caring for a radiation accident victim. Also, no radiation accident victims have been so contaminated that they represented a hazard to providers of health care.

Because of the heightened worldwide awareness of radiation hazards, hospitals have had to augment their capabilities to respond to radiation accidents. Radiologically trained physicians have become focal points of the planning for radiation accidents because of their radiologic expertise. It is important that the emergency response of each hospital be evaluated and tailored to its local requirements and risks. Regional and local health agencies and hospitals should identify one or more physicians who are knowledgeable in biological hazards and the effects of radiation, and who are available on a 24-hr basis for advice in the event of a radiation scare, incident, or accident.

Many fundamental questions must be evaluated and answered when formulating a plan, including the following.

(1) What radiation accidents are likely to occur?

(2) What capabilities and resources are available that could be used in an emergency?

(3) What emergency procedures will be followed?
(4) What is the organization that will respond, and what authority and responsibility does it have?
(5) Will this emergency organization be able to obtain off-site assistance?
(6) What arrangements have been negotiated with fire, police, and emergency medical services?
(7) Do adequate communications exist between the Emergency Command Center and the various elements of the system?

Other questions that must be evaluated and answered pertain to equipment, facilities, and resources, as follows.

(1) Is equipment available to assess an accident, and is it reliable under adverse weather or terrain conditions?
(2) Is the equipment regularly inventoried and tested with both normal power and emergency power?
(3) Has health physics consultation been obtained to develop personnel monitoring?
(4) Have the requirements for triage, first aid, decontamination, and bioassays been integrated into a plan?
(5) Are continuous training and exercises conducted under simulated accident conditions with rigorous critiques?

The majority of radiation accident victims involving personal injury or contamination can be adequately treated at the local hospital if the staff has had some training in radiation accident management. Several excellent references are available to provide systematic guidance in developing a radiation emergency plan (2–5). Many of the requirements identified as the *Nuclear Weapon Accident Response Procedures (NARP) Manual* (see Chapter 17 of the present volume) are directly transferable to the civilian sector. Organization and prior planning are the essential requirements.

III. MANAGEMENT OF RADIATION ACCIDENTS

Life-threatening doses of accidental acute whole-body irradiation (WBI) have been so infrequent that treatment strategies must be devised from the few actual cases, experimental animal studies, and analogous states of clinical disease. Regardless of the type of exposure or contamination, the initial response to a radiation-injury patient is the life-saving treatment of nonradiation injuries. The first actions must be the standard emergency medical procedures: ensure that an airway exists, the patient is ventilated, adequate perfusion is maintained, and life-threatening hemorrhage is stopped. Once the patient is stabilized, decontamination may be instituted. Where feasible, decontamination may be performed simultaneously with treatment. External and internal decontamination were addressed in Chapters 13 and 17.

Since radiation exposure does not represent an emergency, there is no justification for hasty diagnostic and therapeutic procedures without careful consideration. It is unlikely that dose estimates will be immediately available; consequently, statements on prognosis should be extremely guarded. Proper care can significantly impact the survival of patients receiving high doses of WBI in radiation accidents. Although the number of accidental irradiations has been small, the effectiveness of therapy for bone marrow depression has been demonstrated in other diseases (6). It must be stressed that therapy should never be instituted on the basis of physical estimates of dose. Therapy must be predicated on the clinical signs, symptoms, and laboratory data of the radiation victim (7). In evaluating acute accidental irradiations, the hematologic perturbations are not only the most life-threatening effects, but also the best biologic index of exposure dose (8). The degree of injury is generally proportional to radiation dose, with only moderate biological variation.

Persons exposed to less than 10 cGy will require no special medical therapy. Those persons who have received less than 10 cGy may show minimal physical signs but will almost certainly recover. If the dose is in doubt, they may be kept under clinical surveillance with mild sedation, as necessary, until the dose is confirmed. Persons who have received acute WBI of greater than 100 cGr may require specialized care; the remainder of this discussion concentrates on them.

The severity of the irradiation may be estimated from prodromal effects, hematologic changes, cytogenetics, film badges, dosimetry mock-ups, and whole-body counting in the event of neutron exposures (9). The use of the clinical and biological indicators will prove to be the most valuable. The presence or absence of nausea and vomiting may allow separation into fatal and nonfatal categories. Vomiting within 3 hr of exposure suggests a high radiation exposure. Prompt, explosive, bloody diarrhea most likely indicates a lethal exposure. An early lymphocyte decrease is the single best indicator of radiation injury, while the levels of platelets and granulocytes are the most important guides for later therapy. Severe trauma (burns or soft tissue) may also cause a lymphopenia. Consequently, lymphocyte counts may be unreliable in combined injuries (10). If the lymphocyte count is greater than 1500/mm^3 after 48 hr, no significant irradiation has occurred. On the other hand, a decrease to 1000 lymphocytes/mm^3 within 24 hr and 500 lymphocytes/mm^3 within 48 hr represents a very severe exposure. Cytogenetic evaluation of cultured lymphocytes allows dose estimates to as low as 11 cGy (11). Cytogenetic evaluation is a very useful test that should be obtained in all accidental irradiations to corroborate exposure. It is obvious that if a significant exposure has occurred, lymphocytes may be gone within 48 hr. In addition to a requirement for cytogenetic dosimetry is the even greater requirement for histocompatibility typing and storage for later mixed leukocyte cultures. The lymphocytes for this must be obtained immediately.

Once the patient has been resuscitated (if necessary), monitored for radioactivity, and decontaminated, and after the baseline blood work has been drawn, there should be time to attend to the patient's apprehension. Patient apprehension will be significant because of the presence of the many consultants and accident investigators and the patient's ignorance and misperception of radiation. The patient's apprehension may grow because of his sense of isolation resulting from protective environments to prevent infection and his lack of contact with hospital staff and visitors for fear he is radioactive. In potentially lethal exposures, an empathetic physician with control over visitors can significantly help the patient. A knowledgeable physician can also provide much support by answering questions about long-term effects, fertility, and early mortality. For patients with the gastrointestinal syndrome, good nursing care, sedation, antiemetics, fluids, and electrolytes plus large quantities of compassion can be offered. Although this may not alter survival, it will provide emotional support to the victim and his family.

In paralethal exposures, granulocytes will initially increase as a result of an inflammatory release of granulocytes by the bone marrow. The granulocytes are not directly damaged like the stem cells. Consequently, neutropenia develops only as granulocyte reserves are depleted. In paralethal exposures, neutropenia will occur within 4 weeks, while in supralethal exposures, it may occur within 1 week. Platelets are similarly affected by paralethal exposures and slightly lag the response of granulocytes, but reach their nadir at about the same time. Concomitant with leukopenia and thrombocytopenia is a general decrease in resistance to infection. There is a multifactorial manifestation of lymphopenia, leukopenia, thrombocytopenia, impaired antibody production, decreased intracellular bactericidal activity, and damage to skin and intestinal mucosa, which facilitates sepsis (10,12). The primary determinant of survival will be how well the tremendous susceptibility to infection and subsequent bleeding is managed. If the patient can be kept alive for 5–6 weeks, marrow recovery will have begun. Any military hospital with an active hematology–oncology service will have significant experience in handling bone marrow depression.

Several characteristics of radiation accident victims with the hematopoietic syndrome differentiate them from oncology patients, and portend a more favorable outcome for the same degree of bone marrow depression. Among these are usually general good health and nutrition, normal bone marrow, normal bacterial flora, and no presensitization to antibiotics at the time of exposure. The salient characteristics of infections in irradiated patients that are similar to those in immunosuppressed individuals include the presence of nonlethal organisms, rapidity of infectious process, lack of inflammatory process, and presence of nosocomial infections (13). A protective environment can diminish colonization and may reduce the incidence of infection, but the use of standard hospital rooms, gowns, gloves, and masks is of little help for patients with granulocytopenia (14). Other possibilities include using laminar airflow rooms, washing hands carefully, eliminating from the diet green, leafy vegetables (which tend to be colonized with gram-negative bacilli), and doing environmental surveillance cultures (15,16). Most consultants will recommend prophylactic oral antimicrobials for radiation victims when granulocyte counts fall below $1000/mm^3$, although this may increase colonization of the patient with organisms resistant to antimicrobial drugs. Nonabsorbable antibiotics, such as a combination of gentamicin, vancomycin, and nystatin, can decrease infection but may be poorly tolerated by the patient. Prophylactic use of oral trimethoprim sulfamethoxazole and nystatin has decreased the incidence of fever and bacteremia as much as gentamicin and nystatin, and they are better tolerated (17,18).

A new fever or suspected sepsis should be treated with intravenous antimicrobials without waiting for culture results. Most of these infections will be due to *Escherichia coli, Klebsiella pneumoniae, Pseudomonas aeruginosa,* and *Staphylococcus aureus* (19,20). The choice of drugs should be determined by the prevalent organisms and drug susceptibilities unique to each hospital. The use of a semisynthetic penicillin (such as carbenicillin or ticarcillin) plus an aminoglycoside (such as tobramycin, gentamicin, or amikacin) has been recommended (21,22). In patients with suspected fungal infections and refractory fever despite treatment, amphotericin B is recommended, although recent evidence shows a high incidence of pulmonary reactions in patients also receiving granulocyte transfusions (23). Most experts recommend that antimicrobial therapy continue until the granulocyte count is greater than $500/mm^3$, or about 2 weeks, if the patient is afebrile for 5 days and evidence of infection is absent (24).

Granulocyte and platelet transfusions may be valuable during the bone marrow depression (25). Typing of lymphocytes (HLA) of the patient and family members should have been completed immediately after irradiation. Potential donors of platelets and granulocytes should be identified, and plans made for their availability. Several studies have shown that granulocyte transfusions have been helpful in patients with profound granulocytopenia or with persistent or progressive signs of infection (21,25,26). In the absence of a rising granulocyte count, it is unlikely that there will be complete resolution of radiation sepsis. In this setting, granulocyte transfusions may be not only useful but also clearly indicated. Numerous studies have shown the usefulness of granulocyte transfusion for a patient with progressive infection and persistent fever; it is effective therapy especially with a focal infection (perianal abscess). More recently, experience with patients receiving WBI has raised doubts about the use of granulocytes in marrow aplasia because of the risk of cytomegalovirus infections (27).

Concomitant with the neutropenia, the radiation victim will develop thrombocytopenia. Tremendous experience has been gained in treating thrombocytopenia in states of bone marrow failure secondary to chemotherapy or disease. Platelet transfusions are indicated for thrombocytopenia to prevent or treat bleeding (28). Spontaneous bleeding usually does not occur until platelet counts are less than $20,000/mm^3$, although some patients will hemorrhage with platelet counts of $50,000/mm^3$ (29). It is usually desirable to give platelets before an active hemorrhage. Prophylactic transfusions of platelets have been shown to be effective in decreasing morbidity and mortality in patients with platelet counts less than $20,000/mm^3$ (30). Prophylaxis is recommended with 4–8 U of platelets 2–3 times weekly. For active bleeding, surgery, petechiae, closed-head trauma, or microscopic hematuria, it is anticipated that the platelet transfusion should be given at or below 50,000 $platelets/mm^3$. The life span of transfused platelets is approximately 8–10 days (31). Many factors will influence this life span, including drugs, bleeding, fever, infection, disseminated intravascular coagulation, and storage. Platelet transfusions from multiple random donors are associated with transfusion reactions, and subsequent random donor transfusions may be refractory. It is therefore desirable to use single or a few nonrelated HLA-matched donors for platelet transfusions (31). This is especially true in any contemplated bone marrow transplantation.

In supralethally irradiated patients, bone marrow transplantation may be the only therapeutic modality that can offer hope for survival by correcting the manifestations of hemorrhage and infection. A lethally irradiated accident victim who had a twin brother was saved by bone marrow transplantation after 600 cGy of total-body exposure (32). Because transplanted bone marrow takes about 2 weeks to correct the pancytopenia, the decision to transplant should not be delayed past the first week postirradiation. Recent improvements in the survival of nongenetic HLA-matched transplants in cases of refractory aplastic anemia or acute leukemia offer some encouragement for supralethally irradiated patients (33). However, this therapy is available only at specialized treatment centers.

IV. BONE MARROW TRANSPLANTATION

The effects of radiation on various cells of the lymphohemopoietic system were discussed in Chapter 6. The general theme was the effect of radiation on the immunocompetency of the individual. The level of immunocompetency after irradiation was shown to be dependent on the level of survival of the functional immunoresponsive cells and also the pluripotent stem cell of the bone marrow. At doses above the $LD_{50/30}$, severe marrow aplasia is observed, and death occurs as the result of infection. Lorenz and co-workers (34) were the first to show, in mice and guinea pigs, that severe marrow aplasia and eventual death could be overcome by the infusion of syngeneic bone marrow after exposure to radiation. Thus, the concept of bone marrow transplantation and pluripotent stem-cell replacement was initiated.

Since these first experiments, bone marrow transplantation has been evaluated extensively in both experimental and clinical applications. Experimental transplantation has been shown to be successful in the animal models if the donor is syngeneic to the recipient. However, bone marrow transplantation from an allogeneic donor (marrow obtained from a donor not related to the recipient) leads to a later death of the animal as the result of a "secondary disease." This secondary disease is graft-versus-host (GVH) disease. This complication with respect to allogeneic bone marrow transplantation has been shown to be alleviated by meeting specific genetic requirements. These genetic requirements have been shown to be centered around a specific chromosomal region, the human major histocompatibility complex (HLA). The HLA region consists of three gene loci (*A, B, D/DR*) that code for serologically determined antigens and lymphocyte-culture-determined antigens. Successful transplantation and reduced GVH disease have been associated with the complete matching of all elements in the region. In fact, in the first successful human transplants, the donor was an identical twin of the recipient. In spite of genetic matching, GVH disease continues to be observed, and has been attributed to the contamination of immunocompetent donor T lymphocytes in the marrow. Methods have been developed to eliminate the contaminating T cells from the donor marrow, and drugs have been developed to reduce the clinical severity of GVH disease. Thus, bone marrow transplantation is currently the recommended therapy for bone marrow and immunodeficiency diseases (e.g., aplastic anemia and severe combined immunodeficiency). In addition, its use combined with chemotherapy and radiation-therapy regimens has gained wide application. In the following sections, the use of bone marrow transplantation, autologous and allogeneic, will be discussed as potential treatments in rescuing lethally irradiated individuals.

A. Autologous Bone Marrow Transplantation

Autologous bone marrow transplantation is the transplantation of an individual with his/her own bone marrow after irradiation. Although the use of autologous bone marrow transplantation is limited in scope, with some foresight and planning, it can be used effectively. Possible recipients for autologous bone marrow transplantation are individuals who work in high-risk radiation areas, such as personnel working at a nuclear reactor or astronauts on prolonged space missions. In these cases, bone marrow or an alternative source of pluripotent stem cells (stem cells from the peripheral blood) (35) must be harvested and cryopreserved before any radiation exposure. Similar techniques have rescued patients who have undergone extensive chemotherapy and radiation therapy in the treatment of various malignancies (33).

Cryopreservation studies of the bone marrow have shown that most stem cells and progenitor cells survive the freezing, storage, and thawing phases of the procedures (36). The bone marrow of dogs was removed, prepared for freezing in 10% dimethylsulfoxide plus media, and stored for 2–5 months. The marrow was then transplanted into the lethally irradiated (1000 cGy of cobalt-60) autologous recipients (37). After storage for 2 months, the marrow was shown to be as good as fresh marrow in reconstituting the hemopoietic system. The marrow stored for 5 months decreased slightly in viability, but it was capable of reconstituting the hemopoietic system in a manner not significantly different from that of fresh marrow.

Advances in cell separation techniques (counterflow centrifugation–elutriation, density gradients, and monoclonal antibodies) have effectively reduced the contaminating mature cells, red blood cells (RBC), and granulocytes, which complicate the efficiency of cryopreservation (38,39). These techniques provide a more purified marrow population, which is enriched with stem and progenitor cells. In the human system, this enriched population has been effectively stored in 5-ml vials ($1–4 \times 10^7$ cells/ml) at $-170°C$. Autologous bone marrow reconstitution using the cryopreserved cells requires $1–2 \times 10^8$ nucleated cells/kg (36,38). Individuals transplanted with the cryopreserved marrow have shown recovery of all hemopoietic cell lines, and the level of immunocompetency returns to near normal values within 30–60 days. The feasible length of time for potential storage of these cells is uncertain beyond 6 months, and no additional data are available to make any conclusions with respect to the longer storage.

Several important concepts in the potential transplantation of cryopreserved autologous bone marrow must be considered. First, radiation is immunosuppressive, and a certain length of time will be required for the ''critical cells'' of the integrated immunoresponse to fully recover. Autologous transplantation of cryopreserved marrow would supplement these cells with functional immune responsive cells, thus reducing the level and length of time of immunosuppression. In addition, these transplanted cells would most likely enhance the recovery of the surviving cells. Furthermore, the radiation dose received does not have to be lethal or near lethal to merit the use of autologous bone marrow transplantation (40). Thus, although autologous bone marrow transplantation is limited in overall application, it appears to have a significant role in the potential treatment of high-radiation-risk workers.

B. Allogeneic Bone Marrow Transplantation

Allogeneic bone marrow transplantation is the transplantation of bone marrow from one individual (donor) to another individual (host). This type of transplantation is more complex than an autologous one, and it carries a high level of risk. Following is a discussion of the factors involved in allogeneic bone marrow transplantation and its potential application in the treatment of radiation victims.

Animal models have been used extensively to understand the factors involved in allogeneic bone marrow transplantation. The consequence of transplanting a tissue (e.g., skin and bone marrow) into a nonrelated individual is that either the host cells will recognize this tissue as foreign and a host-versus-graft (HVG) reaction will occur, or that the grafted cells, in the case of bone marrow, will react to the host as being foreign and a GVH reaction will occur. Skin graft studies using the mouse model and its extensive genetic base have shown that there must be a level of immunosuppression before the transplanted skin will engraft. WBI and/or cyclophosphamide were used to induce the required level of immunosuppression. This same protocol was applied in the mouse model for allogeneic bone marrow transplantation, and after WBI, the transplanted allogeneic marrow was shown to engraft. Thus the HVG reaction was overcome using this preparative regimen. However, in spite of the observed engraftment, GVH disease developed about 30 days posttransplantation, which resulted in 100% fatality. The clinical symptoms of GVH disease are associated with skin pathologies (e.g., rashes and scaling), gastrointestinal tract disruption (e.g., diarrhea), and liver diseases. In addition to these problems, the recovering immune system appears to be further compromised by the GVH reaction. Therefore, the patient experiences increased susceptibility to infection (e.g., bacterial, fungal, and viral), and the result is death. So the GVH reaction has limited the application of bone marrow transplantation (33) until answers can be ascertained concerning the mechanism, prevention, and treatment of GVH disease.

The genetic development of inbred mouse strains and the development of an outbred model, the dog, have allowed breakthroughs in understanding some of the reasons why a GVH reaction occurs. The dog has been used effectively to demonstrate that histocompatibility testing could predict the outcome of marrow grafts (41). In fact, it was shown that when canine littermates were matched at the major histocompatibility complex, the survival after bone marrow transplantation was better than in the transplantation of unmatched animals. Despite this level of matching, approximately 50% of the dogs died as a result of the GVH reaction. As a result, it was suggested that the minor histocompatibility differences also play a part in fatal GVH disease. These minor histocompatibility differences were found to be attributed to the lymphocyte-culture-determined antigens (42). In humans, the HLA is composed of three loci, *A, B, D/DR,* which give rise to numerous serologically determined and lymphocyte-culture-determined antigens. Complete matching of those antigens at the A and B loci is essential in overcoming the major incompatibility barriers. The importance of matching at the minor histocompatibility regions is still being investigated, and these elements pose continual questions as to their significance in the GVH reaction. Thus, histocompatibility must exist between the donor and the host before any successful bone marrow transplant will occur. As demonstrated by the canine data, the degree of genetic matching predicts the probability of survival, which in turn is directly related to the severity of the GVH reaction. A level of matching of the minor histocompatibility antigens is required, although it is still unknown. Despite these efforts in matching, 35–50% of the patients develop significant acute GVH disease, and 45% of the long-term survivors experience chronic GVH disease (43).

Since GVH disease is a primary consequence of allogeneic BMT, various treatment regimens posttransplantation have been evaluated in animal models and in clinical trials to prevent GVH disease. Studies using intermittent prophylactic treatment with either methotrexate, cyclophosphamide, or cyclosporine for the first 100 days after transplantation have not prevented acute GVH disease. Thirty-five to fifty percent of the patients still develop significant GVH disease in spite of the treatment (44). A combination treatment of methotrexate and cyclosporine used in dog studies showed a significant improvement in survival by reducing the incidence and severity of GVH disease (45). Clinical studies of this combination therapy in human patients are ongoing, and results have not been reported. However, prophylactic treatment regimens have not been shown, as yet, to prevent the GVH reaction.

In mice studies, the mature immunocompetent T lymphocytes of donor bone marrow origin were identified as being involved in the GVH reaction (42). The involvement of other cell populations (the exact cell type of which is uncertain) is likely. In pathology of the GVH reaction, it is observed that donor T lymphocytes are sensitized against the host's epithelial cells of the skin, gastrointestinal tract, and liver. The lymphocytes are found to specifically invade these tissues. Methods have been developed to specifically eliminate these immunocompetent T lymphocytes from the donor marrow *in vitro* before infusion. These procedures include the use of several anti-T lymphocyte-directed monoclonal antibodies, either coupled to the toxic ricin molecule (46) or incubated with complement (47,48), to lyse the monoclonal antibody-reacting lymphocytes before transplantation. Another procedure removes the T lymphocyte from the bone marrow by the combined techniques of rosette formation with sheep red blood cells and agglutination with soybean agglutinin (49). Finally, one technique makes use of the differences in physical size between lymphocytes and hematopoietic progenitor cells. DeWitte *et al.* (50) have been able to isolate a marrow inoculum that is greater than 98% free from contaminating lymphocytes, by using counterflow centrifugation–elutriation. These techniques all appear to be promising in reducing the GVH reaction by a method of T cell depletion of the donor marrow in combination with drug prophylaxis posttransplantation. However, it is too early to draw any conclusions concerning these results.

Bone marrow transplantation has become a treatment of choice for a variety of disease states (43). However, strict requirements must be met before transplantation is deemed feasible. As mentioned above, the donor must be carefully selected or else the treatment may be worse than the original state. The GVH reaction, the actual mechanism of which is still uncertain, limits the application of this technique to the most sophisticated and well-equipped hospitals.

V. LOCAL RADIATION INJURY

As mentioned in the introduction, the most frequent cause of radiation injury is a lost industrial radiographic source. Many of the reported exposures to radiographic sources have involved serious injury to the skin and extremities. Radiation injury to the skin is the best known and most studied response to radiation in humans. Strandqvist (51) quantified skin damage as a function of dose and fractionation. Fifty percent of the people will experience epilation after exposure to 300 R (roentgens), erythema after 600 R, dry desquamation after 1000 R, and wet desquamation after 2000 R (52). (The roentgen is a unit of radiation exposure in air that is not included in the new SI units.) The erythema secondary to radiation-source exposure is similar to a first-degree thermal burn, such as a severe sunburn, and no therapy is necessary. Wet desquamation is a transepidermal injury similar to a second-degree thermal burn. Pain and itching occur soon after the exposure, with subsequent erythema and blister formation. The magnitude of medical intervention depends on the size, severity, and location of the lesions. Radionecrosis occurs with exposures greater than 2000 cGy. These effects are similar to severe chemical burns; they are accompanied by excruciating pain, which must be treated.

Regardless of the type of injury, the histologic result is obliterative endarteritis with resultant ischemia, necrosis, and infection (53). The most important question in obtaining a patient history about a suspicious burn with no specific etiology is, Did you burn yourself? If the patient has no recollection of touching a hot object and if he is a radiation worker or has had access to a shiny metal object, then serious consideration should be given to a radiation burn. Initial management should be conservative and nonoperative. The lesions are progressive with continuing ischemia, fibrosis, chronic ulceration, infection, necrosis, and poor wound healing (54). This progression makes it almost impossible to predict the ultimate level of viable tissue. The conservative approach requires large doses of narcotics (55). An internationally famous anesthesiologist received 85,000 cGy over several weeks to his hands from repeated use of a malfunctioning fluoroscope (56). The pain was so excruciating that he lost 54 lbs in 3 months. He emphasizes, from a personal and professional perspective, the importance of salvaging every possible bit of hand function. It is essential to preserve tissue and joint motion and to avoid contractures. A combination of splinting and physical therapy can usually preserve the range of movement and avoid contractures.

Massive radiation injuries to the extremities may require amputation. In 1967, a man received a midline-tissue dose of 600 cGy from a 300–800-KeV X-ray Van de Graaff accelerator (57). Within 20 min he received 5900 cGy to the hands and 2740 cGy to the feet. This individual had a twin brother who provided an HLA-identical bone marrow transplant graft, enabling him to survive the hematopoietic phase of injury previously described. Over the next 22 months, he sustained 11 operative procedures to amputate gangrenous extremities or revise previous stumps; the final amputation sites were at the midforearm and above the knee bilaterally. Review of this accident and others shows that the magnitude and extent of tissue damage are not readily appreciated until severe radionecrosis occurs, necessitating radical surgery. In severe cases of radionecrosis, it is recommended that large neurovascular pedicle grafts be provided (58). If no surgery is necessary, chronic effects of stiffness, atrophy, and radiodermatitis may ensue. The extremities may exhibit exquisite sensitivity to pressure, heat, and cold (59). Subsequent skin cancer must be anticipated. If dermatitis becomes severe, subsequent grafting may be necessary. Because local radiation injuries are progressive and the clinical course obscure, the therapy and rehabilitation must be carefully planned, and organized efforts must be made to comfort and support the radiation victim and his family.

VI. DECONTAMINATION AND DECORPORATION AFTER EXTERNAL CONTAMINATION BY RADIOACTIVE MATERIALS

The widely accepted use of radionuclides in industry, research, nuclear power, and medicine, in addition to the more frequent transport of these materials, has also increased the likelihood of exposure to workers and bystanders. Because of the greater probability of exposure, it is not enough that only medical facilities located with or adjacent to facilities using radionuclides are trained in the decontamination and decorporation of radionuclides. Decontamination and decorporation have been previously discussed, but several points deserve amplification.

VII. SKIN DECONTAMINATION

The contamination of skin with radionuclides is almost never immediately life threatening. As in every other aspect of radiation accident management, the serious medical problems have priority over decontamination. The primary objective of skin decontamination is to remove as much radionuclide as possible in order to reduce the surface dose rate and to decrease entry into the body. Decontamination also increases the accuracy of whole-body counts of incorporated radionuclide burdens. Zealous decontamination is discouraged, because it may increase the percutaneous absorption (60). Fear among health care providers about radiation exposure must be allayed. The simple removal of a victim's clothing can remove as much as 70–80% of the contamination. No human exposure to date has presented significant risk to the personnel giving assistance. Also, the principles of time, distance, and shielding (discussed in the previous chapter) can reduce any potential radiation exposure to the attending personnel. Personnel participating in decontamination should wear protective clothing, including surgical gowns, gloves, shoe and head covers, and aprons. Health physics monitoring may indicate the need for additional protective gear. Clothing, personal effects, and biologic samples from swabs of the nares, aural canal, and mouth should be placed in plastic bags and glass-stopped tubes with proper identification for later analysis.

The first priority of surface contamination should be open wounds. Since these areas may allow the rapid incorporation of radionuclides, they should be copiously irrigated with physiological saline for several minutes. If contamination persists, gentle surgical debridement may be necessary. Experiments with plutonium oxides have shown that translocation to regional lymph nodes occurs within a few minutes to several hours. After 1 month, the absorbed concentration has been found to be 60% of the implanted dose (61). For this reason, contaminated wounds must receive first decontamination priority. If the radionuclide is plutonium or other alpha emitters for which diethylenetriamine pentacetic acid (DTPA) is an effective chelating agent, treatment should begin immediately. An effective irrigating solution for americium or plutonium contamination is 1 g CaDTPA and 10 ml of 2% lidocaine in 100 ml of normal saline (62). However, Jolly and collaborators recommend DTPA by the intravenous, aerosol, inhalation, and oral routes as well as by irrigation. They also recommend venous occlusion of the extremity during decontamination. If an extremity is so severely contaminated that it cannot be decontaminated adequately, a decision may be required of whether or not to amputate. Amputation should be seriously contemplated only when the injury to the extremity is so severe that it precludes functional recovery, or when the contamination burden is to great that severe radionecrosis will occur. The best conservative advice is still, decontaminate, but do not mutilate.

After contaminated wounds have been treated, other areas can be decontaminated. The eyes, ears, nose, mouth, areas adjacent to uncontaminated wounds, and remaining skin surface should be decontaminated. Gentle, frequent irrigation and suction of the eyes and ears should be sufficient to decontaminate them. Decontamination of the mouth is important because of possible incorporation. The mouth should be irrigated with suction while the head is dependent. A nasogastric tube should be inserted and aspirated for analysis. If radionuclides have been ingested, lavage and decorporation therapy should be started. Decontamination of the skin usually requires only soap and warm water with gentle scrubbing (use of hot water is contraindicated because of subsequent vasodilation). If more aggressive decontamination is necessary, a mixture of half cornmeal and Tide (detergent) has been shown to be very effective. Hair can usually be decontaminated with soap and water. If this is inadequate, the scalp should be clipped rather than shaved to avoid disruption of the skin barrier. A thorough review of internal contamination is covered in Chapter 13.

TABLE 18-2

PERCEPTION OF RISK BY DIFFERENT GROUPS IN ORDER OF RANK[a]

Activity and deaths per year (estimation)	Group		
	League of Women Voters	College students	Business and professional club members
Smoking (150,000)	4	3	4
Alcoholic beverages (100,000)	6	7	5
Motor vehicles (50,000)	2	5	3
Handguns (17,500)	3	2	1
Electric power (14,000)	18	19	19
Motorcycles (3,000)	5	6	2
Swimming (3,000)	19	30	17
Surgery (2,800)	10	11	9
X Rays (2,300)	22	17	24
Railroads (1,950)	24	23	20
General (private) aviation (1,300)	7	15	11
Large construction (1,000)	12	14	13
Bicycles (1,000)	16	24	14
Hunting (800)	13	18	10
Home appliances (200)	29	27	27
Fire fighting (195)	11	10	6
Police work (160)	8	8	7
Contraceptives (150)	20	9	22
Commercial aviation (130)	17	16	18
Nuclear power (100)	1	1	8

[a]Source: Sinclair, W. K. Effects of low level radiation and comparative risk. *Radiology* **136,** 1-9 (1981).

TABLE 18-3

GUIDE FOR HOSPITALS INVOLVED IN A NUCLEAR ACCIDENT

(1) Establish communication with Nuclear Power Plant, Civil Defense, State Bureau of Radiological Health, United States Nuclear Regulatory Commission
(2) Activate an environmental-monitoring program
(3) Mobilize personnel-monitoring devices
(4) Locate spare portable radiation detectors
(5) Maintain an hourly log of events
(6) Designate an emergency telephone line for inquiries
(7) Implement a medical radiation emergency plan
(8) Establish procedure for interviews by media
(9) Schedule radiation information updates
(10) Review sample, whole-body, or thyroid counting capabilities
(11) Identify sources of information
(12) Review hospital's evacuation plan
(13) Distribute radiation information packets

TABLE 18-4

GOVERNMENTAL AGENCIES PROVIDING RADIOLOGIC ASSISTANCE

Brookhaven Area Office Upton, New York 11973 (516) 282-2200	Chicago Operations Office 9800 South Cass Avenue Argonne, Illinois 60439 (312) 972-4800 (Duty hours) or 972-5731 (Off hours)
Oak Ridge Operations Office P.O. Box E Oak Ridge, Tennessee 37830 (615) 576-1005 or 525-7885	Idaho Operations Office P.O. Box 2108 Idaho Falls, Idaho 83401 (208) 526-1515
Savannah River Operations Office P.O. Box A Aiken, South Carolina 29801 (803) 725-3333	San Francisco Operations Office 1333 Broadway Oakland, California 94612 (415) 273-4237
Albuquerque Operations Office P.O. Box 5400 Albuquerque, New Mexico 87115 (505) 844-4667	Richland Operations Office P.O. Box 550 Richland, Washington 99352 (509) 376-7381

Tables 18-2 through 18-4 are included to facilitate emergency planning and discussions of radiologic hazards. Definitive assistance can be provided by the Department of Energy (DOE) Radiologic Assistance offices in Table 18-4.

REFERENCES

1. Personal communication from Karl Hubner, Radiation Emergency Accident Center/Training Site, Oak Ridge, Tennessee.
2. Hubner, K. F., and Fry, S. A., eds., *The Medical Basis for Radiation Accident Preparedness*. Elsevier/North-Holland, New York, 1980.
3. Nenot, J. C., Lushbaugh, C. C., and Lincoln, T. A. *Manual on Early Medical Treatment of Possible Radiation Injury*, Saf. Ser. No. 47. IAEA, Vienna, International Atomic Energy Agency, 1978 (USA-UNIPUB, 345 Park Avenue South, New York, New York 10010).
4. Leonard, R. B., and Ricks, R. C. Emergency department radiation accident protocol. *Ann. Emerg. Med.* **9,** 462–466 (1980).
5. Roessler, E. C., Price, D. G., and Spencer, P. S. A hospital plan for emergency handling of radiation accident cases. *Health Physics in the Healing Arts,* 73-8029. Food and Drug Admin., Dept. of Health, Education and Welfare, Washington, D.C., 1973.
6. Andrews, G. A., Balish, E., Edwards, C. L., Knisely, R. M., and Lushbaugh, C. C. Possibilities for improved treatment of persons exposed in radiation accidents. In *Handling of Radiation Accidents,* pp. 119–135. IAEA, Vienna, 1969.
7. Bond, V. P., Fliedner, T. M., and Cronkite, E. P. Evaluation of management of the heavily irradiated individual. *J. Nucl. Med.* **1,** 221–238 (1960).
8. Andrews, G. A. Radiation accidents and their management. *Radiat. Res., Suppl.* **7,** 390–397 (1967).
9. International Atomic Energy Agency. *Personnel Dosimetry for Radiation Accidents.* IAEA, Vienna, 1965.
10. Conklin, J. J., Walker, R. I., and Hirsch, E. F. Current concepts in the management of radiation injuries and associated trauma. *Surg., Gynecol. Obstet.* **156,** 809–829 (1983).
11. Brewen, J. G., Preston, R. D., and Littlefield, L. G. Radiation-induced human chromosome observation yields following an accidental whole-body exposure to ^{60}Co X-rays. *Radiat. Res.* **49,** 647–656 (1972).
12. Walker, R. I., and Porvaznik, M. Association of leukopenia and intestinal permeability with radiation-induced sensitivity to endotoxin. *Life Sci.* **23,** 2315–2322 (1978).
13. Andrews, G. A. Medical management of accidental total body irradiation. In *The Medical Basis for Radiation Accident Preparedness* (K. F. Hubner and S. A. Fry, eds.). Elsevier/North-Holland, New York, 1980.
14. Nauseef, W. N., and Maki, D. G. A study of the value of simple protective isolation in patients with granulocytopenia. *N. Engl. J. Med.* **304,** 448–453 (1981).
15. Schimpf, S. C., Greene, W. H., Young, V. M., *et al.* Infection prevention in acute non-lymphocytic leukemia. *Ann. Intern. Med.* **82,** 351–358 (1975).
16. Remington, J. S., and Schimpf, S. C. Please don't eat the salad. *N. Engl. J. Med.* **304,** 433–435 (1981).
17. Gurwith, M. J., Brunton, J. L., Lonk, B. A., Harding, G. K. M., and Ronald, A. R. A prospective controlled investigation of prophylactic time-thoprim/sulfamethoxazole in hospitalized granulocytopenic patients. *Am. J. Med.* **66,** 248–256 (1979).

18. Wade, J. C., Schimpf, S. C., Hargadon, M. T., Fortner, C. L., Young, V. M., and Wiernik, P. H. A comparison of trimethoprimsulfamethoxazole plus nystatin with gentamicin plus nystatin in the prevention of infections in acute leukemia. *N. Engl. J. Med.* **304,** 1057–1062 (1981).
19. Schimpf, S. C. Therapy of infection in patients with granulocytopenia. *Med. Clin. North Am.* **61,** 1101–1118 (1977).
20. Sotman, S. B., Schimpf, S. C., and Young, V. M. *Staphylococcus aureus* bacteremia in patients with acute leukemia. *Am. J. Med.* **69,** 814–818 (1980).
21. Keating, M. J., Bodey, G. P., Valdivieso, M., and Rodriquez, V. A randomized comparative trial of three aminoglycosides—comparison of continuous infusions of gentamicin amikacin and sisomicin combined with carbenicillin in the treatment of infections of neutropenic patients with malignancies. *Medicine (Baltimore)* **58,** 159–170 (1979).
22. Love, L. J., Schimpf, S. C., Schiffer, C. A., and Wiernik, P. H. Improved prognosis for granulocytopenic patients with gram-negative bacteremia. *Am. J. Med.* **68,** 643–648 (1980).
23. Wright, D. G., Rochichaud, K. P., Pizzo, P. A., and Drisseroth, A. B. Lethal pulmonary reaction associated with combined use of ampholericin B and leukocyte transfusions. *N. Engl. J. Med.* **304,** 1185–1189 (1981).
24. Pizzo, P. A., Robichaud, K. J., Gill, F. A., Witebesky, F. G., *et al.* Duration of empiric antibiotic therapy in granulocytopenic patients with cancer. *Am. J. Med.* **67,** 194–200 (1979).
25. Schiffer, C. A. Principle of granulocyte transfusion therapy. *Med. Clin. North Am.* **61,** 1119–1131 (1977).
26. Higby, D. J., and Burnett, D. Granulocyte transfusions: Current status. *Blood* **55,** 2–8 (1980).
27. Hershman, J., Meyers, J. D., Thomas, E. D., Buckner, C. D., and Clift, R. The effect of granulocytic transfusions of the incidence of cytomegalovirus infection after allogeneic marrow transplantation. *Ann. Intern. Med.* **96,** 149–152 (1982).
28. Hawk, J. C., and Koepke, J. A. Platelet transfusions. *Clin. Haematol.* **5,** 69–79 (1976).
29. Caydos, L., Freireich, E. J., and Mantel, N. Quantitative relation between platelet count and hemorrhage in patients with acute leukemia. *N. Engl. J. Med.* **266,** 905–909 (1962).
30. Higby, D. J., Cohen, E., and Holland, J. F. The prophylactic treatment of thrombocytopenia leukemic patients with platelets: A double blind study. *Transfusion (Philadelphia)* **14,** 440–446 (1974).
31. Aisner, J. Platelet transfusion therapy. *Med. Clin. North Am.* **61,** 1133–1145 (1977).
32. Thomas, G. E., and Ald, N. The diagnosis and management of accidental radiation injury. *J. Occup. Med.* **8,** 421–447 (1969).
33. Thomas, E. C., Storb, R., Clift, R. A., Fefeo, A., Johnson, F. L., Neiman, D. E., Garner, U. G., Glucksberg, H., and Buchner, C. D. Bone marrow transplantation I and II. *N. Engl. J. Med.* **292,** 832–843, 895–902 (1975).
34. Lorenz, E., Uphoff, D., Reid, T. R., *et al.* Modification of irradiation injury in mice and guinea pigs by bone marrow injections. *J. Natl. Cancer Inst. (U.S.)* **12,** 197–201 (1951).
35. Epstein, R. B., and Sarpel, S. C. Circulating hematopoietic stem cells. In *Biology of Bone Marrow Transplantation* (R. P. Gale and C. F. Fox, eds.), pp. 405–417. Academic Press, New York, 1980.
36. Zander, A. R., and Spitzer, G. In *Autologous Transplantation in Clinical Bone Marrow Transplantation* (U. G. Blame and L. D. Detz, eds.), pp. 331–358. Churchill-Livingtone, Edinburgh and London, 1983.
37. Gorin, N. C., Herzig, G., Ball, M. I., and Graw, R. G. Long term preservation of bone marrow and stem cell pools in dogs. *Blood* **51,** 257 (1978).
38. Wells, J. R., Ho, W. G., Gale, R. P., and Cline, M. J. Hematopoietic stem cell procurement separation and cryopreservation. In *Biology of Bone Marrow Transplantation* (R. P. Gale and C. F. Fox, eds.), pp. 27–137. Academic Press, New York, 1980.

39. Wells, J. R., Sullivan, A., and Cline, M. J. A technique for the separation and cryopreservation of myeloid stem cells from human bone marrow cryobiology. *Cryobiology* **16,** 201–210 (1979).
40. Balner, H. *Bone Marrow Transplantation and other Treatment after Radiation Injury.* Martinus Nijhoff, The Hague, 1977.
41. Epstein, R. B., Storb, R., and Ragde, H. Cytotoxic typing antisera for marrow grafting in littermate dogs. *Transplantation* **6,** 45–58 (1968).
42. Tigelaar, R. E., and Asofsky, R. Graft vs host reactivity of mouse thymocytes: Effect of cortisone pretreatment of donors. *J. Immunol.* **110,** 567–574 (1973).
43. Thomas, E. D., Clift, R. A., and Storb, R. Indications for marrow transplantation. *Annu. Rev. Med.* **35,** 1–9 (1984).
44. Deeg, H. J., and Storb, R. Graft versus host disease: Pathophysiological and clinical aspects. *Annu. Rev. Med.* **35,** 11–24 (1984).
45. Deeg, H. J., Storb, R., Weiden, D. L., Raff, R. F., Sale, G. E., *et al.* Cyclosporin A and methoxiexate in canine marrow transplantation: Engraftment, graft versus host disease and induction of tolerance. *Transplantation (Philadelphia)* **34,** 30–35 (1982).
46. Vallera, D., Ash, R., Zanzani, E., Neville, D., and Youle, R. Anti-T-cell reagents for human bone marrow transplantation: Ricin linked to three monoclonal antibodies. *Science* **222,** 512–515 (1983).
47. Renherz, E. L., Geha, R., Rappeport, J. M., Wilson, M., Penta, A. C., *et al.* Reconstitution after transplantation with T-lymphocyte depleted HLA hypro type-mismatched bone marrow for severe combined immunodeficiency. *Proc. Natl. Acad. Sci. U.S.A.* **79,** 6047–6051 (1982).
48. Sharp, T. G., Sachs, D. H., Fauci, A. S., Messerschmidt, G. L., and Rosenberg, S. A. T-cell depletion of human bone marrow using monoclonal antibody and complement-mediated lysis. *Transplantation (Philadelphia)* **35,** 112–120 (1983).
49. Reisner, Y., Kapoor, N., Kirkpatrick, D., Pollock, M. S., Cunningham-Randles, S., *et al.* Transplantation for severe combined immunodeficiency with HLA-A, B, D, DR incompatible parental marrow cells fractionated by soy bean agglutinin and sheep red blood cells. *Blood* **61,** 341–348 (1983).
50. DeWitte, T., Raymahers, R., Plas, A., Koekman, E., Wessels, H., and Haanen, C. Bone marrow repopulation capacity after transplantation of lymphocyte-depleted allogeneic bone marrow using counterflow centrifugation. *Transplantation (Philadelphia)* **37,** 151–155 (1984).
51. Strandqvist, M. Studien uber die Kumulative Wirkung der Rontgenstrahlen bei Fraktioniervung. *Acta Radiol.* **55,** Suppl. 1, 1–300 (1944).
52. Langham, W. H., ed. *Radiobiological factors in manned space flight,* Report of the Space Radiation Study Panel of the Life Sciences Committee, NRC Publ 1487. Space Science Board, National Academy of Sciences, National Research Council, Washington, D.C., 1967.
53. Casarett, G. W. *Radiation Histopathology,* Vol. II, pp. 37–50. CRC Press, Boca Raton, Florida, 1980.
54. White, D. C. *An Atlas of Radiation Histopathology,* TID 26676. U.S. Energy Research and Development Administration, Washington, D.C., 1975.
55. Ariyans and Krizek, T. J. Radiation effect, biological and surgical considerations. In *Reconstructive Plastic Surgery* (J. H. Converse, ed.). Saunders, Philadelphia, Pennsylvania, 1977.
56. Smith, S. M. Subjective experiences during a 32-year period after resurfacing of hands for severe and acute radiation burns. *Plast. Reconstr. Surg.* **51,** 23–26 (1973).
57. Schenck, R. R., and Bribirh, M. V. Four-extremity radiation necrosis. *Arch. Surg. (Chicago)* **100,** 729–734 (1970).
58. Stern, P. J. Surgical approaches to radiation injuries of the hand. In *The Medical Basis for Radiation Accident Preparedness* (K. F. Hubner and S. A. Fry, eds.). Elsevier/North-Holland, New York, 1980.

59. Cramer, L. M., Waite, J. H., Edcomb, J. H., Powell, C. C., *et al.* Burn following accidental exposure to high energy radiation. *Ann. Surg.* **149,** 286–293 (1959).
60. Voelz, G. L., ed. *Management of Persons Accidentally Contaminated with Radionuclides,* Rep. No. 65. National Council on Radiation Protection and Measurements, Washington, D.C., 1980.
61. Bestline, R. W., Watters, R. L., and Lebel, J. L. A study of translocation dynamics of plutonium and americium from simulated puncture wounds in beagle dogs. *Health Phys.* **22,** 829 (1972).
62. Jolly, L., McClearen, H. A., Poda, G. A., and Walke, W. P. Treatment and evaluation of a plutonium-238 nitrate contaminated puncture wound. *Health Phys.* **23,** 331–341 (1972).

CHAPTER 19

Medical Operations in Nuclear War

PETER H. MYERS*

Armed Forces Radiobiology Research Institute
Bethesda, Maryland 20814-5145

The use of nuclear weapons in war, or even their potential use, will cause military personnel to change their operations. Medical units on the battlefield are no exception. They will have to train and deploy in ways that will minimize the effects of nuclear weapon detonations on their personnel and equipment. The medical units must be prepared to deal with significant levels of attrition of personnel and equipment and be able to rapidly return to high levels of battlefield effectiveness. Finally, the medical units must remain capable of operating effectively in areas of the battlefield that have been affected by nuclear weapon detonations.

I. PLANNING

It is necessary to take certain steps before a nuclear weapon detonation in order to maintain a medical unit's effectiveness after the detonation. At least during times of increased tension, the medical unit should be located in protective shelters away from "nuclear-suitable" targets such as nuclear-capable units, command posts, communications centers, and special ammunition supply points. This location will decrease the likelihood of a nuclear weapon detonating closely enough to the medical unit to severely affect it. Although this remoteness will increase the transport distance for a patient, it will also enable the medical unit to survive and remain effective. The unit should be located in an area that will provide protection from the effects of a nuclear weapon detonation, such as caves or tunnels. Other terrain features that provide varying degrees of protection are ravines, ditches, culverts, and road overpasses. The protective shelters should be colocated with the medical unit's PMMNC (prepositioned medical material for

*Present address: Department of the Army, Office of the Surgeon General, Washington, D. C. 20310-2300.

nuclear casualties) package. This may mean that medical units will have at least two operational sites. During peacetime, the units might be located close to the units they support, but at the first indication of hostility (perhaps through intelligence sources), they should relocate to their preestablished protective shelters.

Basements of buildings can be considered likely positions for medical units, but the structural strength of a building must be determined before making that decision. Steel-reinforced-concrete buildings should be able to withstand blast effects from 100-kiloton (kt) weapons at distances greater than about 1.25 statute miles from ground zero. A below-ground location also provides increased protection from thermal and radiation effects. Occupants of basements would receive only about one-third of the initial nuclear radiation that unprotected persons would receive from a nominal-yield nuclear detonation. The same occupants would be completely protected from the flash effects of thermal radiation emitted from the detonation.

If the medical unit deploys under tentage with plans to remain there for some time, a sandbag wall should be established at the immediate perimeter of the tentage. The sandbags serve several purposes: they protect the tent and its occupants against initial nuclear weapon effects, and they provide protection against residual nuclear radiation if the position is affected by fallout. Accumulation of fallout on the tops of tents may be a relatively intense source of ionizing radiation exposure for the tent occupants. To avoid this hazard, tarpaulins (or tent liners) should be placed over the tops of the tents when the position is established. If fallout accumulates on the tent tops, the tarpaulins can be rolled off the tents, thus containing the fallout within the rolled tarpaulins. The tarpaulins can then be transported a safe distance away from the occupied area.

Medical unit personnel should receive comprehensive, factual training on nuclear weapon effects and, in particular, on the acute and chronic biomedical effects of nuclear radiation exposure. Such training should minimize the frequency and severity of adverse psychological responses to potential or actual nuclear weapon detonations. Information should be provided to medical unit personnel on the procedures to follow to enhance their individual survivability before, during, and after a nuclear detonation. An excellent source for this information is Field Circular 50-15, Nuclear Weapons Effects Mitigation Techniques, United States Army Combined Arms Combat Development Activity, Fort Leavenworth, Kansas 66027 (March 1984).

Alternative battlefield positions should be identified to which the medical unit can be easily relocated if the conditions at the existing position worsen (e.g., intense fallout arrives). Equipment and supplies should be stored in areas or configurations that provide protection from nuclear weapon effects. Items for treating patients (bandages, sheets, sterile water, etc.) should be stored under secure shelter to prevent them from becoming unusable by exposure to nuclear weapon effects.

Any medical treatment equipment that is electrically powered should be designed to resist electromagnetic pulse (EMP) effects (described in Chapter 21) if the equipment is critical or if it cannot be replaced in time to prevent the mission accomplishment from falling below a prescribed critical level. Besides this built-in protection against EMP, a number of things can be done in the field to help protect electrical equipment against EMP. For instance, nonessential electrical equipment should not be connected to an electrical circuit; cable and wire runs should be kept as short as possible and without loops.

Unexposed radiographic film should be stored in an area that provides shielding from penetrating ionizing radiation, primarily X radiations and gamma radiations. Otherwise, the medical unit's ability to provide effective treatment may be severely impaired by the absence of a radiographic capability. Battlefield medical units may have to place their unexposed film in a niche cut into the wall of a below-ground shelter. Ideally, the niche will be at least 3 ft below ground level, where it will remain dry.

Procedures for replacing personnel and supplies must be well conceived and practiced to perfection. Medical units should have on hand those items that will assist in preventing contamination by fallout and those items used for decontamination.

II. CASUALTIES

Nuclear battlefield modeling indicates that it is highly probable that a large number of soldiers will be affected by a nuclear weapon detonation in a relatively short time span. It is projected that 10–20% of United States/NATO ground forces will become casualties following a nuclear attack by the Warsaw Pact, even if the force were expecting the attack. In order for the force to regain its fighting effectiveness (so that the war will not be lost), it will have to reorganize and reconstitute. Because replacements from the continental United States cannot be expected for some time after the onset of aggression, the Medical Department will play a primary role in providing replacements to the tactical unit commander who is in the midst of attempting to reassemble an effective fighting force. Therefore, medical treatment will have to be conducted with orientation toward returning soldiers to duty, rather than retaining them in treatment facilities for the length of time preferred by practitioners during peacetime. Combatants who might be affected by this treatment orientation are those with relatively minor injuries: those who can help themselves (first aid), those who can care for themselves after initial medical treatment, and those cared for by untrained personnel (buddy aid). Some of these minor injuries might be minor lacerations (which may or may not require suturing), minor burns (less than 20% of body area affected by second-degree burns), or minor fractures (after splinting or casting). Exposures to ionizing radiation may also be included, when the individual has passed through the period of prodromal symptoms but not yet lapsed into the manifest phase of the radiation sickness syndrome.

Even with medical evacuation from the battlefield, the treatment resources probably will not be able to provide care for all the casualties. Therefore, two facets of mass casualty care must be actively practiced: first aid (buddy care) and triage. Medical personnel must ensure that combat units understand that the medical care they provide for themselves (buddy care) may be the only care they will receive for some time. The medical personnel must conduct comprehensive first-aid training and ensure that they have adequate stores of first-aid materials. For times when stores are inadequate, combat units must understand how materiel and equipment common to the battlefield can be adapted to emergency first-aid use.

Nuclear detonations release energy in three basic ways, which can cause casualties requiring medical treatment: blast energy, thermal energy, and ionizing radiation energy. (These phenomena are discussed in Chapters 2, 3, and 4.) Blast energy is manifest in two ways: as overpressure waves and as dynamic winds. The overpressure waves, passing through an emplacement, cause a pressure wave to be transmitted through the bodies of the emplacement's occupants. This pressure wave can be quite severe. Its greatest effect is on tissue junctions and organs that contain air spaces. The lungs and gastrointestinal tract are particularly vulnerable to injuries by overpressure. The resulting tissue disruptions can lead to air emboli or hemorrhage, either of which can be rapidly fatal. Injuries from overpressure may range from those that are relatively minor to those expected to result in fatality (1.0–3.5 atmospheres of overpressure). Patients with only overpressure injuries will comprise a very small part of the patient load, because it is likely that other more severe injuries will also be present.

Dynamic winds are responsible for the majority of injuries that occur by translation of a person with subsequent decelerative tumbling or by impact with a solid, unyielding object. Injuries also result from translation of debris and the subsequent impact of missiles with persons, producing either penetrating wounds or blunt trauma. The types of injuries to be expected by these means are concussion, fracture, laceration, and perforation. These injuries will occur farther from the detonation than will overpressure injuries, and they will place greater demands on battlefield medical units. Because dynamic injuries to the head and thorax have a high probability of immediate fatality, the blast injuries seen most often for treatment will be fractures, lacerations, and perforations of the extremities.

The thermal energy released by a nuclear weapon detonation can cause skin injuries or eye injuries. Skin burns can be caused by either direct exposure to the thermal impulse or exposure to the flames of an object ignited by the thermal impulse. Although thermal radiation is a relatively far-reaching effect (especially for high-yield weapons), the probability is relatively low that soldiers will be in the direct line of sight of a nuclear detonation and receive thermal impulses of sufficient intensity to cause second-degree burns. The probability is higher that soldiers will be burned by a flaming environment (forests, tents, etc.), but the variables of environmental flammability are too great to allow prediction of incidence or severity. But it can be said that burn injuries will add significantly to the patient load in treatment facilities that receive patients flowing from a nuclear battlefield. The complications and logistical requirements of providing treatment to soldiers with burn injuries could place great demands on battlefield medical units. Eye injury from thermal radiation will not be a significant problem to battlefield medical units (although other units may be significantly affected). Flashblindness is a very transient phenomenon, with afterimages lasting only 15–30 min. Retinal burns will be a relatively uncommon phenomenon, for which little if any medical treatment is needed.

Ionizing radiation exposures may be received as either acute (less than 24 hr) or chronic; single or multiple; internal or external; and as neutron, gamma, beta, or alpha doses. Persons who are relatively close to a nuclear detonation but shielded from blast and thermal radiation effects (in closed vehicles or shelters) will receive pure, single, acute, external neutron and gamma exposures. Persons operating in a fallout zone for longer than 24 hr will receive single, chronic, external gamma exposures. If bare skin comes into contact with fallout, beta and gamma exposures (from fission fragments) will be received. If fallout particles enter the body, internal exposure will result. Internal exposure can be considered medically unimportant since the probability of significant internal exposure is relatively low. This is because soldiers can be trained to avoid internal contamination, gastric biological elimination of contaminants works in favor of the soldier, and traditional wound debridement and lavage eliminate much of the contamination that may have entered wounds. (Fallout is discussed in more detail in Chapter 3.)

Estimates of the incidence, severity, time of onset, and duration of radiation sickness symptoms as functions of a single acute gamma dose and time postexposure are provided in Tables 9-1–9-8 (in Chapter 9). The $LD_{50/60}$ under these exposure conditions would be approximately 425 cGy (free in air). If the exposure conditions were the same except that the exposure is a protracted one (more than 24 hr), then the percentage of a population exhibiting the specified symptoms would be less, the specified symptoms would be less severe, the symptoms would occur at later times, and the duration of the symptoms would be shorter. Also, the $LD_{50/60}$ would be greater. However, if the radiation exposure is accompanied by some other trauma, the chances for an individual's survival will decline. Evidence has been collected to suggest that, in combined-injury cases, radiation exposures above 300 cGy (free in air) markedly affect the potential for survival, and may shift the casualties into the expectant category.

III. COMBINED INJURY

Analyses of the expected battlefield situations have resulted in the belief that a high proportion of the casualties presenting to battlefield medical treatment facilities will be suffering combined injuries. Further, most of those suffering combined injuries will have received burns along with exposures to ionizing radiation. Unfortunately, the survival of these patients is adversely affected by complications brought on by the treatment. Burn injuries offer multiple portals of entry for infection. Both burns and exposure to ionizing radiation decrease one's immunity to infectious microorganisms. Previous chapters discuss the existing experimental evidence for the increased lethality that is expected in combined-injury cases.

IV. PATIENT MANAGEMENT

Even though the prospect of survival declines for combined-injury cases, patient management should ensure that the conventional injuries are not neglected, if the patient load is manageable. The issue becomes less clear when a mass-casualty situation requires triage, because the triage system is confused by the issue of dosimetry. Currently, not every soldier carries a dosimeter. Two dosimeters (IM-93s or IM-185s) are carried per platoon (40 soldiers), and an average exposure is assigned to the group based on the readings of the two dosimeters. Under this system, it is difficult to be confident that the average exposure even approximates the exposure received by any one member of the platoon. The degree of this uncertainty often would prohibit individual medical officers from using the platoon's average exposure as an aid in conducting triage. Also, the IM-93 tactical dosimeter does not accurately respond to the prompt ionizing radiation that is emitted directly from a nuclear weapon detonation. It is hoped that in the near future, much of the difficulty of determining the ionizing radiation dose received by an individual will be reduced. By 1986, the United States Army hopes to procure and issue to each soldier on future battlefields a dosimeter (DT-236) that will be responsive to prompt neutron and gamma radiations as well as fallout radiations. This will assist the triage officer in identifying those patients who might have a better chance of responding to treatment, based on the extent of their radiation injury.

Some uncertainty will still exist regarding the accuracy of the dosimeter (e.g., orientation of the dosimeter to the source), and this uncertainty requires that primary attention be paid to conventional injuries. However, medical treatment personnel may use another mechanism in an effort to measure the ionizing radiation exposure of an individual; that mechanism is noting the development of the patient's symptoms. Table 12-1 is a suggested scheme for assessing in a gross manner the severity of an exposure received by a patient (previously described in Chapter 12). The use of such a scheme must be used with caution, because some of the symptoms listed can be psychosomatic. The nuclear battelfield will be a highly stressful environment, and some stress-related symptoms may be commonplace. Nausea, vomiting, and diarrhea may occur in soldiers suffering stress reactions, which may cloud the identification of soldiers who received only mild exposures. But it is important that the symptoms of soldiers who received moderate-to-severe exposures are not confused with stress-reaction symptoms. Finally,

TABLE 19-1

SCHEME FOR TRIAGE[a]

Serial	Starting priority	Final priority		
		150 cGy	150–450 cGy	450 cGy
1	Radiation only	T3	T2	T4
2	T1	T1	T1	T4
3	T2	T2	T4	T4
4	T3	T3	T4	T4
5	T4	T4	T4	T4

[a]T1 (Immediate treatment group) includes those requiring immediate life-saving surgery. Procedures should not be time consuming and should concern only those with a high chance of survival (e.g., respiratory obstruction, accessible hemorrhage, and emergency amputations). T2 (Delayed treatment group) includes those needing major surgery, but where conditions permit delay without unduly endangering life. Sustaining treatment, such as intravenous fluids, antibiotics, splinting, catheterization, and relief of pain, may be required in this group (e.g., fractured limbs, spinal injuries, and uncomplicated burns). T3 (Minimal treatment group) includes those with relatively minor injuries who can effectively care for themselves or be helped by untrained personnel (e.g., minor lacerations and fractures). T4 (Expectant treatment group) includes those with serious or multiple injuries requiring intensive treatment or with a poor chance of survival (e.g., severe head and spinal injuries, widespread burns, or large dosages of radiation). These patients receive appropriate supportive treatment compatible with resources, which includes large doses of analgesics if available.

since the development of some of these symptoms requires the passage of time, this mechanism (i.e., noting the development of a patient's symptoms) may not be usable in making decisions about immediate treatment. If greater credence can be placed on dosimeter readings (perhaps once a dosimeter is provided to each individual), then schemes devised to use ionizing radiation doses assigned to individuals could assist in conducting second-stage triage (first-stage triage being based on conventional injuries alone). A scheme that has been reviewed and adopted for use by the NATO signatorees appears in Table 19-1.

V. ADVICE TO THE TACTICAL COMMANDER

The tactical commander will be looking to his medical staff officers for advice on a particularly critical issue: the capability of his force to fight. The radiation exposure status (RES) category system (see Table 19-2) is a mechanism for helping the tactical commander maintain the fighting capability of his force despite the exposure of his troops to ionizing radiation.

If a commander's force has accumulated sufficient ionizing radiation exposure to place the personnel in a particular RES category, then the commander must restrict his troops to involvements in which additional radiation exposure is not expected, unless he is willing to accept the next higher risk to his command. With the understanding that biological recovery does take place, the question asked of the command physician is, When can a combat unit's RES category be downgraded? That is, when has the ionizing radiation injury been repaired to the extent that the commander can reasonably risk the presence of his troops in an area in which they might receive additional radiation exposure? This is not a trivial question. If the command physician underestimates the unit's recovery, he places an unnecessary constraint on the tactical unit commander in that he may not employ his unit in an operation in which it may accumulate any additional

TABLE 19-2

RADIATION EXPOSURE STATUS (RES) CATEGORY SYSTEM

RES	Dose (cGy)	Casualties (percentage)	Nuisance symptoms (percentage)	Risk
0	0	0	0	None
1	0–70	1	2.5	Negligible
2	70–150	2.5	5	Moderate
3	150–	5	Undesignated	Severe

radiation exposure that might exceed the specified acceptable risk. If the command physician overestimates the unit's recovery, the tactical commander might then plan a critical mission into an area affected by fallout, depending on his forces being able to maintain a certain degree of fighting effectiveness even while accumulating some amount of additional radiation exposure. If the biological damage that resulted from the initial exposure had not been repaired to the degree expected by the command physician, the subsequent exposure might result in an amplified, deleterious biological response from the unit. That amplified response could very well result in the fighting effectiveness of the force being degraded to below that required for the successful outcome of the mission. Too little is known about recovery from ionizing radiation injury to permit a straightforward prediction of what an individual's or a combat unit's "effective remaining dose" might be at some time following a particular ionizing radiation exposure. It is expected that the supporting medical unit will test exposed personnel for depressed biological activity, which may indicate a lower magnitude of initial radiation damage. Tables 9-1–9-8 (in Chapter 9) may be of some use when making this determination.

Also, a rule of thumb may be applied to those who have received ionizing radiation exposures to determine their residual biological damage. The rule of thumb indicates that 90% of the total biological damage done by the ionizing radiation exposure will be repaired within 30 days after the exposure. This rule is based on the premise that 10% of the *remaining* damage, which resulted from a single exposure, will be repaired per day. For instance, if an individual had received a 100-cGy exposure on day 0, after 1 day had passed (or on day 1), 10% of the biological damage done by the 100-cGy exposure would have been repaired, resulting in a *remaining* effective dose of 90 cGy on day 1. On day 2, 10% of the remaining damage would be repaired (10% of 90 cGy is 9 cGy recovery), resulting in a *remaining* effective dose of 81 cGy (90 cGy − 9 cGy = 81 cGy). This sort of rule results in an exponentially declining function which, if taken through a 30-day period, arrives at a prediction of 90% recovery within that period. Some have criticized this estimation of 10% recovery of the residual damage per day as being optimistic. They maintain that the correct estimation should be closer to 2.5% recovery of the residual damage per day. If 2.5% recovery per day of the residual damage is taken through a 30-day period, the prediction of biological recovery would be closer to 50% recovery from the total biological damage within that period.

A scheme has been proposed by the British for determining a remaining effective dose at some time after an initial exposure or even multiple exposures, although it has not yet been ratified by the United States. In the scheme, the operational exposure dose (OED) is determined by

$$\text{OED} = \text{AAD} - 200 - 15t \text{ (cGy, free in air)}$$

where AAD is the accumulated air dose and t is the number of days postexposure. This equation is being proposed for use in situations in which the accumulated exposures are low-dose-rate, survivable, protracted, gamma-only exposures. Also, in applying this formula to units having received multiple exposures, the 200-cGy figures may be subtracted only once. There are at least two difficulties associated with using this formula in deriving residual effective dose:

(1) The proponents of the formula agree that the formula can be used only for low-dose-rate, gamma-only exposures. On nuclear battlefields, combatants could certainly receive moderate, *prompt* ionizing radiation exposures from the detonation of nuclear weapons. With existing dosimeters (IM-93 and IM-185), these exposures cannot be discerned from fallout exposures, and so the formula could not and should not be used.

(2) Even if there were a way to ensure that the only readings received on a dosimeter were from low-dose-rate, gamma-only exposures, the formula has another fault: it is based on the analysis of information about human-exposure lethality. That being the case, if one were to apply the formula to determine the remaining effective dose for a group of combatants in order to return them to a fight in which they would receive an additional ionizing radiation exposure, then the use of the formula would guarantee that the combatants would not die from the additional exposure. But it would not guarantee that the combatants would not suffer radiation sickness symptoms, which could very well degrade their combat performance to the extent that the battle would be lost.

The concept is a valuable one; there is no argument that this sort of tool is badly needed and is one that would demand great use by the Medical Department. An attempt is being made to derive a similar formula based on the radiation sickness symptoms that would result from either prompt or protracted exposures.

If exposures are able to be maintained below 150 cGy, the overall effectiveness of the combat unit will not be significantly degraded. However, if the exposures become large (as may happen if an aggressor uses nuclear weapons), then the tactical commander must be advised of his force's capability to continue the fight. Figures 19-1 and 19-2 provide an estimate of the combat effectiveness of combat units as functions of acute dose and time postexposure. These figures have been developed from subhuman primate studies at the Armed Forces Radiobiology Research Institute (for times less than 60 min postexposure) and from an assessment of how symptoms of radiation sickness will affect the performance of combat tasks (for times longer than 60 min postexposure). The prediction associated with persons identified as being "combat effective" is that they will be suffering radiation sickness symptoms of such a nature that they will be able to maintain at least 75% of their preexposure performance level. Those predicted

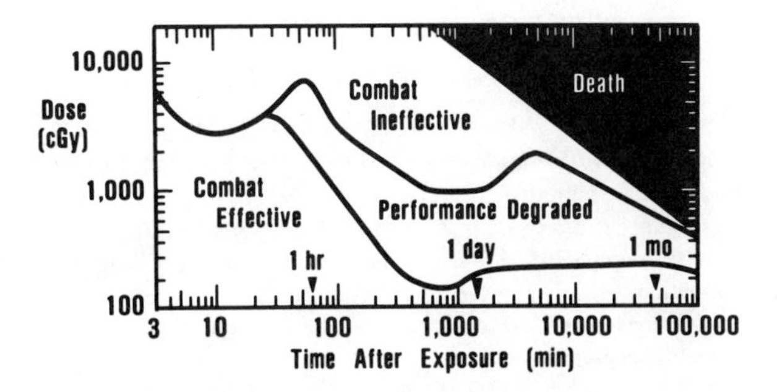

FIG. 19-1 Expected response to radiation for physically demanding tasks.

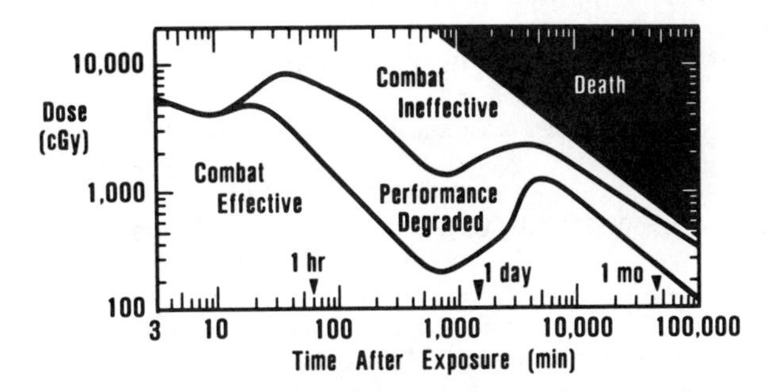

FIG. 19-2 Expected response to radiation for physically undemanding tasks.

as being "performance degraded" could be operating at a performance level between 25 and 75% of their preexposure performance. Those predicted as being "combat ineffective" should be considered as performing their tasks at 25% (at best) of their preexposure performance level. Of course, these predictions are based on combatants suffering from only one stressor, that being ionizing radiation exposures. The Medical Department has always had the responsibility of providing the tactical commander with assessment of the performance capabilities of his command. The command physician has had to assess a multitude of stressors that might affect the fighting effectiveness of the force: malnutrition, continuous duty, endemic disease, fatigue, sleeplessness, general poor health, etc. It is clear that ionizing radiation exposures will now have to be considered together with how these other stressors might affect the total performance capability of the force.

Another facet of nuclear warfare that has the potential of impacting the performance capacity of the force greater than any other is the psychological response of the force required to participate in nuclear warfare. This issue is covered in detail in Chapter 15.

SUGGESTED READINGS

Baum, S. J., Anno, G. H., Young, R. W., and Withers, H. R. *Symptomatology of the Acute-radiation Effects in Humans for Doses from 0.5 to 3.0 Gy,* PSR Note 581. Pacific-Sierra Research Corporation, Los Angeles, California, 1983.

Kaman Tempo. *Symposium on Biological Effects of Nuclear Weapons, Proceedings—Technology Liaison Group 3,* Spec. Rep. 198. Kaman Tempo, Santa Barbara, California, 1982 (Foreign Government Information).

North Atlantic Treaty Organization. *Medical Effects of Ionizing Radiation on Personnel,* STANAG 2866. NATO, Brussels, Belgium, 1976.

North Atlantic Treaty Organization. *NATO Handbook for the Medical Aspects of NBC Defensive Operations,* AMedP-6. NATO, Brussels, Belgium, 1976.

North Atlantic Treaty Organization. *NATO Handbook on the Concept of Medical Support in NBC Environments,* AMedP-7 (A). NATO, Brussels, Belgium, 1978.

North Atlantic Treaty Organization. *Commander's Guide on Nuclear Radiation Exposure of Groups,* STANAG 2083 (NBC), 4th ed. NATO, Brussels, Belgium, 1981.

U.S. Army Combat Developments Command. *Personnel Risk and Casualty Criteria for Nuclear Weapons Effects (U).* Institute of Nuclear Studies, USACDC, Fort Bliss, Texas, 1971 (Confidential).

U.S. Army Combined Arms Combat Development Activity. *Nuclear Weapons Effects Mitigation Techniques,* Field Circ. 50-15. USACACDA, Fort Leavenworth, Kansas, 1984.

U.S. Army Nuclear Agency. Addendum to *Personnel Risk and Casualty Criteria for Nuclear Weapons Effects.* USANA, Fort Bliss, Texas, 1976.

U.S. Army Nuclear and Chemical Agency. *Combined Injury from Nuclear Weapons Effects,* Jt. DNA-USANCA Study. USANCA, Springfield, Virginia, 1979.

U.S. Department of the Army. *Nuclear Handbook for Medical Service Personnel,* TM 8-215. USDA, Washington, D.C., 1969.

U.S. Department of the Army. *The Effects of Nuclear Weapons,* DA Pam 50-3. Headquarters, USDA, Washington, D.C., 1977.

U.S. Department of the Army. *Staff Officers' Field Manual, Nuclear Weapons Employment Doctrine and Procedures,* FM 101-31-1. USDA, Washington, D.C., 1977.

Vineberg, R. *Human Factors in Tactical Nuclear Combat,* Tech. Rep. 65-2. Human Resources Office, George Washington University, Washington, D.C., 1965.

CHAPTER 20

Low-Level Effects

ROBERT T. DEVINE* AND RAYMOND L. CHAPUT**

Armed Forces Radiobiology Research Institute
Bethesda, Maryland 20814-5145

The acute effects of radiation on humans at dose levels above 100 rads of low-LET (linear energy transfer) radiation are reasonably well understood in terms of classical radiobiology (1,2). The acute effects of high-LET radiation are not as well known with respect to dose, but they are qualitatively the same as low-LET radiation. The late effects of radiation are common to both high and low doses. These somatic effects occur in an irradiated individual, whereas genetic effects occur in the descendants of an exposed individual. The somatic effects are cataracts, changes in growth and development, nonspecific life shortening, and cancer. All but nonspecific life shortening have been shown to occur in humans. The genetic effects of radiation are manifest by increased frequency of mutation, but they have not yet been demonstrated in the descendants of persons exposed to the high doses at Hiroshima and Nagasaki. The quantitative estimates for human genetic injury are extrapolated from animal models.

The quantitative relationship between dose and effect for low doses and low dose rates is inferred by extrapolation from data on high dose rate. The extrapolations used in providing protection criteria for occupational exposure represent the most conservative assumptions rather than the application of well-understood models of effects. Estimates of effects of low-LET radiation are the result of consensus (3–5). The method of extension of high-dose data to low-dose effects of high-LET radiation is the subject of controversy. It is questionable to use any of the data for assigning causation of cancer to low-level radiation, rather than to any other of the countless suspected carcinogens to which the population is exposed.

*Present address: Science and Technology Biomedical Effects Directorate, Defense Nuclear Agency, Washington, D.C. 20305.

**Present address: Biomedical and CBR Defense Exploratory Technology, Office of Naval Technology, Washington, D.C. 22217-5000.

I. LATE EFFECTS OF RADIATION

The following summarizes information in a report of the Committee on the Biological Effects of Ionizing Radiation (1). For many years, it has been known that large doses of radiation increase the probability of late effects, such as lens opacification, shortening of life span, leukemia, and other types of cancer (neoplasms). Irradiation does not appear to produce any uniquely new disease; all observed late effects are those normally seen in an aging population. Thus, irradiation increases one's chance of developing a disease that the nonirradiated person already has a definite probability of developing.

The probability of late effects of radiation markedly differs from the probability of early effects. If an individual is exposed to a large dose of radiation, it is certain that he will show severe cell depletion in the bone marrow and peripheral blood, and he may die if the dose is high enough. However, the same large dose will increase only the *probability* of his developing late effects. Thus no precise predictions are possible with respect to late effects in a given individual, and one can deal only in terms of probability. The principal concern following exposure at low doses and low dose rates is a possible increase in the incidence of cancer.

The probability of late effects increases with dose, dose rate, and LET. These observations are similar to those for acute effects. Late effects have a very long latent period. The time between the irradiation and the appearance of an effect may be years.

A. Somatic Effects

1. Cataracts

The term "cataract" implies a vision-impairing opacity of the normally transparent lens of the eye, varying from small lesions to extensive opacification that results in total blindness. It is well known that cataracts may be caused by exposure of the lens to ionizing radiations, such as X rays, gamma rays, beta particles, and neutrons. Lens opacification has been observed in patients exposed therapeutically to X radiations and gamma radiations, in the survivors at Hiroshima who were within 1000 m of the center of the explosion, and in physicists accidentally exposed to neutrons during the operation of cyclotrons.

In many cases, radiation-induced cataracts can be differentiated from cataracts due to other causes. The cataract due to ionizing radiation is characterized initially by a dot in the posterior pole of the lens. As it enlarges, small granules and vacuoles appear around it. As it continues to grow, the cataract develops a relatively clear center, shaped much like a doughnut.

For humans, the threshold dose for cataractogenesis appears to be several hundred rads of acute-dose, low-LET radiation. Lower doses may produce some damage, but it is not clinically significant. For doses of 250–650 rads, the latent period is about 8 years; at doses of 650–1150 rads, it is reduced to 4 years. Additional increases in dose result in shorter latent periods.

While information appears to indicate the existence of a "practical threshold" for the induction of vision-impairing lens opacification in man by low-LET radiation, data on humans do not allow the assumption of a practical threshold for neutron irradiations. However, animal studies have revealed a relative biological effectiveness (RBE) for fast neutrons of about 10 at high doses of hundreds of rads to above 50 or more at doses of fractions of rads. This increase in RBE for neutrons is due to the relative ineffectiveness of gamma rays or X rays at low doses in producing cataracts.

2. Growth and Development

At exposure levels in the tens of rads or more, serious effects on growth and development have been demonstrated in man. Japanese children exposed to atomic bomb radiations showed slight retardation of growth and maturation, as did the Marshallese children who had been exposed to fallout radiation. The massive irradiation of the growth centers in bone inhibits growth and shortens bones, especially in young animals and children.

It is generally conceded that effects on growth and development are not a concern in adults after exposure to radiation. This is not true for the prenatal child. As many as 24 types of major abnormalities have been reported among children irradiated with relatively large doses *in utero*. Injuries to the central nervous system, the eye, and skeletal development were the most frequent. Mental retardation and reduced head size have been seen in Japanese children exposed during the first few months of embryonic life to doses in excess of 50–100 rads, and the data suggest that these effects may occasionally occur as the result of lower doses. Work with animals has shown that changes could occur after doses as low as 25 rads.

3. Nonspecific Life Shortening

Although a prominent feature of radiation injury is the ability of cells, tissues, and organs to recover from sublethal injury, changes induced at the cellular or higher level may lead to late effects. Irreversible injury, which does not produce a recognizable specific disease, might be expected to decrease the vitality or reserve capacity of these cells and organs in which it occurs. Presumably, this loss of vitality is somewhat similar to the loss in normal aging, or it may add to the degenerative changes of aging that determine the length of life. Thus, following sufficient exposure to radiation, a recognizable shortening of life span may occur, either in association with increased or earlier incidence of specific disease, or from more general causes that are similar (at least superficially) to premature aging.

Evidence from recent studies of human and animal populations indicates that life shortening can be totally accounted for by the increased numbers of cancers in the low-to-medium dose range. At lower doses, acute effects are minimal.

4. Dose–Effect Relationships for Cancer Induction

Cancers in animals and humans after irradiation are well documented. However, it is still not possible to accurately deduce from animal studies the frequency of cancer in man at low doses.

At high doses the early somatic effects appear, and definitive statements can be made about the relationships of dose and effect. The late effects are only a probability; they may never occur. If a late effect does occur, it may not be the result of treatment, since all late effects occur "naturally." Consequently, in studying the late effects of ionizing radiation, a direct approach is often not feasible because of the extremely low incidence of somatic effects at the very low doses of interest.

In making toxicological evaluations, it may be necessary to deal with dose–effect relationships in which groups of persons have been or are exposed to different doses of a toxic agent. Then a determination can be made of the percentage of exposed individuals who respond by developing a given effect over a stated period of time. From such studies, response curves are obtained.

The classical "S-shaped" dose–effect curve is well known in pharmacology

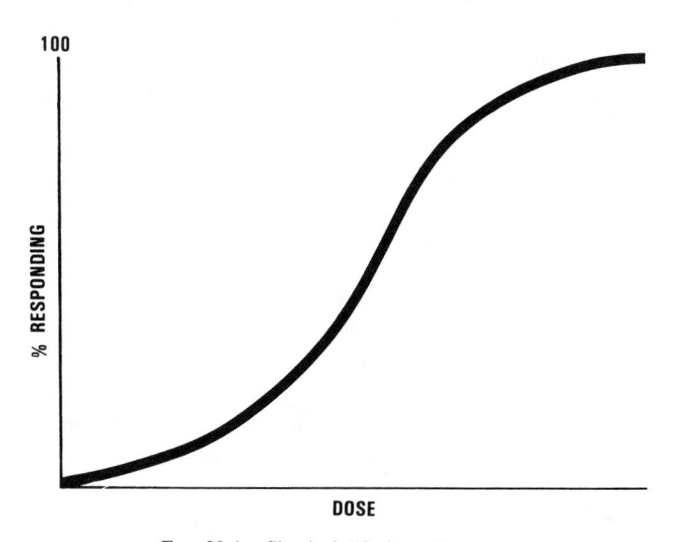

FIG. 20-1 Classical "S-shaped" curve.

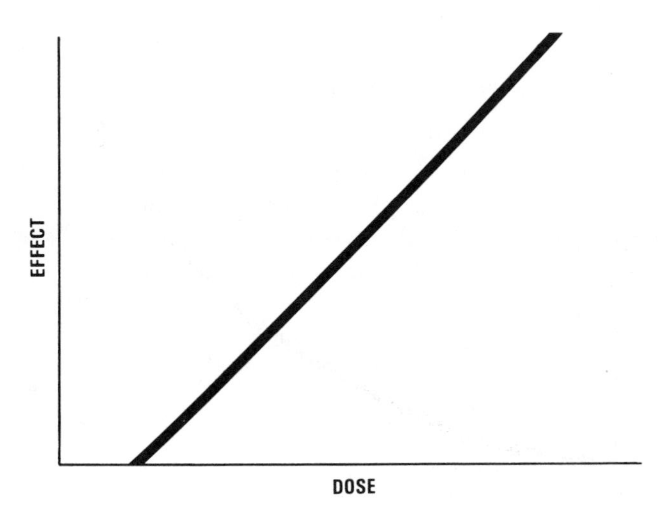

FIG. 20-2 Linear–threshold curve.

and toxicology (Fig. 20-1). Such a curve might be obtained by giving different groups of persons (e.g., 100 per group) increasing numbers of aspirin pills and determining the percentage in each group showing a particular effect. If each individual in a group is given one pill, probably one will show the effect. As dose increases, the percentage showing an effect will also increase. Eventually, at very high doses, "all" in a group of 100 would show the effect. The S-shaped or sigmoidol dose–effect curve is obtained from plotting the percentage of each group that manifests the effect versus the dose received by each individual in that group. The shape of the lower part of this S-shaped curve is not well established; however, it has been commonly assumed that there is a threshold below which no individual would respond, even if the group tested was much larger than the 100 in each of our hypothetical groups.

A linear–threshold curve (Fig. 20-2) is a variation of the S-shaped curve. In such a response pattern, the effect per unit dose is constant from the threshold dose. This type of response has been observed with toxic agents.

The nonlinear–nonthreshold curve (Fig. 20-3) is similar to the nonlinear curve described above except that a threshold dose does not appear to exist. Based on recent data, the incidence of cancer after irradiation appears to follow this pattern. At low doses, cancer incidence is proportional to dose; at higher doses, it is proportional to dose squared. This curve thereby reflects a quadratic relationship between cancer incidence and radiation. Although cancer incidence at low doses is extremely small, there is no threshold dose. The response of many biological systems follows this form, and the curve has been recommended by two national and international committees attempting to resolve the problem associated with

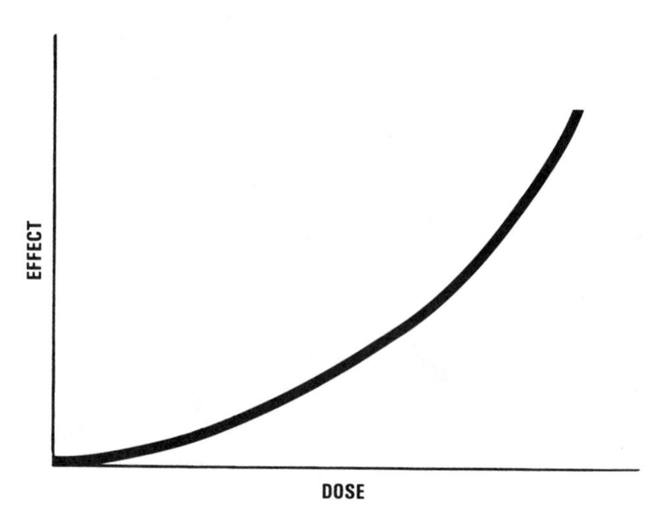

FIG. 20-3 Nonlinear-nonthreshold curve.

cancer induction at low radiation doses. They are the Committee on Biological Effects of Ionizing Radiation (BEIR) of the United States National Academy of Sciences and the United Nations Scientific Committee on the Effects of Atomic Radiation (UNSCEAR).

Since great uncertainties exist in the human and animal data obtained at low radiation doses, the BEIR and UNSCEAR committees resort to the linear–nonthreshold curve (Fig. 20-4) to obtain estimates of cancer incidence at doses close to or at background levels. If such a curve applies, there is of course no "safe" dose below which no effect on the population can be presumed; i.e., there is a chance, however small, that some exposed person will show the effect. This interpretation, applied to the aspirin example, leads to the expectation that in a very large population given even a very small fraction of one pill, a few individuals would show the particular effect. If a harmful or toxic agent is everywhere in the environment and exposes a very large population (such as that of the United States), then even a very small percentage effect would mean a large absolute number of affected persons. (Consider that the number of Americans killed per year in auto accidents is approximately 0.025%. Yet this percentage of 200 million Americans is the approximately 50,000 killed per year.) Hence, by using the linear curve, extrapolation results in a conservative estimate and will result in an overestimate of the risk from cancer.

The types of cancer now generally considered to increase in incidence following exposure at high doses and high dose rates are myelocytic leukemia, cancer of the female breast, lung cancer, thyroid cancer, cancer of the bowel, and possibly others. The data are sparse but are sufficiently quantitative to allow rather

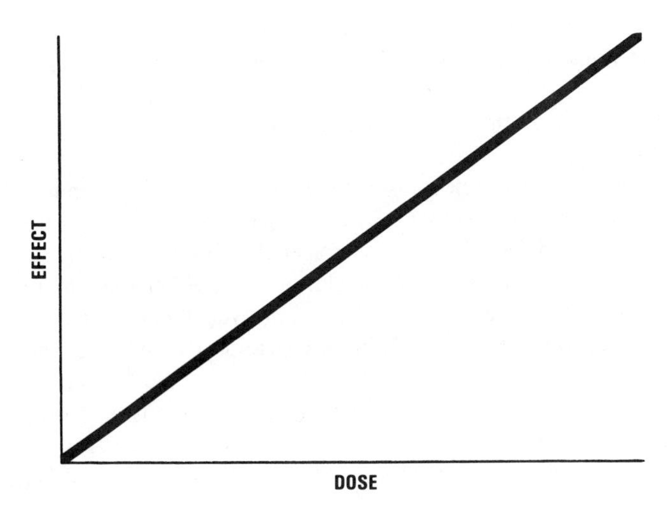

FIG. 20-4 Linear–nonthreshold curve.

accurate predictions of what will happen in an irradiated population exposed to high doses at high dose rates. For instance, it is possible to say that, in a large population exposed to 100 rads, the incidence per year of these different types of cancers will increase over the normal incidence by some calculated percentage.

At the level of a millirem to a few rems per year, no clear-cut direct observations can be made regarding the incidence of cancer. No documented cases of leukemia or cancer in the adult human can be shown definitely to have resulted from exposure at these low doses and low dose rates. This does not mean that the incidence did not increase or that such an increase is not important. The main reason for the lack of data is that the incidence is low and therefore, if it exists, it is extremely difficult to demonstrate.

Some of the most complete data on the induction of cancer in humans by external radiation concern the increased incidence of myelocytic leukemia in the persons exposed in Hiroshima and Nagasaki.

The risk estimate obtained from these data for radiation-induced leukemia is 46 deaths per 1 million people exposed per rem. The assumption of a linear relationship between dose and incidence would make the number of leukemias the same, whether 10,000 persons were exposed to 100 rads each or 1 million people to 1 rad each.

The human experience with radiation-induced cancers may be summarized as follows:

(1) Skin cancer was observed in physicists, radiologists, and dermatologists who used radiation in early years without regard to safety.

(2) Lung cancer was observed in uranium miners.

(3) Bone tumors were observed in workers who painted clock dials with radium.

(4) Liver cancer occurred in patients who received Thorotrast as a contrasting medium.

(5) Leukemia and other neoplasms were observed among atomic bomb survivors at Hiroshima and Nagasaki. These cancers were also observed among patients who received radiation therapy for ankylosing spondylitis.

(6) Thyroid cancer was observed in children whose enlarged thymuses were treated with radiation. It also was observed in the Marshallese exposed to fallout from nuclear weapon tests and in children whose heads were irradiated to treat tinea capitis infection.

(7) Breast cancer has been observed in patients irradiated for postpartum mastitis.

(8) Leukemia was observed among children prenatally exposed to radiation.

The latent period varies for the different types of cancers, ranging from as low as 5 years for leukemia to more than 27 years for cancers of the pharynx and larynx. A great degree of variability also exists for a specific cancer. For Hiroshima and Nagasaki, leukemias appeared 5–20 years after irradiation, with a mean interval of 14 years. For all other cancers, the median latency was about 25 years.

Examples of Cancers after Irradiation. Leukemia is discussed above for Hiroshima and Nagasaki. If medical data are included, the risk for radiation-induced leukemia becomes 1–3 excess leukemias/10^6 person–years/rad.

The thyroid is an organ of high sensitivity for radiation carcinogenesis. However, the tumors develop slowly and are histologically well differentiated, so they can be removed by surgery or treated by iodine-131. As a result, mortality is low. Incidence is higher in females and in persons under 20 years of age. Risk estimate is 4 cases/10^6 person-years/rad. Thyroid cancers appear within 25 years of irradiation.

Breast cancer has been frequently observed after irradiation. Cases appear 10–30 years after irradiation, and women 30 years old or younger seem to be most susceptible. For atomic bomb survivors who had been exposed to greater than 1 rad to the breast tissue, the absolute risk of cancer per year at 10 years after exposure or after age 30 (whichever occurs later) is about 5 cases/ 10^6 person-years/rad. For women exposed in mastitis examinations, the results are similar in magnitude (where the statistics are good).

Bone cancer is rare in the human, but has been observed in patients and radium-dial painters who received very high doses of radiation. Estimates of risk are very difficult to assess because of the long-term internal exposures associated with these cancers. The best estimate for alpha radiation is 1 bone sarcoma/10^6 person–years/rad alpha dose. In general, the major hazard is from bone-seeking radioisotopes, which concentrate in growing areas of bone and give rise to a large local dose from alpha particles. This risk for low-LET radiation is 20 times lower than for alpha.

Skin cancer has been seen in radiologists and other medical workers during the early years of radiation use when safety standards did not exist. The risk of skin cancer with dose is not known except that a dose exceeding 1000 rads is required. The absence of skin cancer among the atomic bomb survivors supports the view that skin has a low susceptibility for cancer after exposure to radiation.

Lung cancer is an occupational hazard among underground miners exposed to radon gas. Radon is one of many isotopes formed when rocks in the earth decay to stable lead. Radon decays to a solid that is deposited in the lung, and the subsequent decays that occur will result in alpha irradiation to that organ. Lung cancers have also been observed in patients with ankylosing spondylitis who received 400 rads to the lungs during therapy. The latent period appears to be about 40 years. The data for persons exposed to low-LET radiation yields 3 excess cases/10^6 person–years/rad.

Other tumors observed after irradiation include those in the brain, pharynx, stomach, uterus, liver, pelvic organs (such as cervix, ovary, and rectum), salivary glands, and others. Radiation is considered to be potentially carcinogenic for all tissues, but apparently tumors will develop only under specific conditions of irradiation. Certain tissues are definitely more susceptible to cancer than others; examples are the thyroid, bone marrow, breast, and lung.

Prenatally exposed children may have a risk of malignancy during the first 10 years of life, which is a factor of 2 over the risk of similarly irradiated adults.

Total cancer risk is estimated at 70–165 deaths/10^6 exposed persons/rem, depending on the risk projection model. Using these values and assuming a background radiation dose in the United States of 0.1 rem per year, it can be estimated that, in the United States population of about 200 million, 1000–3000 deaths occur each year from background radiation.

B. Genetic Effects of Radiation

Radiation does not produce new and unique genetic effects, but it increases the frequency of harmful traits as a result of gene mutation, chromosome aberration, or chromosome breakage. Gene mutations involve a change in DNA structure: in base composition, in sequence, or both. Normal incidence is about 1% of persons born with an appreciable handicap, including sickle-cell anemia, polydactyly, and Huntington's chorea. Chromosome aberrations occur when a cell contains too few cells or too many cells, exemplified by Down's syndrome. Chromosome breakage results in various recombinations. Such changes are rare naturally, but they are the most frequently observed radiation-induced effect on chromosomes. (These effects were discussed earlier.)

Radiation-produced effects were first observed with the fruit fly *Drosophila*, in which the number of lethal recessive traits produced was linearly related to a dose in the range of 25 rads to thousands of rads. From these data, it was assumed that a linear–nonthreshold type of response occurs for genetic effects, and this was dogma for many years. It was believed that genetic effects were independent of dose and dose rate, and that they were more serious than carcinogenic effects. More recent data have shown that different mutations have different rates of induction. Furthermore, a significant dose–rate effect does exist. In males (who are more radiosensitive than females), repair processes appear to greatly reduce mutations at 0.8–90 rads/min. Below 0.8 rads/min, no further reduction is observed.

Also, the genetic consequences of a radiation dose can be greatly reduced if a time interval is allowed between irradiation and conception. To reduce risks to insignificant levels for man, this period is estimated to be 6 months. The average rate of induction of mutations is about 0.25×10^{-7} per locus per rad. Based on the mouse data, a doubling dose (discussed below) is calculated for man at 20–200 rem.

Estimating Genetic Hazard of Radiation Exposure in Man

The general principles for estimating hazard are that (1) most mutations are harmful, whether spontaneous or induced by radiation; (2) any dose of radiation entails some risk; (3) the number of mutations is linearly proportional to dose; and (4) genetic effects are independent of dose rate.

To gain an appreciation of what genetic mutations and an increase in the rate of mutation would mean in a human population, it is necessary to consider the normal "genetic load" or number of "genetic defectives" in human populations. It is estimated, variously, that somewhere between 5 and 10% of individuals in human populations fail to reproduce or die before they reach maturity because of genetic defects, manifesting themselves as "genetically connected abnormalities." In addition to the human suffering involved, such persons are a burden to society from many aspects, including financial.

In order to estimate the significance of radiation exposure on the mutation rate of a population, the concept of doubling dose is introduced. A doubling dose of radiation is defined as the dose of radiation which, if delivered to generation after generation, would exactly double the mutation rate relative to the spontaneous rate. The doubling dose of radiation for man is not known with precision; it has been estimated, variously, as between 20 and 200 rads. If populations were exposed to the doubling dose, the incidence of mutations would rise detectably. They also would quickly disappear from the population because of the natural processes of selection against individuals who die or fail to reproduce. When a doubling dose is administered repeatedly to generation after generation, mutations quickly double within a few generations and remain at that doubled level until the doses are no longer received. At this point, the incidence of mutations falls back to the normal rate within three generations, just as for a single dose. However, hidden or recessive mutations may not appear until generations later. From this discussion, we can see an argument for discounting, to a degree, the effects of radiation on genetics. Even a doubling dose (certainly a very large dose in terms of medical practice), if delivered to a single generation, produces mutations that are quickly eliminated from the population. Thus, one could argue that even this rather extreme situation, a doubling dose to one generation of the entire population, would not be particularly damaging to the overall population. On the other hand, certain hidden genetic changes could be induced in a population and not become obvious for generations. Thus, in the irradiation of populations, injuries may be induced unknowingly.

A final consideration is offered to help keep the subject in perspective. One can become unduly alarmed with the thought that even small doses of radiation may produce genetic changes in a population. It should be remembered that, in Hiroshima and Nagasaki, many thousands of persons were exposed to large doses of radiation, ranging from tens to hundreds of rads. Yet it has not been possible to detect genetic changes in that population to date. A determined effort was made to examine the criterion thought most likely to yield positive effects, namely, the change in sex ratio of offspring that might be expected in such a population. Irradiation of the male parent should result in a relative decrease in male offspring; irradiation of the female parent should result in a relative decrease in female offspring. It was initially thought that the results were positive, although of marginal statistical significance. But additional work and evaluation showed that the differences are not statistically significant. Also, for some 80 Marshallese exposed to approximately 175 R of radiation, no evidence of genetic effects has been evident in the offspring. These facts, coupled with the tendency for selection against harmful mutations, should allay the fear that large and dramatic changes in a population will occur as a result of genetic effects of exposure.

On the other hand, the subtle nature of such changes should be kept in mind, and the fact that recessive changes may not appear for generations must not be overlooked. Because of this, radiation exposure of the gonads should be kept to a minimum.

Some reports have been published of epidemiological studies of small groups who might have been exposed to radiation at nuclear tests and at naval shipyards (6–8). The results do not indicate a significant increase in cancer rate over similar groups of nonexposed persons. This should not be interpreted as demonstrating a threshold effect, because the sample size was too small to demonstrate a significant change if the consensus standards are regarded valid. The reasons for performing a study in which it can be reasonably estimated that a negative result will be obtained are usually political rather than scientific.

II. CONCLUSIONS

Risk assignments can be made to given practices involving exposure to radiation, because sufficient data are available for the effects of high-dose, low-LET radiation and because sufficient confidence exists in the methods of extrapolation to low doses and low dose rates. The confidence in the extrapolations is based on the fact that the risk is not expected to be overestimated, using the assumptions made (as opposed to the possibility that the extrapolations represent an accurate estimate of the risk). These risk estimates have been applied to the selection of permissible exposure levels, to show that various amounts of radiation involve no greater risk to the worker than the risk expected in another industry that is generally considered safe.

The setting of standards for protection from exposure to low levels of ionizing radiation is made by expert committees at the national and international levels. In this process, social factors as well as scientific factors are weighed. It is difficult to understand how given levels of permissible dose can be arrived at if only scientific data are considered. Recently these committees have been more specific in describing both the scientific and the social criteria in standard setting.

Data on low-level effects may be applied when assigning a "probability of causation" to a certain exposure of radiation. This has become a prominent method for arriving at an equitable award for damages caused by such exposure (8–10). The generation of these tables requires as many (if not more) social and political considerations as does the setting up of protection criteria. It is impossible to extract a purely scientific conclusion solely from the protection standards and other legal decisions.

Sufficient information exists on low-LET radiation that safety standards for exposure can be rationally (if not scientifically) agreed upon. This is not the case for high-LET radiation. Our earlier standards for neutrons were based on the bombings of Japan and the supposed difference between the amounts of neutrons in the blasts. Reevaluation of the dosimetry has nullified the difference, resulting in the loss of almost the entire human data base for exposure to neutrons. Few data exist for humans exposed to other high-LET particles, but much discussion is found (11).

The acceptance of radiation as only one of the causes of cancer and genetic damage, among the many causes to which the population is exposed, is a rational method of dealing with this hazard. The current undue emphasis on radiation has resulted in inappropriate allocations of money and manpower, which would be better spent on mitigating the sources of harm that result in higher risk to the population.

REFERENCES

1. Committee on the Biological Effects of Ionizing Radiation. *The Effects on Populations of Exposure to Low Levels of Ionizing Radiation.* National Academy Press, Washington, D.C., 1980.
2. United Nations Scientific Committee on the Effects of Atomic Radiation. *Ionizing Radiation: Sources and Biological Effects,* 1982 Report to the General Assembly. United Nations, New York, 1982.
3. International Commission on Radiological Protection. *Recommendations of the International Commission on Radiological Protection,* Ann. ICRP 1(3). ICRP, Washington, D.C., 1977.
4. National Council on Radiation Protection and Measurements. *Basic Radiation Protection Criteria,* Rep. No. 39. NCRPM, Washington, D.C., 1971.
5. National Council on Radiation Protection and Measurements. *Review of the Current State of Radiation Protection Philosophy,* Rep. No. 43. NCRPM, Washington, D.C., 1975.
6. Stern, F., *et al.* A case control study of leukemia at a nuclear naval shipyard. *Lancet* (in press).
7. Caldwell, G. G., Kelley, D., Zack, M., Falk, H., and Heath, C. W., Jr. Mortality and cancer frequency among military nuclear test (Smoky) participants, 1957 through 1979. *J. Am. Med. Assoc.* **250**(5), 620–624.
8. Robinette, D., Jablon, S., and Preston, T. *Studies of Participants in Nuclear Tests.* National Research Council, Washington, D.C., 1985 (in press).
9. Ad Hoc Working Group to Develop Radioepidemiological Tables. Report to the National Institutes of Health, NIH Publ. No. 85-2748. National Institutes of Health, Bethesda, Maryland, 1985.
10. Oversight Committee on Radioepidemiologic Tables. National Research Council. *Assigned Share for Radiation as a Cause of Cancer: Review of Assumptions and Methods for Radioepidemiologic Tables, December 1984.* National Academy Press, Washington, D.C., 1984.
11. Straume, T. A radiobiological basis for setting neutron radiation safety standards. *Health Phys.* **49**(5), 883–896 (1985).

Index

B

C

D

G

H

I

K

L

M

Q

R

U

V

W

X

Y